COSMIC DNA AT THE ORIGIN

A Hyperdimension
before the Big Bang:
The Infinite Spiral Staircase Theory

Foreword by physicist John Brandenburg, Ph.D.

CHRIS H. HARDY, PH.D.

COPYRIGHTS

Artwork layout by Chris H. Hardy and Vincent Winter.
(Vincent Winter Associés, www.vwaparis.com).

Cover design by Vincent Winter, based on a Hubble telescope photo of the *Egg nebula* (source: hubblesite.org/ Public domain).

Author’s blog at
http://cosmic-dna.blogspot.fr

"When a cosmologist perceives how the laws and principles of the cosmos begin to fit together, how they are intertwined, how they display a symmetry that ancient mythologies reserved for their gods, [...] then he or she perceives pure, unadulterated beauty.

"The scientific concept of creation encompasses no less a sense of wonder: We are awed by the ultimate simplicity and power of the creativity of physical nature—and by its beauty at all scales.

"[The universe] evolution was written in its beginnings—in its cosmic DNA, if you will."

George Smoot, Nobel Prize in Physics.

ACKNOWLEGMENTS

I dedicate this Infinite Spiral Staircase theory and vision to the scientists, artists, and seers, who are going to access and explore the 5th dimension in their minds and make it a new dimension of consciousness for humankind, in sync with other civilizations in the galaxy.

My heartfelt thanks to physicists John Brandenburg and Massimo Teodorani for their enthusiastic response regarding this ISS theory and for our in-depth discussions.

My deep appreciation goes to several scientists whose work and theories have illumined my research, notably George Smoot, Roger Penrose, Stephen Hawking, Gerard 't Hooft, Lisa Randall, Jack Sarfatti, Brian Josephson, Karl Pribram, Igor and Grichka Bogdanov, Dean Radin, Stephan Schwartz, and Ervin Laszlo.

I'm grateful to the colleagues and friends who have given me their relentless support, Leslie Allan Combs, Linda Dennard, Fred Abraham, Vincent Winter, Martine Winter, Carole Sedillot, Sidney Tegbo, Didier Dufresnoy, Jacques Donnars. My friendly and appreciative thanks to Vincent Winter, who has given me a robust helping hand in terms of layout and images, and—I must confess—the expert management of quite a few digital catastrophes along the eventful process of bringing this book to completion.

CONTENTS

LIST OF FIGURES

FOREWORD

Let the readers prepare themselves for an epic intellectual journey down the spiral staircase of time and space, to its very cosmic origins. Chris Hardy has shown herself to be an original thinker of historic brilliance in unifying the metaphysical requirements of quantum mechanics with the properties of the primordial singularity required by General Relativity, all in the first instant of creation. The results of this conceptual unification are original, profound, and world-view altering.

Physics is normally about measurements, and metaphysics is about meanings, but both must come together when Schrodinger's cylinder is opened, and where the cat is discovered to be alive or dead.

Obviously, the Cosmic Cat was alive and sprang forth purring! This fact of the conscious mind collapsing the wave function, has always been understood as mandating a role of consciousness in quantum mechanics. Some have even extended this logically to ask the role of consciousness in the Big Bang. They ask, if the Big Bang was in fact the collapse of a cosmic wave function, then what consciousness caused this collapse, and what did it learn? Now Chris Hardy has found this act of consciousness: it was an evolution in an extra dimension of the spacetime manifold.

Chris Hardy follows in the footsteps of many scientists in invoking a hyperdimensional cosmos, but where the ISS theory blazes a new path is in including consciousness and meaning in this hyper-dimensionality and further—to me as scientist, most significantly—by hypothesizing a signature of this new degree of freedom, a truly ground-breaking concept.

The existence of this 'dimension of significance and meaning' or 'syg-metadimension' changes the evolution of the universe from the vacuum, like the existence of any new degree of freedom changes the evolution of any system. However, in this case the existence of this hyperdimension of consciousness is fundamental to the evolution of the universe we experience now, all the way from its earliest moments before the Big Bang.

Chris Hardy has identified the signature of this new degree of freedom, the 'sign of its existence,' as a primordial field of collective cosmic consciousness, imprinted in all forms and structures of matter now found in the universe, from the particles to the stars, in a compact and curled-up hyperdimension. She has also traced this signature back to the earliest phases of the primordial universe, dominating its physics well before the Big Bang, even before space, time, and the first particles came into being. Curiously, the imprint Chris Hardy has found is hidden in the types of physics most often studied. But this fact makes her discovery all the more remarkable and brilliant. She has deduced the existence of this cosmic metadimension of consciousness from her two decades long research that posited semantic or syg fields, an extra dimension of minds and all complex systems.

The post Big Bang physics is now well known from precise measurements of the abundance of the primordial elements and from the WMAP data of the Cosmic Background Radiation showing minute fluctuations and ripples in this microwave relic radiation lingering about 270 000 years after the Big Bang but issued from the first seconds. This data gives the astonishing result that, except for some isotopes of lithium, everything that is measured requires that "the-physics-we-know," was well established by the first second of the universe's existence, and that before this was an "inflationary phase" of spectacular growth. However, we know that even before this epoch that sees the emergence of space, time and the first particles, this "physics-we-know" breaks down and another physics entirely, takes over. This pre-Big Bang physics, dealing with information and something else than matter and matter-energy is the Big Mystery, the open question. So the challenge that systems theorist Chris Hardy has taken up, was to find the dynamics of the process that both sparked the universe, expanded it, and then determined its present physics. The imprint of that process, the actual signature of this extra degree of freedom, that surrounds and permeates us even now: she has found it.

The signature of the cosmic semantic field, this hyperdimension, is a dynamical geometric pattern, self-generating and imbued with consciousness, and it suggests that the origin of the universe was not a random event, but an act of conscious logic. That is, the universe did not begin with a violent and random explosion, this is only the physics we presently understand. What the reader will see, laid out in compelling logic, is that the cosmos began with an *idea*.

So read on my friend, and drink deeply of the fine wine Chris Hardy has now poured into your glass and offered you. It is a new and memorable wine, an excellent vintage. The world, the whole cosmos, will never look the same once you have tasted it.

John E. Brandenburg, Ph.D.,
Plasma physicist and senior propulsion scientist at Orbital Technologies Corporation, Madison, Wisconsin. Author of *Beyond Einstein's Unified Field, Life and Death on Mars*.

INTRODUCTION

When I started to tackle the dimension of consciousness at the origin, I had no idea that it would take me into a full blown hyperdimensional theory of the universe—one that addresses not only the beginning of our universe, but also the 'deep reality' of all particles and matter-systems, and of course of all individual consciousnesses.

For the last twenty years, I've been elaborating a theory of nonlocal consciousness, that is, consciousness operating on its own meaningful or *semantic dimension.* This Semantic Fields Theory (SFT) posed that consciousness was indeed an energy in and of itself, of an unknown nature, that I called *syg energy*; this energy was creating connections at a distance spontaneously, and therefore it was operating in a nonlocal way, neither bound by space and time, nor to the laws of EM fields. In SFT, I used the frameworks of chaos theory, systems theory and neural nets in order to account for the mind as a dynamical network—a *semantic field*—an intelligent system able to learn and to reorganize itself according to its own intention and free-will. This network, in sync with body and emotions, is able to connect all levels of the 'mind-body-psyche' system between them, all the way down to the neuronal level. The mind as a semantic field is also able to spontaneously connect and exchange information at a distance with other minds, other semantic fields.

Moreover, all the living and complex systems in the environment (and even what we thought was 'inert matter' such as a lake or a mountain), have eco-semantic fields (in short, eco-fields). The whole of nature and the universe is thus bathing in a consciousness dimension fueled by syg energy.

In *Networks of Meaning*, I had analyzed in depth the architecture and network-dynamics of the human semantic field (and the connective syg energy) to make of SFT an enlarged cognitive theory, modeling also the mind-world interactions. While postulating the energy of consciousness, syg energy, as nonlocal, I had sorted out that it operates through bonding, resonance, and sympathy with others and the environment.

The present *Infinite Spiral Staircase* theory is building on this previous research on the *semantic dimension*, but reaches out to fathom the universe at its very beginning. I hypothesize that the semantic dimension and syg energy existed in the universe right at its origin, long before the Big Bang, as a hyperdimension that is still pervading the universe.

The revolutionary thesis of this book is that entirely different forces prevail at the origin of the universe; specific laws of organization and dynamics exist, wildly different from the quantum and then electro-magnetic laws that will appear later in our 4D universe. Indeed, there exists an energy threshold called the *Planck scale*, and it's only after the Planck time, when the universe was an infinitesimal fraction of a second old, that the first energy-particles could be born in the Higgs field. Physicists infer that before this Planck scale, there were no particles and therefore no matter (nor was there any electro-magnetic force, strong force, weak force), and that no time and no space had formed yet. And nevertheless the primordial universe had a gigantic energy, despite the fact that we are still, at Planck scale, an immense time before the Big Bang or inflation phase.

So that, if (1) there was an immense *energy*, and if (2) there were no particles and *no matter*, then it had to be a *non-matter energy*. And given (3) that there were no time and no space yet, then I thought I knew what this energy was: it was consciousness-as-energy—the very *syg energy* that I had posited and researched in depth since more than two decades.

However, as I finally understood it, syg energy was not the only force organizing the primeval universe: two other forces were involved in shaping its dynamics and architecture. All three were woven into a triune braid forming one hyperdimension (the 5th dimension) that was encompassing the matter-universe (the classical 3+1D spacetime of Einstein). The triune hyperdimension consisted of three enmeshed 'metadimensions' or 'metadims':

- Syg metadim (syg for letter S sigma in Greek): the semantic dimension operating with syg energy (or S).
- A hyperspace created by number Pi creating a circle out of the point/center, called Center metadim (or C).
- A hypertime created by a cyclical rotation in a spiral based on the phi ratio (the golden proportion) and that I call Rhythm metadim (or R).

The three metadims form the *CSR hyperdimension* (Center-Syg-Rhythm). The hyperspace and hypertime are entwined with *Syg metadim* (the semantic dimension) to create the *triune braid* that takes the shape of a near-infinite spiral staircase issuing from the point of origin.

Metadim Center creates the primordial spiral of the origin by incrementing at a blinding speed the radius of the universe by the phi ratio and drawing the next quarter circle or 'bow' of the golden spiral.

Metadim Rhythm jumps in frequency at each bow and vibrates, thus creating a staircase of frequencies. Each bow ejects, like an arrow, a virtual particle called a *sygon,* prefiguring the strings-particles appearing after Planck scale.

Thus we have a dynamic and self-propelled geometry—a topology driven by a specific rhythm in a spiraling progression, and the whole staircase is in-formed and inspired by a cosmic field of information and a conscious collective intelligence (Syg metadim).

The infinite spiral staircase (or ISS) thus contains a near-infinite number of distinct frequencies, like musical tones, that are also the frequencies of the virtual particles: the sygons. But these frequencies, because they are entwined in a braid with Syg metadim, are loaded with meaning and alive active information.

This ISS theory hypothesizes that the *infinite spiral staircase* is containing, on its infinite number of bow-frequencies, the information about all the systems and civilizations that evolved in parent universes and all possibilities of evolution of our universe: it is a *Cosmic DNA*.

So, the dynamical topology of the early universe issuing from the point of origin shows two numbers, Pi and Phi, enmeshed in the cosmic consciousness of the self-creating universe. Thereafter, in our 13.7 billion years old universe, evolving in spacetime, all information about its evolution is inscribed in a synchronistic and instantaneous score on some of the ISS bows, thus creating an ongoing evolving melody for each given system (such as a solar system or humanity on Earth).

How I came to envision the Infinite Spiral Staircase of the origin

So how did I myself make the leap from a theory of semantic fields in cognitive sciences to a theory of deep reality before the Big Bang, as a hyperdimension of all-that-is—of all particles, matter, and systems, including living beings and minds?

There are a few reasons, let me share some with you.

The first and foremost is the deep intertwining of mind with matter, that I had posed and developed at the very beginning of my research on SFT (in the early nineties). I've always been convinced, mostly by my capacity to 'see' the syg-energy fields on individuals, groups, and specifically objects and sacred buildings (as I discuss it in *The Sacred Network*) that consciousness is an energy in itself. It's not an information carried by a carrier wave (such as conversations on a cell phone's microwaves), but a specific type of energy that is itself finely tuned and imbued with meaning and thoughts. Syg energy is thus a concept similar to psychologist Carl Jung's *psychic energy* and to physicist David Bohm's *active information*. Syg energy, I posed, pervades the universe and influences the organization of matter, of systems, and the world at large. And yet, all minds use it and activate it, and it is itself modulated and colored by the psychological state, emotions, thoughts, values, beliefs, etc. In brief, syg energy is colored by the actual state of the person's semantic field.

So, if mind and matter/energy are so entwined that an energy of consciousness exists and is able to organize complex systems, then what kind of scientists could tackle the theory? Physicists who may know very little about the dynamics of consciousness? Or cognitive and systems scientists who may know very little about physics? Obviously both should try to bridge the gap, as it occurred with the famous collaboration of physicist Wolfgang Pauli and psychologist Carl

Jung, both eminent theorists in their respective fields. The researchers willing to elaborate a candidate theory should propose foundations and dynamics across dimensions, but the onus is on them to sort out the pertinent forces and dynamics within the consciousness domain.

Thus, the crucial problem that such a hyperdimensional theory has to solve is to explain how and why meaning, intention, and free will are real forces able to influence events in the world. It is not enough to state that desire and intention (or need, or focus, etc.) will bend random events toward the desired state, as has been experimentally proven, or else that mind and matter are entangled. What we need a theory to explain is why such psychological factors (such as intention or meditation) are having an influence on events, and on what foundations and dynamics lies the mind-matter entanglement. As far as I'm concerned, in the mid-nineties I've posed the processes and parameters of this entanglement in the mind-body-psyche system, and as well between a mind and the environment's eco-fields, and since then I have strenuously kept on extending this SFT theory. As highlighted by Jung, Bohm, and Brian Josephson, meaning (as an 'I' or *1st person* perspective) is central to this mind-matter entanglement. This is why I call 'semantic' (or syg) the energy of consciousness, from the Greek root referring to 'giving meaning to something.' Syg energy is what allows us to live in a meaningful environment.

Second, psi research up to now is our best source of data showing (through robust double or triple blind experiments) that mind can indeed influence biological matter and the distribution of randomness as well—to make it conform to the subjects' intention. And as I did my doctoral thesis on psi research linked to altered states (such as dreams, meditation, relaxation, etc.) and worked to prepare it at a psi research lab called the Psychophysical Research Laboratories, in Princeton, New jersey, I certainly had a good grasp of these data.

Third, my book *The Sacred Network* focused on the real life experiences showing consciousness (1) as creating syg-energy fields, with an analysis of their parameters, structures, psychological correlations, etc., and (2) as allowing a tinkering with space and time (consciousness as nonlocal). The whole third part of *The Sacred Network* was devoted to 'Space and Time Singularities,' that is, exceptional collective semantic fields or Telhar fields, creating a shared consciousness that showed a clear-cut tinkering with either 3D space or the linear time of classical physics. All the arguments and discussions were directly based on objective and factual evidence from real-life experiences (a phenomenology approach). I believe I demonstrated clearly, to minds free of preconceived ideas, that a part of our consciousness operates indeed beyond space and time, that it behaves as a still mysterious new type of energy and is sometimes observed infringing on, or overriding, the classical Einsteinian spacetime framework.

Let me dare say that these experiences are invaluable in regard to the present ISS theory. On the one hand, they make the link with a part of humanity pursuing an age-old quest for a 'know thyself' and an inner and spiritual path of knowledge, to which part I also belong. And on the other hand they bring an objective demonstration (albeit not a proof) of the reality of consciousness-as-energy. Indeed, they bring many cues as to the workings and dynamics of this syg energy in our minds and in the world. Syg energy is in an 'exalted' or 'excited' state in the semantic fields of people having peak experiences, as well as in sacred places and objects, and thus becomes 'visible' for the sensitive, appearing as syg-energy structures such as a sphere, a torus, or a rod, in a domain of frequencies usually outside the range of visible light.

Fourth, the real trigger was the work I did for my book called "Jung's Prediction," primarily focused on *synchronicities*—that is, meaningful coincidences—a phenomena discovered and studied by the renown psychologist Carl Jung in the middle of the twentieth century. Early in his research on synchronicity, Jung entertained a correspondence with physicist and Nobel laureate Wolfgang Pauli, one of the pioneers of Quantum Mechanics (QM). Jung and Pauli both had the strong conviction that there existed a dimension of *deep reality* in which mind and matter were merged. In my view, it pointed to consciousness being an energy and having physical properties, an energy enabling its interaction with matter. Such deep interactions would allow synchronicities and nonlocal processes to happen, and would be also the dimension in which dwells a higher consciousness in human beings—their Self (atman, soul). The deep reality as a CSR hyperdimension is the core subject of this present book: not only is it a semantic level of organization in all the living and in matter systems, but it resides also at the subquantum scale.

I was practically finished with the writing of this book on Jung (in November 2010), when I had a breakthrough and conceived three 'meta-dimensions;' I then wrote a whole new chapter while the ideas were still crystallizing about how these metadims worked and how they constituted a new manifold, which I called Center-Syg-Rhythm. The creative spur was so sudden and so intense that I kept writing new stuff in the book up to a few hours before sending the belated file to my French publisher. At the end of the chapter, I used metadim Rhythm to explain the entanglement of particles irrespective of distance (as shown in the famous EPR paradox experiments that proved the instantaneous nonlocal entanglement between particles. And still, I had no inkling of the ISS yet.

Finally, when I got involved in books and data on the Big Bang and the pre-space, I came to understand that consciousness must have existed at the very birth of our universe, before the emergence of spacetime, because I knew and had theorized that syg energy was operating on parameters and dynamics independent from spacetime.

And that gave me a head start as well as a definite momentum to try to fathom how consciousness-as-energy (syg energy) could exist and operate before the Big Bang. And I dove enthusiastically into more research. As I was finally writing a synthesis of what I had learned (the blueprint of the actual chapter 1), I suddenly had a grand insight that unfolded itself at a stunning speed—a conceptual, architectural, visual and musical understanding of the Infinite Spiral Staircase as the self-organizing architecture and dynamics of the pre-Big Bang universe; it was a topological yet dynamical theory positing the self-actualization of our primordial universe, and its cosmic consciousness.

The core of the whole theory did unfold in about two weeks (October 28 to November 15, 2012) and it included, as a second leap, formalizing the ISS (and thus the Center-Syg-Rhythm manifold) in each particle, system, and mind (the actual chapter 7).

At that point, I had the feeling that everything was in its place; that what I had worked on all along—syg energy, the semantic dimension, and the semantic fields—had now a foundation as a hyperdimension that could later be framed in mathematical physics. It had also a keystone in ontology (the science of Being and of the ultimate questions on the nature of existence): it posed consciousness as *the* universe's fundamental force. Indeed, a force creating order and complexity in the world (or negentropy) is duly observed at work everywhere by scientists: just think for example about our science and technology growing exponentially! The universe is screaming for the theoretical foundations of consciousness as the negentropic force organizing reality, a force that counts for such an immense part of the dynamics of the whole that physics has now to reckon with it.

The new paradigm of the ISS theory: consciousness as part of a hyperdimensional theory of the universe

The ISS theory poses the foundations of a new stand for physics: that of integrating consciousness as a real force and a specific set of dimensions in a hyperdimensional theory of the universe. Among the novel concepts having a bearing on an integral theory of the universe:

- our own input, as individuals, in the cosmic ISS, along a two-way flow of syg energy (simultaneously information and energy);
- not only the spontaneous imprinting of all events and beings in the ISS (the ancient concept of Akasha) but influences received (often as archetypal forms) from it, and our own intentional influences on it;
- the allowance for imperfections, discordant melodies, and roots of conflict at the ISS as source;
- the existence of previous universe-bubbles, along a collar of bubbles (that need not be the only such collar), whose information would be inscribed in the steps (or bows) of our own cosmic ISS, and yet leaving an immense space for novel input;
- the hypothesis that the universe-bubble that immediately preceded our own ended up in a global black hole that, through a X-shape funnel (or hourglass), transformed itself into a white hole launching with enormous energy a new ISS—a new universe;
- the hypothesis that the information of the previous universe-bubble is inscribed in our ISS, in a tiny part of its global system, acting as a cosmic DNA;
- this doesn't preclude numerous such collars of universe-bubbles and accommodates the multiverse concept (but not its random feature);

- the prediction of numerous intelligent civilizations existing in our galaxy and in our universe-bubble, all feeding the cosmic ISS; and the added idea that, all of them being linked to the One-ISS, we share the same ubiquitous Center-Syg-Rhythm hyperdimension; so that, somehow, within the whole bubble, the evolution of conscious beings must be resonant or enmeshed, even if some of these intelligent civilizations have a probability to be millions of years older than ours;
- And lastly, let's realize that we are free and therefore responsible for the future of Earth as a collective planetary semantic field endowed with free will, creativity, and vision.

In the present book, I have kept the development of the core chapters practically untouched, as well as the spontaneous unfolding of concepts, the process of creative insights feeding into one another, launched and fueled by the exquisitely profound dreams of Pauli, the constant back and forth between left-brain logic and right-brain intuitive thinking, the entwining of logic and intuition. Thus I hope that this book, which means for me the sharing of an insight, can also be an exciting ride along a creative spur and may trigger insights and a grasp of the deep reality we are living in and sharing, and of its boundless potentials for experiencing our own Self and exploring our mental capacities.

Provence, France, July 17, 2014

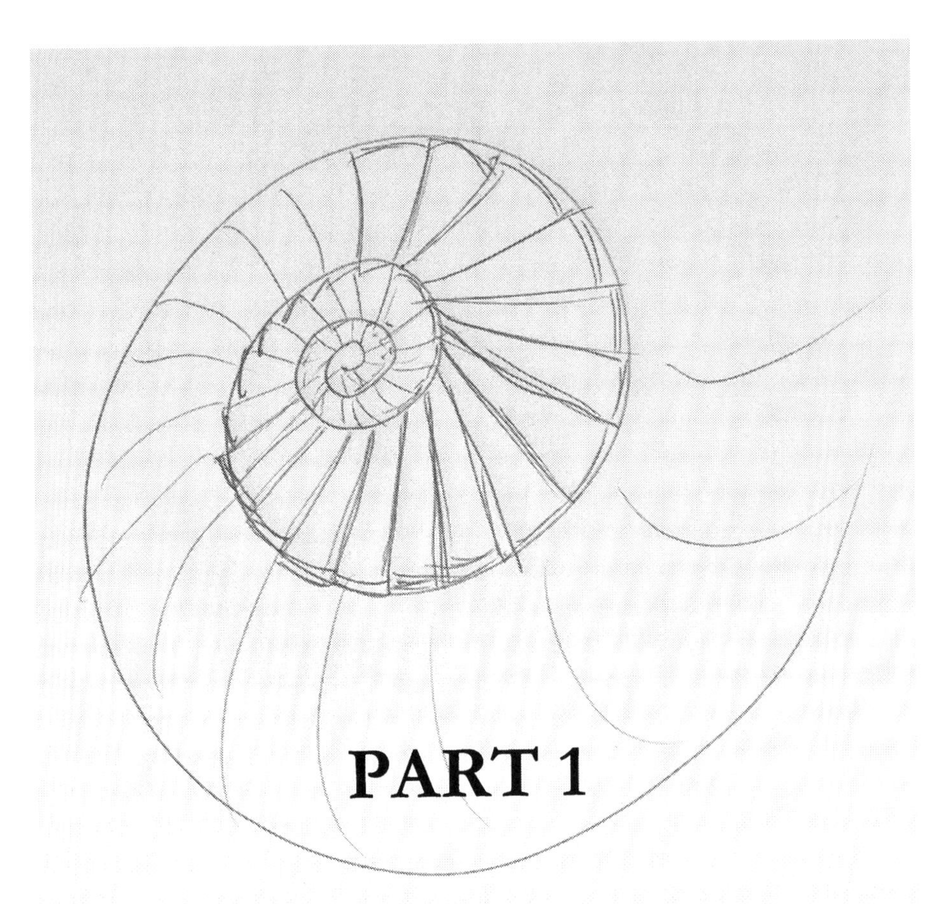

PART 1

THE EDGE OF RESEARCH ON THE BIG BANG AND THE VOID

1
THE POINT-SCALE UNIVERSE AND THE BIG BANG

"Only at the instant of the Big Bang do we see the full power of the hyperspace theory coming into play. This raises the exciting possibility that the hyperspace theory may unlock the secret of the origin of the universe." Michio Kaku (Hyperspace, 27)

The edge of the research in physics reaches, with the point of origin of our universe, a reality that steps altogether out of the domains of both quantum mechanics and Einsteinian relativity physics. A reality that is still largely unknown, and that resides before space and time existed, before particles and matter existed, and beyond causality. The point of origin of our universe—let's call it the X-point—happens an immense 'time' before the Big Bang, and this Big Bang is itself immensely posterior to the quantum scale (or Planck scale) defined by Planck constant. Indeed Planck constant defines the smallest quantity of action—a quantum—and before this scale no particle can exist; then the particles can only jump from one energy state to another, by multiples of this quantum, and this is why the domain of quantum events is discontinuous, that is, discrete. From Planck constant are derived several facets of the universe at that scale, such as its radius and its energy. But it is also a threshold or "cut off" allowing particles and therefore matter to exist. Indeed, there can be no particles before the universe reaches the size of the Planck length of 1.6×10^{-33} centimeter, that is, thirty three orders of magnitude below 1.6

centimeter—one million billion billion billionth of 1.6 centimeter. And this happens when the universe is still in a fraction of a second old, in effect, 5.3×10^{-43} second. The Planck scale is thus an absolute boundary separating quantum physics from subquantum physics; we need some mental acrobatics, or what I like to call an Escherian logic (from the paradoxical geometrics of the famous artist Escher) to fathom that the quantum void (or vacuum) is a cosmological surface and yet it resides also at each virtual particle scale. Consequently it is both infinitely small and extended in the cosmos, as if opening at each point of the whole geometry of spacetime. Furthermore, it is existing both at the universe's point of origin and at the scale of the cosmos, at all times and all places of spacetime. We'll see with the present Infinite Spiral Staircase Theory (ISST) a hypothesis about how a hyperdimensional topology may solve this paradox.

To get back to the point of origin, the stupendous consequence is that before this precise Planck length not only are there no particles, but neither space, nor time, nor matter. Of course it raises the endgame question: if there is no matter and no spacetime before Planck scale, then what exists there? What has produced the immense quasi-explosive and quasi-instantaneous expansion of the universe—the Big Bang or *inflation phase* that happened an infinitesimal fraction of a second after Planck scale, at exactly 10^{-36} second?

Einstein was a proponent of the idea that "hidden variables" (e.g. yet undiscovered deterministic forces) were keeping the universe ordered and were thus counteracting the indeterminacy and probabilistic behavior of quantum systems posed by quantum mechanics (QM). With two colleagues, Podolski and Rosen, Einstein proposed his famous thought experiment now called *EPR paradox*, in order to disprove the QM axiom that particles could remain correlated and entangled at great distance (a distance so great that a signal at light-speed would not have the time to pass from one to the other). To the opposite, he envisioned a substrate below the Planck scale that would act as a hidden force or 'variable' influencing the behavior of quantum events (thus discarding the entanglement concept, that nevertheless was proven in the early 1980s). Einstein further suggested that an *algebraic system* could describe this substrate. According to the Wikipedia article *EPR Paradox,* "If time, space, and energy are secondary features derived from a substrate below the

Planck scale, then Einstein's hypothetical *algebraic system* might resolve the EPR paradox."[1]

An unknown and pervasive force was also posited by physicists David Bohm and his collaborator Basil Hiley who proposed a nonlocal hidden variable, the Quantum Potential, acting as "active information" (and also as a field of form) in a deep 'implicate order' of the universe. Whether in line with the concept of a deterministic universe or not, several other scientists predict that before matter comes into existence, the reality consists of information; this is the case for Igor and Grichka Bogdanov. Some of them propose a digital and computational substrate of the universe at the Planck's scale, such as Edward Fredkin's *Finite Nature Hypothesis*, or Stephen Wolfram.

Other scientists invoke a *field of form* (informational or not) that can be studied and described via its structure or topology. (There are different types of topology in mathematics, such as algebraic topology or geometric topology). Thus, Detlev Buchholz, ex-head of the prestigious chair of theoretical physics at Göttingen University, states: "At the zero scale, spacetime is in a topological state and its 'dynamics' is Euclidian." (cited by *Bogdanov*, 2004, 156) A statement that feeds directly into the Infinite Spiral Staircase theory, in which the topology is the Golden spiral and the dynamic evolution is set by a progression along the Fibonacci sequence. The Bogdanov brothers propose that "at instant Zero, the evolution of the universe (the cold Big Bang) is thus definitely happening in an imaginary time—and not in real time."

At the Planck scale and earlier, the universe is classically considered a point—and a point in physics is called non-dimensional; without dimension, it is nevertheless modeled mathematically as a sphere. We know that the Zero Point or vacuum opens on a near infinite energy. Similarly, in the ISS framework, the domain of information below the Planck scale (acting also as a topological field of form) contains an infinite information, that is, all possible information about the universe past states and about its possible future states.

Stringent physical laws and unconstrained consciousness

This does not imply necessarily that the universe is totally pre-determined by this information-field at its beginning. However, we do have to account for two irreducible sets of facts. First, that about twenty physical constants and variables are so well and so finely adjusted that an infinitesimal modification either way (bigger or smaller) would impede our universe from existing as it is, with solar systems, planets (and therefore Earth), and living organisms. Second, not all reality in the universe is constrained by stringent laws. In total contrast, the mind and consciousness give intelligent beings an immense capacity for creation, innovation, and freedom of choice. And in what we call high-level cognitive processes—that include the emotional life (the psyche) and mental levels (mind), but also states of consciousness and capacities such as volition and intention—the rule for healthy processes is constant change, evolution, and innovation. A repetitive emotional or mental state is at the opposite end of a normal and healthy cognitive process. The optimal mind state is creativity itself, and the optimal psychological states are enthusiasm, love, discovery, opening up to others and the world—all these states thriving on change and evolution.

This is the most paradoxical aspect of the reality of the universe: we have to account for both stringent physical constraints with finely tuned dynamics, and an unconstrained consciousness. Yet innovation, emergence and creativity are not accounted for in classical cognitive theories, and only some partial modeling exists using systems theory and chaos theory. This is why we, the researchers in consciousness studies and cognitive sciences, must take on the task of modeling the global dynamics of consciousness in a way that explains creativity and innovation—something I did with the Semantic Fields Theory.

One thing is sure with the discovery of the subquantum scale, it is that a Theory of Everything has to account not only for matter forces but for a reality beyond matter, and also for consciousness as an unconstrained force able to create and produce new organization and new global orders. In a word, it has to account for consciousness as an organizing force, that is, a negentropic force at work in the universe. Negentropy is the creation of order and it sets the rise in complexity in nature; it is a force that we can observe in action in most complex systems and especially whenever mind, psyche, and consciousness at large are concerned. And in fact we can readily see how a global information and technology civilization like ours can modify the very matter of our planet, in terms of ecological systems, climatic and geological systems, etc. As we are now well aware, our influence is so great that not only it has an effect on the future of our planet but it puts this very future at risk. This is another inescapable reason why a theory of everything cannot limit itself to the matter-only reality of the universe: it is evident that cultures, technology, and science influence the geopolitical choices, and all these semantic forces have an impact on the physical organization of our world. The last facet of reality to take into account is the existence of perfectly random processes—such as the indeterminate events and processes in the quantum domain. As examples of indeterminate processes, we have in nature the radioactive decay, and in technology a perfectly random white noise—the latter widely used in experiments as 'random generators.'

In conclusion, if we follow the logic of the point-scale universe opening on a near infinite energy, then we have to envision that it opens also on a near infinite information. But as the universe is not fully determined (given that consciousness is creative and unconstrained), it would mean that this 'field of information' is not a deterministic process—such as a musical chore that would have to be performed in a perfectly predetermined way. The solution I propose here with the Infinite Spiral Staircase Theory is that this field of information is an immense, in fact quasi-infinite, field of possibilities of evolution of the universe; and this despite the fact that biological and physical structures have been, by previous evolutions, channeled toward their optimal organization.

Can Theories of Everything be complete without integrating consciousness?

Many among the foremost scientists, following Einstein's lead, aim at achieving a *Grand Unified Theory* (or GUT), also called Theory of Everything (TOE); that is, the unification in one single mathematical model of the four basic forces of the universe: strong force, weak force, electromagnetic force, and gravity. In calling this unified theory a TOE, scientists believe that it would amount to the understanding of everything in the universe and consequently it would allow perfect predictions on the future state of the universe and therefore of our planet. Whether or not we blow up the planet Earth, that is!!

This amounts to a kind of Laplacian belief grounded in a fully deterministic and materialistic view of reality, and which, despite its name, is way out of place ('la place' meaning 'the place' in French—pun intended). This belief is shattered to pieces first by the existence of indeterminate and random processes in the quantum domain, by self-organizing processes in complex dynamical systems, and by the existence of newly discovered forces such as dark energy. Moreover, despite being an eminent mathematician and an astronomer to whom we owe a lot, Laplace's belief in a mechanical clockwork-universe does not take into consideration that science is constantly evolving, and at an exponential rate at that, which means that scientists have no idea of what they don't know yet and will know in the future—and this state of affairs applies to all times.

Yet, we won't get a complete description of the universe even if we achieve the unification of the four forces, that is, the blending of quantum physics (the Standard model) with General Relativity (a field theory of gravity). In the actual state of the search for a TOE, the Standard Model of quantum physics unifies three of the four forces: electromagnetism, weak interactions, and strong interactions; only gravity has not been really integrated yet, and its domain of expression starts below Planck threshold and at a scale extremely smaller than the three other forces. Furthermore, a decent physics-only TOE should also accommodate the newly discovered and still enigmatic dark energy and dark matter, and the as mysterious acceleration of the expansion of the universe.

Indeed, in reaching the unification of the four forces, we would only achieve a unified theory of a blind universe—one that accounts for matter and matter energy, and possibly ultimately dead information, but that leaves no possibility for a dynamical meaning-generating force—consciousness—to exist. Such a Blind Forces Theory would not allow any psycho-mental dimension such as scientific knowledge, learning, invention, feeling... not even the most basic psychological and cognitive process such as a meaningful perception (that is, meaningful to the specific person who is perceiving). With a unified theory, we can certainly explain operational laws, interferences, causality, algorithms, quantum indeterminacy and chance events—but neither a mind like yours or mine, nor the origin of these laws. Yet, obviously, mind, spirituality, self-consciousness are in us and with us at all times, as the very essence of self—of our feeling of being alive as an 'I' who feels, thinks, acts, at any single moment of our waking state, what cognitive sciences call the "1st person perspective." And even in the sleep state, in our dreams, we see that our unconscious is involved in complex mental processes such as recalling, feeling, behaving, and even thinking. Surely, we cannot, in our theories of the universe, have a constant and crucial part of our reality, the psycho-mental processes, be accounted for solely as a Creator God who would be out-of-the-universe, out-of-Man, and acting solely at the no-time of the origin, meaning at about 13.79 billion years in the past, in an infinitesimal fraction of the first second, and yet an immense 'time' before the threshold of Planck.

All in all, whatever the way we formalize it in terms of physics, the psycho-mental reality has to be accounted for at any moment of time or pre-time in the evolution of the universe. Yet, consciousness has not been accounted in physics terms—neither in the Einsteinian physics (of 3+1D), nor in the quantum dimension (of 3D of particles + 3D of wave), nor in the latest 10D or 11D string theory, M-theory, supersymmetry and superstring theories. In this perspective, an irrefutable logic leads us to assume that only in positing new dimensions of reality can we account for it. A likely solution would then be a hyperdimensional universe, with some added dimensions being metadimensions of consciousness.

To those who still harbor the conviction that physics knows all that exists in the universe and therefore that postulating new types of dimensions is downright foolish, I remind them that physics is actually confronted to the gravest of situations: in 1998, physicists inferred the existence of a new gigantic force—dark energy—that is driving the acceleration of the expansion of the universe (this acceleration, enormous, was at the time just discovered by the COBE astronomical observations). As of March 2013, the results of measurements and mapping by the satellite PLANCK have established with great accuracy that dark energy makes up 68.3% of the total energy in our universe! Furthermore, it has been determined that our good old matter itself (all the particles, atoms, stars and galaxies) constitutes only about 5 percent of all energy in the universe. And moreover, the remaining quarter of this total energy is composed of *dark matter*, another still mysterious state of the 'physical' reality that had been predicted by

Einstein and De Sitter, another theoretical physicist. With dark energy, we have, in my opinion, a new leap in the dematerialization of physics. This is where 'matter' stands right now—and as well a physics based only on matter and matter particles: a bare 5% of what physics should know and account for. About the other 95% we know next to nothing, apart from the very definite fact that they exist and the strength of their attractive (dark matter) or repulsive (dark energy) pressure radiation. However we know that with dark energy, we are dealing with an 'energy' because it is precisely what produces the acceleration of the expansion of the universe. Thus the nature of 95% of the stuff of reality, *95% of the energy* in our own universe, is still a perfect enigma to our physics. Furthermore, many physicists agree that the physics before Planck scale, at the subquantum scale, is a totally different order of physics. Says for example John Brandenburg "The fact that physics must change at some small size to make the electron mass finite has been found to be necessary in analysis of the strong and weak forces also, so it must be physical. *Somewhere in the deep subatomic scale, physics changes."* (*Beyond Einstein: GEM Unification*, 119.) The least we can do, in such humbling situation, is to be open to new ideas and to weigh and consider new theories in this light.

End of the 'blind forces paradigm' in physics

In being face to face with the Planck threshold, physics does much more than reaching the ultimate limit of the matter universe: in fact it reaches the limit of its own materialistic paradigm. The Planck scale sets the smallest packet of energy existing in our universe—one quantum—and it was posited in 1900 by the genius physicist Max Planck, thus setting the stage for quantum physics. This smallest packet of energy is $6.6{\times}10^{-34}$ J·s (that is 0. and then 33 zeros after the dot before reaching the number 66—thus 66 million (6) billion (9) billion (9) billionth (9) of a Joule/second). This number, called Planck's constant (***h***), also sets the time of the emergence of energy particles at $5.3{\times}10^{-43}$ second—an infinitesimal fraction of a second after the instant zero. It also posits the size of the universe at this very instant of Planck. This size is extremely small and yet it is very far from zero; the universe is a sphere with a diameter the size of Planck length of exactly

1.616×10^{-33} centimeters (the 'length of Planck' is the distance covered by light during a Planck instant). At the Planck's scale, the universe is a sphere of pure energy, yet it has a mass of 22 micrograms—and given the infinitesimal size of the universe, this is an enormous mass about 10 billion of billion times heavier than that of a proton! More importantly, it has a frequency—which is at about 10^{43} hertz (Planck frequency, precisely $1.85 \times 10^{43}\ s^{-1}$), which means that the point-scale universe vibrates about 18 million billion billion billion billion times in one single second. As for the energy of the Planck sphere, it had been calculated to be 10^{19} GeV (giga or million electron-volts) and its force (Planck force) is 10^{40} tons. The Planck length and time mark the onset of space and time (3+1d) and of the first particles (viewed as strings in string theory). Then, an infinitesimal fraction of a second after Planck's threshold, at 10^{-36} second, happens the sudden explosive expansion of the universe called the *inflation phase* (the big-bang) in which its size will be multiplied 10^{50} times. Yet, the size of the egg-universe is so small, and the forces it houses so huge, that it stretches our imagination to try to figure its reality. And nevertheless, the Planck scale stands very far from the instant zero.

The four forces of the universe

The crucial point, for us, is that it's only at Planck's scale that the spacetime manifold appears, a split second before the Big Bang sets the explosive expansion of the baby universe. Temperatures are so high, to the order of millions of billion of degrees, that we can hardly fathom them. In a timely procession, the four major forces of the universe get into effect. Gravity before Planck's threshold, and then the weak force (radioactivity), the electromagnetic force, and finally the strong force (nuclear force). Gravity is believed to be the only force to exist before the 3+1d space and time are born, the only one coming into effect before the universe reaches Planck scale; its value is 10^{-39}. Then the weak force, radioactivity, has a value of 1/1million or 10^{-6}; the electromagnetic force is 1/137; and finally the strong force, which is the nuclear force, the glue of particles, has a value of 1.

Starting at Planck's scale, quantum mechanics, string theory, and Relativity theory allow us to model the blitz expansion of the universe, the emergence of the first energy particles—Higgs bosons within the field of Higgs, a quark-gluon plasma existing between 10^{-12} and 10^{-10}

second. The Higgs field is understood by most physicists as containing as many particles as anti-particles, and thus exhibiting supersymmetry.

At 10^{-4} second, the massless gluons exert a *strong force* on the *up* and *down* quarks (the first and lightest quarks) that then form protons and then neutrons, each having three quarks in a sort of unbreakable connection. The *up* and *down* quarks are among the six types of quarks including also *top, bottom, strange, and charm,* and their respective anti-particles with a negative charge. The way quarks aggregate by three is thus: an *up, up, down*, combination (u, u, d) gives a proton, while a *down, down, up* combination (d, d, u) gives a neutron. Thus protons, neutrons—up and down quarks—and electrons are all that is needed to form all the existing atoms. And then the remaining lonely quarks are annihilated. At 10^{2} seconds (about one and a half minute after instant zero) the radiation era ends and the matter era begins; all protons and neutrons of our actual universe have already been created. Starting at minute 3 until minute 20, protons and neutrons will combine via a process of nuclear fusion, thus forming the nuclei of atoms (a process called *nucleosynthesis*). Then the first simple atoms will form when electrons start bonding with these nuclei and orbiting them, mostly atoms of hydrogen (the lightest atoms, comprising one proton, and thus of atomic weight 1), deuterium, and helium-4. It is breathtaking to realize that the nucleosynthesis took only 17 minutes and that all hydrogen still existing at our present time in our whole universe has been formed within twenty minutes after the Big Bang. (Let's note, though, that the precision of these infinitesimal instants keep being accrued at nearly a year pace with new calculations and new measuring devices, and let's be aware that slight variations may exist between the different teams' results.)

At the same time as the *nucleosynthesis* phase, matter will start separating from radiation (a process called the *photons decoupling*). A first decoupling of the neutrinos happened within the first second. Then the photons, which up to now kept popping up from the field or ocean of Higgs, only to be smashed to pieces instantly reabsorbed by the plasma field, are suddenly able, with a colder plasma, to thrust themselves free from it and start on a cosmic journey. This wave of photons will light up the universe. It is its relic (or afterglow) image that we detect now at about 370,000 years after the Big Bang, and that forms the Cosmic Microwave Background (CMB). The photons, set on a course, will slowly get cooler (they are now at 2.7 degrees above the absolute zero) and their frequency will progressively decrease (while their wavelength increases) from the gamma rays of the origin, to the low microwave frequency they exhibit now. My view is that these photons, through their travel, are themselves creating space and time as they go along. (See Fig. 1.1)

Fig. 1.1. Cosmic Microwave Background (CMB) or relic radiation. (Credit: NASA/WMAP Science Team. Public domain. http://jca.umbc.edu/~george/images/cosmology/)

After this first shot of light, the baby universe enters the 'dark ages' phase where it will be totally obscure, until the first galaxies and stars are formed around 500 million years after the Big Bang; in November 2012, Adi Zitrin, a Hubble scholar at the California Institute of Technology Pasadena, detected a primordial grouping of stars about 420 million years after the Big Bang , called MACS0647-JD. (NASA news releases 12-397 & 14-283, on www.nasa.gov/press/2014).

Our galaxy the Milky Way is believed to have formed about 8.8 billion years ago, and given our sun is a late generation sun, our solar system has, in its formation, used matter particles and gases from previous solar systems.

In the process of the formation of galaxies, dark matter is an essential component believed to have speeded up the process by billions of years. Dark matter is impervious to the electromagnetic force and radiation: it doesn't absorb or emit photons and thus is not subjected to radiation pressure. Consequently, it may have coalesced much earlier than the luminous matter subjected to the electromagnetic (EM) force, and the decrease of its force as an inverse function of the square of the distance (called in short EM's 'inverse square law').

Here is the scenario scientists consider at the present the most likely: luminous matter is attracted to dark matter and coalesces into a star-like aggregate that then collapses into forming a black hole. The black hole rotates on itself and attracts matter in its flat accretion disk that, while doing so, becomes more and more massive and rotates progressively faster. From its gaseous disk, hydrogen and dust will, in their turn, coalesce into forming stars that will increase in mass over billion years and be subjected to the gravitational force of other stellar bodies in their galaxy.

Thus, from an infinitesimal sphere (the Planck sphere), the universe will, after a wild and explosive inflation phase happening at a fraction of its first second, slowly expand as a nearly flat disk: actually WMAP has shown it is not perfectly flat but has a slight curvature, and therefore it may be modeled mathematically as a sphere. Let's note, though, that General Relativity models spacetime as a Riemann's sphere (curvature of space) and has been proven correct in many predictions. However, quantum theory (both in the Standard Model and the superstring M-Theory) is set in a flat space and has also led to many correct predictions.

According to the Bogdanov, the evolution in time of this virtual sphere takes the form of the *light cone*. Hermann Minkowski, a great German mathematician, had been the professor of Einstein in Berlin. He became absorbed in the study of the newly born Special Relativity Theory and made a breakthrough in 1908 by modeling the light cone. (See Figure 1.2.)

Inside the light cone is our (3D+1D) spacetime; an event (or particle) situated at the center of the hourglass has a straight 'worldline' extending unto the future (top of figure) and extending also in the past (represented by the bottom cone). As a cosmological representation, like the one elaborated by Igor and Grichka Bogdanov, the universe is evolving from the tip of the cone—the point of origin, the singularity—toward the future at the open large end of the cone (our actual time 13.7 billions years later).

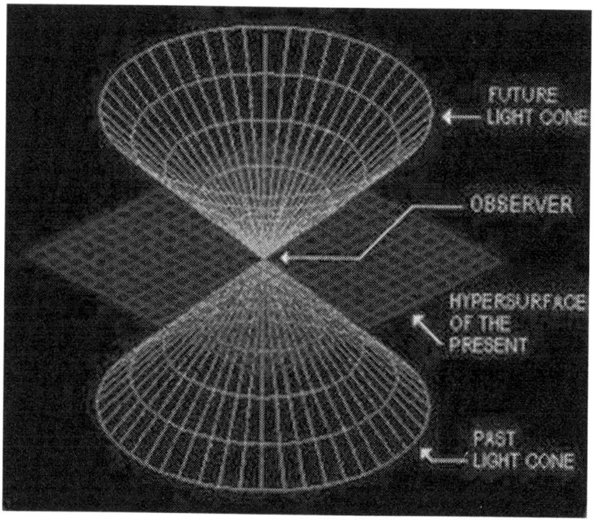

Fig. 1.2. Minkowski's Light Cone. Our whole spacetime stands within the future and the past cones, showing some of the possible straight paths (past-future) of the worldlines of events happening at point O (Observer). The 'Elsewhere' region is outside of the double cone. (Public domain.)

However, since the discovery in 1998 of the acceleration of the expansion of the universe, the representation shows the cone curved toward the outside since about 6.3 billion years ago, and also during the initial gigantic inflation phase; in between, it presents a huge lessening of the rate of expansion just after the afterglow light-pattern (the CMB) and then a regular expansion (near flat line) until this expansion rate accelerates again. The Bogdanov, using the universe timeline cone, call the point of origin "Zero point;" that's a bit confusing. We know that the "zero point" is the vacuum or so-called "false vacuum," referring to the fluctuation fields at the quantum scale (as opposed to the "real vacuum" that is the ground state of quantum systems). Let's remark then, for the sake of clarity, that the Point of Origin or X-Point I'm talking about is an immense time before Planck scale and the vacuum. This is why I felt the origin should have a distinct name to differentiate it from the Zero Point Fluctuations.

A rocky road toward recognizing the Big Bang

The point of origin (or X-point) is what is called in physics a singularity, just as the virtual center of black holes, and the maths of singularity (called the *Theorems of Singularity*) were conceived by two brilliant theorists, Stephen Hawking and Roger Penrose. In 1970, they demonstrated mathematically that there existed a singularity at the origin of the universe, something that the detection and mapping of the cosmic microwave background (CMB) had already confirmed. Yet, not only are there scientists (even if rare) who still refuse the reality of a point of origin, but the road from the hypothesis of a Big Bang up to its astronomical observation in 1964 (with the CMB) has been full of obstacles. (Let's note that we associate now the Big Bang with the inflation phase, and not with the much earlier point of origin.)

The first theorist to have deduced from calculations that there had been an origin and an evolution of the universe, was the genius mathematician Bernhard Riemann, professor in the famous University of Göttingen (where will happen the most advanced discoveries in mathematical physics of the times). Riemann's name is associated to his 1854 mathematical modeling of the curved space, the *Riemann sphere*. He developed with it the model of a spherical and finite universe. Yet, while Riemann's maths of the hypersphere (at 3 dimensions) paved the way for Einstein's curved spacetime, Riemann (contrary to Einstein's concept of an eternal and fixated universe) had predicted that the radius of this sphere had increased with time. (I will be inspired, here, by the excellent presentations of the evolution of research given by the Bogdanov (*Le visage de Dieu)*, by Greene (*The Elegant Universe)*, Kaku (*Hyperspace*) and Brandenburg (*Beyond Einstein*). As we will see, the breakthroughs kept shifting from astronomical observations to theorists' calculations, hypotheses, and models. Let's remember that Albert Einstein developed his theory of Relativity in two stages, Special Relativity in 1905 and General Relativity (GR) in 1916-17; as soon as 1905, with both the photon explanation of the photoelectric effect (for which he received the Nobel Prize in 1921) and his Special Relativity, he was a recognized genius, the first among physicists and a towering figure in the famous Solvay conferences, the one whose judgment on any matter of physics will be taken for gospel. But for Einstein and all scientists at that time, the universe was eternal, immobile, infinite, and of course it had had no beginning.

In 1912, the astronomer Vesto Slipher, in Arizona, made a stupendous observation, that a dozen nebulae were speeding away from our solar system at the incredible speed of one million and a half kilometers per hour (932,000 miles per hour)! But the presentation of his discovery at the American Society of Astronomy in 1914 was not understood and subsequently it fell into oblivion.

Then in 1917, a Dutch astronomer, Willem de Sitter, stated that the GR equations do not prohibit the possibility of an expanding universe and he modeled a dynamical universe, in constant movement. By doing so, he just angered at first Einstein who wrote in substance to him that to admit such possibilities seemed non-sensical. (A few years later, Einstein and De Sitter will sign seminal articles together.)

A breakthrough was made that same year 1917 by an eminent mathematician from the university of Saint-Petersburg, Alexander Friedmann, who deduced from the equations of the newly born General Relativity that the universe had had a beginning and was then a point. He was 34 years old at the time but he had been an early genius: while still in college and at age 17, he had already published an article in the prestigious *Annals of Mathematics* of Germany, the most advanced journal in the field of mathematics, directed by David Hilbert, an outstanding mathematician who did the maths of a curved space (called *Hilbert Space*). Friedmann wrote his seminal article in 1922 in the most important journal of physics—*Zeitschrift für Physik*—, and a year later he published a book called *The world as space and time*. To reach his conclusion, he had taken out of the equations of General Relativity a late addition introduced by Einstein in order to maintain the universe as fixated, eternal, and infinite—which was the cosmological constant.

In several mathematicians' eyes—including Friedmann's own thesis director the great Paul Ehrenfest, who was a friend of Einstein—this constant was a mistake (Einstein was a genius in physics, but not in mathematics). They checked the maths of Friedmann and had no other option than to recognize he was right. His conclusion was that billions of years ago the universe had been wholly condensed in a single point (without dimension or volume), and that the radius of this sphere had kept increasing. Thus was posited by Friedmann not only the beginning of the universe as a non-dimensional point, but simultaneously its subsequent expansion in time.

On reading the published article, Einstein, already crowned by the Nobel Prize since 1921, got into a furor; he started to crumple and trample the article and then wrote in anger to the journal, denying the validity of Friedmann's mathematical physics argument. After which, the father of Relativity Theory got four decades of respite, until the discovery of the relic radiation in 1964.

A few years later, the Belgium mathematician and priest Georges Lemaître got to the inescapable conclusion that the universe cannot be static. He developed the concept of a "primitive atom" and the model of a spherical universe in expansion and published his findings in 1927. However, Lemaître thought that this initial atom had fragmented in smaller and smaller pieces (while it's the complete opposite). As was to be expected, Lemaître failed to convince Einstein that the universe had had a beginning when he met him at the famous Solvay Conference of that same year (these international conferences brought together the greatest physicists and mathematicians of the time generally every three years in Brussels). And on top of it, he got a nasty remark from Einstein: "Your calculations are correct, but your physical insight is abominable."

The next move in the saga of the Big Bang came from the observations of the eminent astronomer Edwin Hubble, working in the best observatory of the time, that of Mount Wilson in California, and of his assistant Milton Humason, an extraordinary self-made astronomer. In 1924, the scientific world still believed there was only our galaxy, the Milky Way, in our universe. But in 1925, Hubble shatters the idea to pieces, announcing that his and Humason's observations showed the existence of millions of galaxies, possibly billions. The discovery becomes official when Hubble writes a first article on it in 1931; and in a second one published in 1949, Hubble and Humason posit the expansion of the universe and come up with what became known as *Hubble's Law* (but in fact posited earlier by Lemaître), that the velocity to which a galaxy recedes from us increases with its distance from the earth, as an effect of the universe's expansion. In fact, it is widely admitted now that it is space itself that is expanding—its metric—rather than galaxies speeding away as if moving within a fixated space, yet the law gives correct predictions. However this expansion of the fabric of space itself doesn't seem to account for the enormous velocity of a huge dark current stirring whole clusters of galaxies, called *dark flow*, and whose velocity is immensely greater than outside of its flow, notwithstanding the fact that it is at the same distance from earth. (We'll see it in detail in chapter 2.)

It's only in 1933 that Einstein will go visit Hubble on Mount Wilson's observatory and, being shown the evidence from observations, will accept that the universe had had a beginning, and therefore, that space and time were also born with it. Yet the establishment's recognition of this reality will come only in 1964.

Now the research saga shifts again to the theorists, with a student of Friedmann in St. Petersburg, George Gamow, who was able with his wife, also a physicist, to escape the USSR and settle in Boston, where he got a professor tenure at George Washington University. Gamow had the brilliant insight of an extremely hot phase at the beginning, but it is his own student in doctoral thesis, Ralph Alpher, the genius mathematician, who in 1948 conceived, modeled and did the maths of the nucleosynthesis (the formation of the nuclei of atoms) of all the hydrogen and helium in the universe in only the first five minutes of the initial point-sphere, that is, in the extremely dense and hot initial phase. In their book *The Face of God*, the Bogdanov brothers bring to light how Gamow appropriated shamelessly the results of his protégé Alpher and went as far as using his great clout to publish a near copy of Alpher's seminal article exposing his results *before* his student, in the same journal *Nature*.

Alpher's article, *Evolution of the Universe* (published on November 13, 1948) was moreover predicting the background relic radiation in astounding details: Alpher had understood that the photons had been glued to the primeval plasma and could only pull out from it violently when the temperature got down to 3000 degrees. That was essentially correct. Alpher had also calculated that a relic radiation should be detected about 300,000 years after the Big Bang, and the latest PLANCK estimate is 370,000 years, and the photons had undeniably separated from the field of Higgs violently. Alpher's comprehension was nothing less than prodigious!

Ralph Alpher raises a wave of enthusiasm when he passes his thesis in April 1948, just two weeks after the publication of the famous article on nucleosynthesis *The origin of chemical elements*, nicknamed 'Alpha, Beta Gamma' from the initials of the three authors: Alpher, Bethe, & Gamow, that reproduce the first three letters of the Greek alphabet. Hans Bethe was already famous before he got the Nobel, because he had been one of the great physicists of the Manhattan Project, geared toward building the bomb and headed by Robert Oppenheimer. Alpher and his all time colleague and friend Robert Herman were thus persuaded that the initial dense and hot phase and the subsequent radiation of photons had left a trace, a diffuse light now extended to the whole cosmos, that is, a diffuse light that should be detected in all directions from earth. They try to get a research project going but fail because the opposition to the idea of a point of origin is too powerful in the establishment science. It is led by Fred Hoyle, the renowned British astronomer from Cambridge who later was to become the president of the Royal Astronomical Society, and also by Arthur Eddington, another eminent astronomer, director of the Cambridge Observatory. Ironically, this opposition also had a great impact into the saga: the very name 'Big Bang' was spurted out derisively by Fred Hoyle during a BBC interview in 1949.

Despite Hubble's observations, Hoyle practically killed the very idea of an expanding universe for another 15 years. The indisputable discovery of 'the trace' was made by two young radio-astronomers, Arno Penzias and Robert Wilson, in 1964-65. While working for Bell company in New Jersey and intent on repairing an antenna disrupted by a strange and constant noise, Penzias and Wilson came to understand that they had detected the relic radiation from the Big Bang. The strange parasitic signal was perturbing the communication with one of Bell's satellite, and they kept hearing it whatever the direction they pointed the antenna to. The two radio-astronomers spent a year recording the signal and making precise measurements, in their attempt to find the source of the noise among all the possible radiations from earth or from the cosmos; but it fitted none of the known sources of noise, and furthermore, it was more clear in the wavelength of 7.5 centimeters and seemed to originate from a body about 3° above the absolute zero.

Meanwhile, the astrophysicists Jim Peebles and Robert Dicke of Princeton University were looking for a sort of relic radiation left over from when the universe was immensely more hot and dense than our sun, along the prediction of Alpher and Gamow, and they had just finished developing a sensitive microwave detector. That's when

Penzias and Wilson, who were just given the information, got in contact with them, and a team headed by Dicke came to the Bell's lab to meet them and discuss their data. Dicke recognized the microwave radiation they were searching: Penzias and Wilson had precisely registered the *cosmological background light*—what we call now the *cosmic microwave background* radiation (or CMB). (Arno Penzias and Robert Wilson got the Nobel prize in 1978 for their discovery.)

This relic radiation is the remnant of the photons being set free from the plasma and streaming freely through space. This 'decoupling' of the photons happened at 1.40 minute after the Big Bang according to some physicists, and much later according to others. In 2008, the analysis of the WMAP data led to the discovery that neutrinos had also decoupled from the plasma, this by the first second (the temperature was then 10 billion Kelvin) thus creating their own 'cosmic neutrino background.' About 370,000 years after the Big bang, where the cosmic microwave background lingers (the photon relic radiation), the universe was then a thousand times smaller, and yet its density was a billion times higher. Thus, in May 1965, the news shook the world that it was now proven that the universe had indeed originated from a Big Bang. And five years later Hawking and Penrose came up with the Theorems of Singularity, which posed the mathematical physics of the pre-Planck era, together with that of black holes.

A new phase of discoveries was ushered by the satellites and instruments conceived to observe this background radiation in the microwave wavelength. The first one, COBE (for Cosmic Background Explorer), launched in 1989, was conceived by George Smoot, John Mather, and Mike Hauser (at Lawrence Berkeley Laboratory and NASA's Goddard) and was put in an orbit at 900 kilometers (560 miles) from earth. It brought at first some information via an infra-red detector (Mather's FIRAS), that the relic radiation had a blackbody spectrum and was in a state of thermal equilibrium. But in fact, it is only very near to such a state, because infinitesimal variations (or anisotropies) in temperature were evidenced in spring 1992 by Smoot's microwave detector (called DMR, Differential Microwave Radiometer). This detector was able to detect variations to the order of ten millionth of a degree. On the stupendous photos these variations appeared as dots of different colors, the red for the hotter regions, the blue for the coolest ones. These anisotropies were called "wrinkles" by Smoot, in his book *Wrinkles in Time*. (See Fig. 1.1, p. 31)

The second observation satellite, WMAP, was conceived by Charles Bennett and launched in 2001. (The W in WMAP stands for Wilkinson, from one of the scientists who, together with Robert Dicke and Jim Peebles in Princeton, had predicted the relic radiation.) Says the Wikipedia article on WMAP: "The study found that 95% of the early universe is composed of dark matter and dark energy, the curvature of space is less than 0.4 percent of 'flat' and the universe emerged from the cosmic Dark Ages 'about 400 million years' after the Big Bang." This Dark Ages period, showing a black background starting after the tempest of photons, is due to the cosmic neutrino background. It means that "it took over half a billion years for the first stars to reionize the universe," that is, for light to shine again.[2]

With the third satellite, PLANCK, launched in 2009, the international research shifts to ESA (European Space Agency) in Toulouse, France, in collaboration with NASA; the satellite was conceived by Jean-Michel Lamarre and the project director is Jean-Loup Puget, both French scientists. PLANCK has a ten times higher sensitivity than WMAP. On March 21, 2013, the all-sky map of the CMB was released. It brought a greater precision to the earlier findings.

Staggering numbers

In the inflation scenario, inflation (the Big Bang itself) happened at 10^{-36} of the first second and after Planck threshold); the universe's size was one hundred million times smaller that today. Then the size of the universe increased 10^{50} times in an infinitesimal fraction of a second, at a speed estimated by some physicists (such as Alan Guth and Andrei Linde in their inflation theory) to be billion times that of light.

This scenario is actually being questioned in the light of the 2013 results by the LHC scientists who discovered the Higgs boson. At the time (10^{-10}s) the universe was a super hot plasma—the field of Higgs—the first particles (such as the Higgs bosons, the quarks, gluons, etc.) were glued together, and photons and neutrinos were trapped and crushed inside that field until the moment they could tear themselves from it. However, the Higgs field has a high-energy and a stable state, very improbable at this high-energy level, and this fact disagrees with a disordered and quasi-explosive inflation.

According to some physicists, neutrinos free themselves from the plasma (they decouple) during the first second, and the photons decouple at 10^2 seconds. After decoupling, both waves of particles launch their fantastic travel through space—thus *creating space* at the speed of light. When the Higgs field and Higgs bosons appear, the temperature of the point universe reaches one million of a billion degrees. That's about one billion times hotter than our sun which, at its peak temperature in the corona can reach around two million degrees. The sun's temperature is only between 3.700° et 7.800° in its chromosphere (the solar atmosphere), while its surface is at a mere 5.500° of mean temperature. Then, during the 13.79 billion years that spanned from the photons' ejection out of the primeval matter up to now, and while expanding with the universe, the density of the photons decreased, and they cooled down progressively. Their temperature was about 3000° at the time of the decoupling of photons, and reaches now 2.7 degrees above the absolute zero (the absolute zero stands at 0 degree Kelvin, that is, minus 270° Celsius). While cooling down, this light went through the whole spectrum of known EM waves, its frequency decreasing while its wavelength, conversely, kept becoming longer. It thus shifted from the highest frequencies of the gamma rays, to the X rays, to ultra violet, the visible light, and then infra-red, and it has now reached the low frequencies of the microwaves. This decrease of frequency below the visible light window is the reason why the background radiation is now invisible. To the contrary, when this first light was in the window of frequencies and wavelengths of the visible light, then for an observer standing on a virtual planet Earth in the night time zone, the background space surrounding stars would have been illuminated even harder than at daytime with our present sun. Igor and Grichka Bogdanov spell out numbers that are rather bewildering: In our actual universe, there are 10 billion times more photons than matter particles. The number of atoms in our universe is actually estimated to be between 10^{78} and 10^{82}, and the number of galaxies around 300 billion (some suggest 500 billion), each galaxy containing about 400 billion stars, so that the total number of stars could be around 10^{23}. And the photons from the first light make up 96% of all photons reaching us—that is about 400 Big Bang photons by cubic centimeter around us when we walk in the street! This creates the greatest part of the noise we get in between stations on a radio and on our late analog TV screens when there was no image. It is in fact everywhere around us, and even in us; we are bathing into it constantly. The remaining 4% of all photons in our actual environment are the light of stars, created by nuclear fusion in the core of an immense number of yellow stars like our sun.

The expansion of the universe has now been precisely calculated: it nowadays expands about the size of the Milky Way (our galaxy) every five seconds (but in our actual time only, because the expansion is itself in acceleration). So, let us indulge in some calculus: given that there are 86,400 seconds in one day, the universe expands the size of 17,280 Milky Way by day, and over *6.3 million Milky Way per year.* The numbers, as usual, are staggering! The most difficult fact to fathom for us is that this expansion creates space and time with itself. So we are left with an open nagging question: If spacetime expands as a quasi-cone, what is outside the cone? Remember the light cone: Minkowski may have found part of the answer (see Figure 1.2, p. 33). He postulated that in the Elsewhere (outside the cone) time becomes quasi-spatial and space becomes quasi-linear. (We will get back to this fantastic concept.)

Already, Henri Poincaré, twenty-five years older than Einstein (the latter born in 1879) but his contemporary, had come up with the mathematical concept of *imaginary numbers* (or negative numbers). The 'theory of groups' had been elaborated by the genius French mathematician Evariste Galois, who died at twenty years old in a sword duel, in 1832. Anticipating the lethal outcome of this somber sport he didn't mastered, the courageous Galois spent his last night putting down in writing his whole theory. Prolonging this theory of groups, Poincaré as soon as 1890, modeled the universe as a four dimensional spacetime, with three dimensions of space, and one time dimension in imaginary, negative, numbers. To represent it, he invented a *metric*, +++-, thus showing the 3D of space as positive and 1D of time as negative. The Bogdanov will hypothesize in the baby universe a time dimension fluctuating between positive and negative.

To summarize, the greatest discovery of Penzias and Wilson was that the relic of this first light (the CMB) proved that not only the universe had had a beginning but space and time too. Said Penzias: "To be coherent with ourselves, we have to admit that not only was there a creation of matter, but also a creation of space and time." (Cited by I. & G. Bogdanov, 2010, 114.) Then Penzias added that, to many researchers, the Big Bang appeared to be a "sudden creation from nothing"—and many scientists, in awe, will see the hand of God in this creation. These religious overtones created a tempest of their own within the scientific community, many scientists having rejected the idea of a Big Bang because of it, notably two of the greatest astronomers of the time, Fred Hoyle and Arthur Eddington.

In parenthesis, we have here some clear examples of how the greatest scientists and the ones who were the most essential in making science take giant leaps, may nevertheless be blind to some other types of discoveries, or simply to the genius advances of a younger generation. And thus that the condemnation of a potential breakthrough by a crowned figure of science is not to be taken as a sure sign that this novel hypothesis is worthless. Another example was that of Poincaré who, half a century before Einstein, was considered the foremost scientist of his time, having made seminal advances in fundamental mathematics and statistics and who was furthermore the pioneer of several sciences which were to flourish in the late twentieth century—such as chaos theory, and the concept of retrocausality. Also, he proposed the concept of an unconscious mind—that he called the *subconscious ego* ("moi subconscient")—intelligent enough to be able to discover new mathematical functions and to pass the information (i.e. the complete equations) to the conscious mind (See *Science and Method*, Chapter 3, published posthumously in 1913). Poincaré had several instances of such "illuminations" or "intuitions" as he called them, and he made an analysis of the process—way ahead of Carl Jung. And yet, despite being himself a genius precursor in several domains, he violently opposed the concept of a Big Bang and got into about the same fit of anger as Einstein. And since we are at it, let me express, from a long-standing experience, that it is a pitiful and sad sight to see a great white haired scientist to whom we owe a lot, displaying denial and skepticism toward the new advances of science. At least the innovative scientists are not burnt at the stake anymore but their lives and their creative spirit are sometimes smothered by depression after being violently rejected, as it happened to Friedmann. Or else they get so dispirited that they abandon their research, as did the three eminent researchers of the George Washington University,

Gamow, Alpher, and Herman, who had hypothesized the background radiation and had figured the physics of the very hot phase followed by a cooling down and then the decoupling of photons.

To conclude, recent advances made us aware that neither the Big Bang happening after Planck's threshold, nor this cut-off itself were instant zero; there had been another reality before the birth of spacetime, before the first quantum of energy. Some visionary scientists then got focused on this new frontier called by the physicist Jack Sarfatti "subquantum physics."

The progressive dematerialization of the physical reality

The discovery of a domain before and beyond space, time, and matter, is a very exciting development—and another leap in a series that showed a definite tendency toward the 'dematerialization' of physics.

The first such leap took place when the equivalence matter-energy was made, and we started figuring that any 'material' or 'physical' system (from galaxies to planets, to atoms and particles) can be viewed as energy, according to the famous Einstein equation $E=mC^2$ that equates energy (E) to the mass (m) multiplied by the speed of light (C) squared. But there are also massless (or near-massless) particles such as photons and some types of neutrinos that can reach very high levels of energy. The concept of energy is extremely complex and has been also expressed through quantum parameters such as frequency and momentum. So that, here, energy is now disconnected from mass.

A second leap in the 'dematerialization' of physics occurred when we understood that to any wave is associated a particle—or more precisely, that the particle is also a wave. This was achieved *after* the wave nature of light was duly proven in 1803 by Thomas Young's double-slit experiment. It happened in 1905, when Einstein discovered that the photoelectric effect couldn't be explained properly unless invoking a particle that would generate discrete jumps in energy levels. In order to solve the conundrum of the photoelectric effect, Einstein drew on Planck's energy formula that makes energy proportional to frequency. In this way, he viewed light as packets or 'quanta' of energy; thus physicists say that Einstein 'quantized' light: Einstein's "light quanta" were the name given to 'photons' until 1925. Then Louis-Victor de Broglie, in 1924, came up with his famous *de Broglie hypothesis* that not only light but all matter (all particles) also exhibit a wave behavior. Werner Heisenberg, when positing his *uncertainty principle* (early in 1927) posited a *wave-particle duality,* and stated that measuring accurately a specific property (such as the particle's position) entails a loss of accuracy in the complementary property (here the wave momentum or velocity) and he conceived ways to calculate precisely the loss of accuracy. That same year, Niels Bohr (the central figure of the 'orthodox' or 'Copenhagen interpretation' of QM), reacting on Heisenberg's finding, came up with his *Complementarity Principle*: a quantum system exhibits different properties depending on the type of experiment and measuring device, and these quantum properties go by complementary pairs (such as position-momentum). Thus, in order to get the most comprehensive information on a system, all quantum properties highlighted by diverse experiments need to be taken together. Whereas both Heisenberg and Bohr thought this wave-particle duality was due to the *observer effect,* physicists now tend to view it as an intrinsic nature of quantum systems. Now, matter itself could be translated in terms of waves, a definite leap toward dematerialization.

The third leap happened with the experimental evidence, and the confirmation of the discovery of the Higgs boson, on March 14 2013; the stupendous consequence being that mass is not an intrinsic property but just an effect of particles interacting with the Higgs field.

And now, the fourth leap is in the making: we have to envision a reality that precedes the apparition of both matter particles and energy particles and how these are issued from a field of information—a field that would contain all possible information. At this point, we are all in awe. Yet we meet a profound ontological problem: the fact that information in itself needs a mind or consciousness to read it, understand it and act on it. A system of information (such as a CD containing books) is a whole different matter than a mind reading the books saved on this CD. The CD becomes highly active information when a mind reads it, that is, when a mind makes sense of this information. This is to add to the fact that the CD needed a creator of this system of information. Indeed, even if a program can make operations (following internal laws) and then set machines to perform a task, nevertheless some mind had to create the machines and the program in the first place, as well as the laws that define their operations. Thus, it seems unavoidable that the universe in which thrive self-conscious intelligent beings cannot not have consciousness at its core and origin, that is, at the subquantum scale.

Some will argue that consciousness could be an emergent phenomenon springing out of ever increasing complexity. Far from denying self-organized and/or emergent processes, I deem them fundamental in nature, such as the well known tendency toward accrued complexity highlighted by Murray Gell-Mann and Stuart Kauffman. Chaos theory also has shown that most complex systems in nature are 'chaotic' and that they organize themselves internally: they are self-organized. This means that they are able to display novel global orders—thus an emergence of new organization. Of course, minds are the most complex systems and definitely display chaotic behavior (See the theories of Combs, Freeman, Goertzel, Hardy).

In my view, emergence and the rise in information and complexity are the key to a universal force—negentropy—counteracting disorder or entropy; and this negentropic force is also consciousness. Let's put it this way: consciousness couldn't emerge out of a fully determined universe highly constrained by eternal laws, nor could it emerge out of a totally indeterminate and random universe. Why that? Because consciousness is a process of creation of meaning, and this implies a radically different type of force rooted in beingness and an ability to think for oneself and self-reference. It is also a negentropic force, that is, an organizing force creating more and more information, as we see it in full fledge action in the exponential development of science and of cultures. Consciousness and the mind exhibit increasing complexity, just as many complex dynamical systems, and Gell-Mann showed that the increase in complexity was an essential process in nature.

Another line of reasoning is to view the universe as a complex hologram: any part of the universe-hologram will contain the information on the whole, whether in space or time. This was the perspective of physicist David Bohm, and the most important force he posited in this holographic universe was "active information," an organizing force based on meaning. This holographic or 'holonomic' framework was further developed by brain scientist Karl Pribram in the field of the brain sciences. Astonishingly, the Greek philosopher Plotinus, who lived in Alexandria in the third century CE, expressed clearly this conception of the universe as hologram, moreover organized by a central cosmic soul (*anima* in Latin, *psyche* in Greek). He states in the *Fourth Ennead*: "This universe (...) has in itself a *soul* (psyche), who pervades all its parts." And also: "The immaterial [the One] is as a whole in everything." (*Ennead* 6.4) In this hologram-type framework, any emergent process of organization at any point in time should have its root (or meta-force) at the very origin. (We'll get back to Plotinus in Chapter 3.)

Given that intelligence is a specific force, if this force is at work somewhere in the universe's hologram, then it is 'known' by all the facets in all times; then, at the minimum a sort of primeval or "proto-consciousness" has to be at work everywhere (this was first proposed by philosopher David Chalmers in the mid-nineties). The bottom line is, the universe can't disregard and ignore sentience, intelligence and consciousness as powerful triggers of evolution, if it knows about it. And in a holographic universe, it does know about it!

At the present time, several scientists seem to recognize in the *Zero Point Fluctuations* field (ZPF) or in the field of information at the origin, the foremost concept of God; and for several of them, the subscript to such a perspective is that of an omniscient and personalized Creator God. However this subscript is quite a leap! We have already shifted from a (possibly infinite) information field to a self-conscious intelligence (such as in Kauffman, Carr, Chalmers, Hardy), and then we effect a grand leap to a Creator as expressed in monotheistic religions. Furthermore, whatever the scientific and philosophical underpinnings of these theories—either suggesting a highly personal God able to have wills and to give orders, or an impersonal and holistic cosmic consciousness—we are confronted to the open questions of how exactly this immaterial consciousness, and/or divine mind would affect the evolution of our world, and also how this consciousness dimension interacts with matter.

In my view, we cannot just refer to God as if this multivariate concept was saying it all! We have to distinguish the ontological need for a global consciousness pervading the pluriverse, and a projected concept of God shaped by religions—the latter posing a totally distinct entity not only from the matter universe but also from us, humanity—the created. In psychological terms, the personalization of god could be viewed as a projection from our standpoint as intelligent self-conscious persons—the projection of our humanity (intelligence, goals, means of action) into an ideal personality concentrating all qualities to the utmost. This is also the projection of the archetype of the father and authority figure. This was the standpoint of Sigmund Freud, who, in *Moses and Monotheism*, saw in the concept of the monotheistic God the projection of the father figure—as a supreme authority—borne out of a human leader, Moses the Egyptian, who had had a real historical life. According to Freud, the murder of Moses by his people, in rebellion against his iron rule, was not consciously acknowledged and became repressed, an unconscious source of permanent collective guilt. Then this father figure—together with the repressed guilt—was internalized psychologically in the *superego*: in the Freudian system, it is the psychological instance in each person at the root of repression.

To summarize the ontological problem: we do have to account for consciousness in the universe (either pervasive and/or emergent), and for a fundamental force able to engender more order, information, and complexity, that is negentropy. This negentropic and organizing force has to exist not only in the matter universe (or pluriverse), but also at the very beginning, at the quantum and subquantum levels, and even before, in the X-point of the origin. Here is the line of reasoning: if there was only a field of information as a program, and a series of blind rules or forces driving the evolution of the universe, then consciousness would be altogether unnecessary, and why should it evolve out of matter, or exist in the first place? Thus, another way to consider the problem is that, given we exist as intelligent and self-conscious beings, therefore consciousness is an essential and distinct dimension of reality, and it has to be accounted for on its own grounds in any global theory for it to be complete. (We'll see the ontological argument in more detail in chapter 10.)

This notwithstanding, we don't necessarily need to project the figure of God the Creator, a concept bound to be culturally and regionally 'colored' (even within the same religion) and thus with fuzzy parameters. Indeed, there are alternative options and the one that I will pursue in this book is to posit a hyperdimension of consciousness, the *semantic dimension,* existing from the origin and pervading the pluriverse (the ensemble of all universe-bubbles like ours).

When scientists project a Creator God on the Big Bang, they introduce radically different dimensions of reality. If we consider Gödel's second theorem, an operative system of rules or axioms cannot contain internally the proof of its own consistency (or validity)—therefore it is necessarily incomplete. (Hence its name *incompleteness theorem*.) Consequently, any operative system needs a more global system, literally, a *meta*dimension, in order to ground its self-consistency (*meta* in Greek means *beyond*, or *more global*). Thus, we have to posit a meta- or hyper- dimension of global consciousness and sentience (awareness of one's own individuality and needs) at the very least. And we also have to add self-reference (the ability to think and ponder about oneself) to ground the intelligence and self-consciousness exhibited by evolved beings. Nevertheless, we shouldn't indulge in limiting this hyperdimension to fixated preconceptions and antiquated social constructs—such as for example 'only human beings are endowed with consciousness.' In fact, if we posit a hyperdimension of consciousness pervading the universe all the way back to the origin, we have then to figure that all systems, as constituent parts of the cosmos, share in this hyperdimension. That means that all complex self-organized systems, from stellar bodies to living beings, and even to particles, are open on this hyperdimension. We move then toward the view of a field of primeval consciousness—which I call *semantic field*—existing in all systems, including particles. A semantic field is both consciousness in action, as a process, and the memory of a sentient or conscious being (its organization and data). Then we have a more or less complex semantic field and consciousness depending on the complexity of the system, from proto-consciousness in complex systems to self-referent consciousness in human beings.

Up to now, physics has not been able to account directly for consciousness (as dimension, force, or dynamics) in any process or system in the universe. Moreover, nothing in the emerging physics of the zero point and subquantum processes allows us to infer and extrapolate more than: near-infinite energy, field of information, and supersymmetry (a unified substrate at the origin). For now in physics, we don't have a sentient life nor a conscious one, how much less an intelligent, free-will entity, gifted with self-reflection, learning, imagination, and creativity—such as a human being!

Integrating consciousness in physics

However, there are two noteworthy attempts at integrating consciousness into quantum scale physics, both more or less indirect. The first one is by Heisenberg and the 'orthodox' Copenhagen school of physics, who stated that the 'observation' (the observer, the consciousness) introduces a modification in the organization of the quantum systems being measured or observed. Several physicists have argued for the involvement of consciousness in quantum processes, notably Wigner, von Neumann, Stapp, Josephson, and Sarfatti.

However, how consciousness modifies the observed quantum system was not itself integrated within the fundamental equation of quantum physics, the Schrödinger equation: it was only inferred from it. This equation describes superposed quantum waves (or states, constituting the wavefunction's state-space) and their probabilistic collapse into a particle's momentum or position in spacetime coordinates. All the superposed states were 'local' that is, bound to spacetime coordinates. Then David Bohm proposed his Quantum Potential (Q) and revised the equation. According to Basil Hiley, Jack Sarfatti, and Massimo Teodorani, this quantum potential introduces a **nonlocal** term in Schrödinger's equation. The quantum potential is a process by which the state of a quantum system could be influenced or piloted via a "pilot wave" by another distant quantum event, in a nonlocal way. This is what we observe with the entanglement of particles and their still mysterious exchange of information at great distance that Sarfatti keenly calls "signal nonlocality," which we know cannot involve a classical transmission through spacetime.

Furthermore, with the quantum potential, minds at a distance in space or time could now bring changes in the superposed wavefunction, and the collapse doesn't need to occur but is not forbidden altogether: the pilot wave acts as a causal force guiding the state of the system (see Sarfatti 2006, 82-6). (We will get back to it.)

2
VOID AND EXOTIC ENERGY

"Somewhere in the deep subatomic scale, physics changes."
John Brandenburg (Beyond Einstein: GEM Unification, 119)

The more our minds ponder on a reality near the origin that would exist beyond space, time, matter and causality, and the more we face the unavoidable realization that our universe reveals itself as *hyperdimensional*, that is, having more dimensions than just the matter and matter-energy ones. The point is: if, at the origin, it is NOT matter, what else can it be other than information?

A field of information at the origin?

We still have great difficulties to figure an ocean of energy that could have existed before matter and spacetime came into being. The fact is, our concept of energy, in the equations, is still indissolubly linked to various spacetime and physical parameters such as mass, as in the famous Einstein equation $E=mc^2$—where Energy (E) is equal to mass (m) multiplied by the square of the light speed constant (C). And yet, the original energy of the pre-space or sub-Planckian scale—an energy that pervades the "elsewhere" in the words of Teodorani—cannot be formalized using matter and spacetime parameters; it wouldn't do justice to the actual organization processes involved in the pre-space. Indeed, being beyond spacetime, the pre-space should be formalized using its own inherent parameters and dimensions. And this is where laying hypothetical dimensions of the pre-space, such as this ISS theory, is crucial. Of course, if we consider this original information to be of the digital type described by Shannon's Theory of Information, and quantified as bits (as 0 and 1), then what we have is inactive and dead information, one that needs a 'thinker' to not only create it, but also understand it and act with it. At that point we get into what is called in philosophy an 'infinite regress': in order for the world of matter to unfold according to the information or program set at the beginning, we need both a creator of the program and a force to set it into unfolding and performing its commands. Furthermore, this world would be totally predetermined.

An alternative is active or alive information, for example a field of self-organized and complex alive information—in short, consciousness. Several physicists and scientists have hypothesized a field of information at the origin; let's see the main models.

Wolfgang Pauli, Nobel laureate in Physics, while working with psychologist Carl Jung, posited a deep layer of reality underlying the quantum domain, called *deep reality.* In this sub-quantum domain, psyche and matter/energy are so blended as to form a unique 'stuff.'

Moreover, the 'deep reality' is the domain of "acausal" processes such as synchronicities—links and connections between events at great distance based on meaning and independent of space and time. Pauli and Jung started writing a common book on synchronicity, in which the theoretical quantum physicist Pauli showed the symbolic framework underlying Kepler's seminal work in astronomy, and the psychologist Jung stressed the novel physics principle (the acausal synchronicities) of such a deep layer of reality. Namely, Carl Jung clearly pointed to acausal and instantaneous connections based on meaning and consciousness, as well as the 'trans-spatial' and 'trans-temporal' qualities of such connectivity, what we call now *nonlocality*. Both scientists predicted that the scientific understanding of this novel domain would usher a radical and global reformulation of our concepts of spacetime and matter, and similarly of our concepts of psyche and consciousness. When Jung and Pauli looked eagerly for mind-matter blended domains, or a "common language," they came up with (1) numbers as archetypes, and (2) acausal phenomena or synchronicities, or meaningful coincidences. To that Wolfgang Pauli added the spin complementarity and the beta decay.

Physicist David Bohm started the elaboration of a group of theories called Pilot-wave or Bohmian theories, that is actually getting more and more physicists working within its framework, and has two central tenets. The first one poses *nonlocality* as inherent to the quantum interactions domain. David Bohm thus added a nonlocal term to the basic QM Psi wavefunction (Schrödinger equation), inserting a 'pilot wave' that introduces a guiding action at a distance (or Quantum Potential), thus a nonlocal term. Astonishingly, the Psi wavefunction, in Bohm's equation, represents the superposed states not only of the local wavefunction, but of the universe itself, acting as nonlocal guiding potential from the underlying causal substrate or 'implicate order.' Thus the universe, as a hologram, has a guiding influence on all quantum waves and systems. The second tenet of Bohmian physics is *active information* (or meaning-as-process) as being for Bohm the prominent ingredient in the implicate order and in the working of the Quantum Potential. The latter, while still mysterious, being the pilot wave theoretically able to orient and guide the quantum processes, as a 'nonlocal hidden variable,' thus skewing the quantum indeterminacy toward specific states or events. For Bohm, active information being a guiding force and a 'pilot wave' (along the lines developed by French physicist Louis de Broglie), can be thus viewed as an organizing process at work. The only problem posed, in my view, by the Pilot Wave framework, is the strong deterministic and 'one-way' process it posits, because all causal chains affecting the 4D world are determined by the implicate order; and thus it doesn't account for the creative input of individual minds and complex systems (at least in the framework of Bohm and his main collaborator David Hiley). Among the theoretical physicists actually working within the Pilot Wave Theory framework, beyond David Hiley, are Jean-Pierre Vigier, Jack Sarfatti, Massimo Teodorani. (We'll discuss these models in chapter 6). Let's note for clarity that the Bohm-revised Schrödinger equation, as it represents the whole universe with its innumerable possible states being superposed, is not found by physicists to be usable for specific solutions, unless they use the state-space as representing the ensemble of states of a given set or system (instead of that of the universe), and thus bringing the wavefunction to be constrained to this system. The state-space of a wavefunction (the set of superposed states) is normally viewed as 'quantum coherence' and since the quantum system is bound to interact with other quantum waves and systems in the environment, these interferences soon bring a *decoherence*, the disaggregation of the first wavefunction.

Building on David Bohm's Pilot wave theory, Jack Sarfatti posits "*an information-rich giant quantum coherence field* [...] immune to environmental decoherence." This signifies that the information carried by the quantum potential acting on a system's wavefunction will not be lost through the interactions with the environment usually triggering the decoherence. Sarfatti is a strong advocate of what he calls "*signal nonlocality,*" that is, the exchange of information between two particles in an EPR-type experiment. (Sarfatti, 2006, 167) Sarfatti furthers Andrei Sakharov ideas and the "More is different" concept of Philip W. Anderson's (in his famed 1972 *Science* article), and points to levels of emergence and phase transitions in the universe, the vacuum being the locus of such a phase transition.

John Wheeler has also postulated a field of active information able to organize the universe, with its "*it from bit*" principle, in which 'it' stands for any matter system (particle or even spacetime as a whole), and 'bit' for its information: "*It from bit* symbolizes the idea that every item of the physical world has at bottom—*a very deep bottom*, in most instances—an *immaterial source* and explanation; (...) [It means,] in short, that *all things physical are information-theoretic in origin* and that this is a *participatory universe*." (Wheeler, 1990) Wheeler was fathoming this information as being a digital encoding (in bits), but his concept definitely postulated *a layer of information and "meaning"* which, just like Pauli with his "deep reality," he viewed as "a very deep bottom" and "immaterial."

Wheeler was the one to define the strong 'Participatory anthropic principle' (PAP) and he hypothesized that the observers from a very distant future could be the consciousnesses able to in-form the baby universe toward a life-friendly set of variables. Likewise, we in the present, "We are participators in bringing into being not only the near and here but the far away and long ago."

Sarfatti's model, following Wheeler, proposes that, by a feedback loop, we could in-form the primeval universe; he modifies Wheeler's formula to "*it for bit*" and makes it reversible. He views the 'it' as an information system particle-like and point-like, "rolling on the landscape of the BIT pilot wave." Now that's interesting in the sense that Sarfatti (always expect from him a wholly novel and original perspective) translates the universe's wave-function of Bohm, into an information-laden landscape, self-conscious and "mental."

In Sarfatti's view, the universe is not only self-creative and self-organizing, but it also instantiates a two-way influence (as opposed to a unique top-down causal creator agent), meaning that individual consciousnesses may have an influence on the whole, just as postulated in this ISS theory. Says Sarfatti: "Post-quantum theory, *with inner consciousness,* I posit, is when the relation between IT particle system point and the *intrinsically mental BIT* pilot wave landscape is 'two-way' in a self-creative adaptative spontaneously self-organizing feedback-control loop." (2006, 97)

Sarfatti addresses the existence of psi as a fact and a reality that clearly overrides indeterminacy: "Human consciousness represents a strong violation of orthodox quantum theory if one believes the data on remote viewing for example." (94) John Wheeler's and Sarfatti's ideas are that of a conscious and self-observing universe, who would be able to send the information backward to the origin and have an input in the self-creation of our universe. (98)

Thus Sarfatti is a vocal advocate of an inner consciousness universe, with an "information-rich" field that would be the vacuum itself, or strongly linked to it; his theory proposes a process of phase transition from a subquantum domain toward the 'macro-quantum' domain of our 4D world. Emergences (e.g. of matter) would be the effect of a hidden symmetry, as was expounded by Philip W. Anderson in 1972 (Sarfatti, 2006, 197-8). Furthermore, says he: "It looks as though my 'back-action' theory of matter on its pilot quantum wave, which generates consciousness, and my physics/consciousness model predicts *VALIS* in the far future of an open universe, which continues to expand forever. My superluminal theories and cosmology are compatible with Penrose's recently published works."

As VALIS stands for Vast Active Living Intelligence System, we are definitely very near to the ISS core concept of a 'cosmic semantic field' (pervading all times and space of a universe bubble at least). This cosmic syg-field is postulated as a self-organizing collective consciousness in which all individualized or group minds and all complex systems do have an influence on the whole. Nevertheless, I don't endorse Sarfatti's hypothesis of an endless expansion and moreover the ISS theory differs in that this cosmic collective consciousness, as a hyperdimension, exists at all times.

Massimo Teodorani envisions also a field of information at the scale of the universe, that would be linked to the vacuum: "The *energy* that mysteriously springs from the void is the manifestation of a latent world, hyperdimensional and atemporal, *where dwells the consciousness of the universe*." (*Bohm*, 6.1; my translation and emphasis.)

Igor and Grichka Bogdanov pose a field of information before the Big Bang, associated with an imaginary time, or to be more precise, a time dimension oscillating between a positive and a negative value, integrated in an otherwise normal metric with three positive dimensions of space. They propose in *La fin du Hasard* ("The End of Randomness") a scalar field consisting of "a cloud of numbers; an ensemble of numbers," existing before the Big Bang and that would "dilate" spacetime. They write "At instant zero, there is nothing else apart from information. Something purely numerical, yet that 'encodes' with a vertiginous precision all the properties of the universe that are bound to emerge after the Big Bang." (my translation) The way the universe would unfold from this field of information is seen as deterministic, a one-way strict causality.

The systems theorist *Ervin Laszlo*, in his inspiring work *Science and the Akashic Field*, proposes a cosmic field of "in-formation" implying the vacuum, both active information and having an influence on matter systems. Laszlo draws some of the possible interactions (between this A field, the matter universe, and our consciousnesses) from recent discoveries on the ZPF, as well as torsion and resonance waves, but he does not specify the global architecture nor dynamics between these and the A field itself.

In SFT, *semantic (syg) energy* is a negentropic force in action in the universe, allowing each system to self-organize as an entity (with its own individuality) and thus to increase its own complexity. For self-conscious minds, it allows the creation of meaningful interactions with others and the world for a specific subject or observer—an 'I' who instantiates a 1st person's perspective. The creation, self-organization and in-formation of semantic fields is instantiated by a connective dynamics. Similarly, at the collective level, a people is co-creating a meaningful culture and all minds are co-creating the *planetary semantic field* of which each of us is an intelligent and relatively autonomous facet. All living beings and all matter systems have an extra semantic dimension, and the interaction between semantic fields (of all types) is always two-way, each field both influencing and being influenced.

The relative force (and thus influence) of the syg energy is stemming from a set of parameters such as intensity, coherence, recurrence, etc. The key parameter of influence is not the hierarchical level of a semantic field but its syg energy intensity. One member of a community can transform the whole community.

In the ISS framework, this extra semantic dimension becomes the Syg Metadim, part of the triune 5th dimension, curled up in all systems at all scales (thus forming their 'semantic field' (or syg field) and similarly setting a two-way exchange and influence. Top-down processes are postulated to act from the origin toward the actual 4D systems (including minds). Yet they interact with bottom-up processes that set an input from individual minds and systems into the cosmic ISS at the origin, via their personal compact and curled up 5^{th}D, the individual ISS. The universe as a whole is self-organizing and instantiates a two-way influence between systems and the Center-Syg-Rhythm hyperdimension.

Some scientists have invoked God as being the (one-way) causal agency at the instant of creation. Bernard Haisch—an astrophysicist and eminent researcher on the quantum void—has proposed a "*God Theory*" (the title of his 2006 book), saying in substance that the zero-point field (or quantum vacuum) could be understood as God. This was the conclusion to his presentation at a 1998 European conference of the Society for Scientific Exploration in Valencia, where I myself presented a paper on my Semantic Fields Theory. Still filled at the time with the orthodox interpretation of QM, that stated the indeterminacy of quantum events, I asked him at a get-together among presenters after dinner how the Zero Point random fluctuations could ever be related to the paragon of all order. In my view at the time, only a totally different type of energy—such as syg energy—could be an organizing and negentropic force, and it had nothing to do with quantum mechanics, it was a totally different layer of reality. This interaction happened years before my great breakthrough of November 2012 in which I fathomed the model of the ISS driven by the numbers Pi and Phi that I'm presenting in this book. Nevertheless, I still refuse to link the ISS or syg-energy to a personalized, one-way causal, and especially eternally self-similar, deity. A creator God can in no way explain the free consciousness of individuals, unless the deity would share his consciousness capacities with the created—in which case we are back to square one: what in the world is this consciousness stuff that can be shared? It has to be collective!

George Smoot, in his book *Wrinkles in Time* (p. 296) came in his conclusion to evoke an underlying unity and order of the universe, yet allowing complexification, and a Cosmic DNA: "To me the universe seems quite the opposite of pointless. It seems that the more we learn, the more we see how it all fits together—how there is an *underlying unity* to the sea of matter and stars and galaxies that surround us. [...] Nature is as it is not because it is the chance consequence of a random series of meaningless events; quite the opposite. More and more, the universe appears to be as it is because it *must* be that way; its evolution was written in its beginnings—in its cosmic DNA, if you will. There is a clear order to the evolution of the universe, moving from simplicity and symmetry to greater complexity and structure." The Bogdanov brothers remarked about it (in *Le visage de Dieu*, p. 24): "If the cosmic DNA of which Smoot talks in his book does exist, then it will have undoubtedly to be searched for at the origin (...), well before the Big Bang." (my translation).

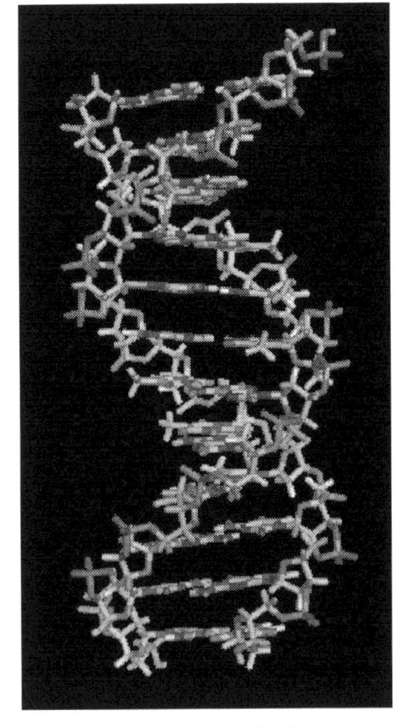

Fig. 2.1. DNA double helix with pairs of bases. (Creative Commons. Zephyris at English language Wikipedia.)

Yet, ontologically, we cannot view this cosmic DNA as a sort of biological 'program'—as we tend nowadays to view the DNA in the living—because then we will meet the same infinite regress, back to the first cause—the intelligence who wrote the program in the first place. I propose to view it instead as the state-space of the complex configuration of the ISS three enmeshed metadims. And in an older universe like ours, the cosmic DNA shows the snapshot of the organizational state of the whole universe (in our 4D matter universe), and nevertheless, this state-space contains also the information of all its possible states in all times.

Zero Point Fluctuations Field (ZPF)

The classical physics definition of the void or vacuum state, is the lowest possible energy or ground state of a system (such as a particle). Far from being empty (as was thought early on), the quantum void is filled with virtual and turbulent waves and particles of a fleeting and transient nature. But the vacuum state can also be assessed in the universe as the 'cosmological vacuum,' whose energy would be the cosmological constant introduced by Einstein (although this is still an object of debate).

Not only do we know now that the zero point field has an immense energy, but scientists have been able to calculate it: its density of energy is close to 10^{108} J/cm^3. That means ten followed by 108 zeros Joules for only one cubic centimeter! According to Massimo Teodorani, this density means that each cubic centimeter of the quantum void has already more energy than all known matter in our universe! This unfathomable quantity of energy, says Teodorani "shows that the world of matter in which we live is but a transitory crystallization in a tridimensional world, of *an energy that comes from elsewhere, an energy that governs the entire universe* and life in it." (*Bohm*, 3.4, my emphasis.) So is not this "elsewhere" precisely the pre-space? Then he cites Bohm saying about it: "The universe is not separated from this *cosmic ocean of energy*, it is a ripple at its surface, a sort of 'excitation zone' in the midst of an incomparably vaster ocean." (My translation and my emphasis.)

Einstein was the one to initiate the idea of a turbulent vacuum filled with EM fields, and he corrected the previous view of Planck that at the absolute zero temperature, there would be no EM fields. Einstein posed instead that the zero point state of the vacuum has to be full of energy, that no absolute void existed at the cosmological level. As the quantum field researchers understand it now, the subatomic, Planck-scale, vacuum has to be in a state of permanent quantum fluctuations (derived from Heisenberg's indeterminacy), hence its name Zero Point Fluctuations. In fact, the closer the ZPF is observed, and the greater the fluctuations, so that the vacuum is like a wild sea of random fluctuations, superdense and quasi liquid, full of transient particles and possibly EM fields.

The energy of the vacuum was put in evidence through the *Casimir effect*. Two plates of metal are set parallel and very close to each other. On the outside of the two plates, the vacuum fluctuations are hitting the plates, but the tiny width of the cavity between the plates will impede all greater wavelengths from the void to penetrate it. And thus, the void inside the cavity will be less dense than the vacuum's density pressure outside. The 'Casimir effect' makes that the two plates get nearer to each other.

Another experiment that showed interactions between particles (such as the electron) and the vacuum was that of the *Lamb Shift* (after Willis Lamb, who got the Nobel prize for this discovery). Lamb used hydrogen atoms and observed that when electrons were leaping to higher energy orbits, it created a tiny shift, different for each orbital, which was due to the interaction with the vacuum. Thus electrons and charged particles were constantly interacting with the vacuum, proving definitely that it was polarizable (as Sakharov and Puthoff postulate it): the vacuum had a texture of charged particles either bearing or responding to EM fields.

Dirac sea: the vacuum as a sea of negative energy

The British physicist Paul Dirac brought a new understanding of the vacuum as an infinite energy, as we conceive it now. He predicted the existence of antimatter back in 1928, with his famous Dirac Equation, that was the first bridge between quantum physics and relativity theory. With it, he postulated that the electron had an anti-particle, positively charged, that was later to be called positron when it was experimentally evidenced by Carl Anderson in 1932. But Dirac went much further with the concept of antimatter, stating that for any particle of any charge, there was an antiparticle of opposite charge. Says the Wikipedia article (Dirac_sea): "for each quantum state possessing a positive energy *E*, there is a corresponding state with energy -*E*."

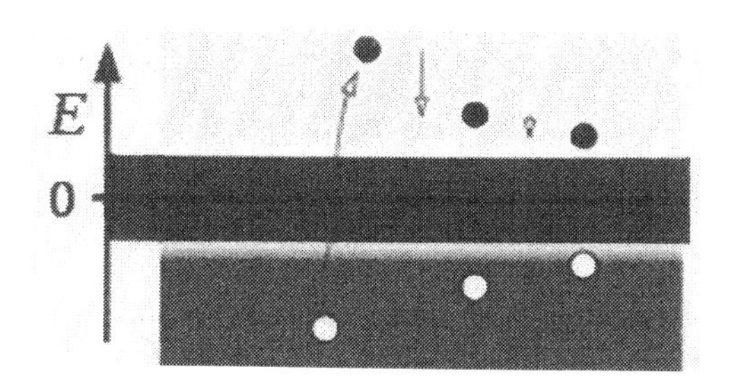

Fig. 2.2. Dirac sea with particles and antiparticles. Massive particles as black dots on pale grey, and antiparticles as white dots on black. Artwork by Incnis Mrsi, public domain, for http://en.wikipedia.org/wiki/Dirac_sea

Then, in the thirties, Dirac postulated that *the vacuum is an infinite sea of negative energy*—now called Dirac sea—in which the electrons and all anti-particles would be like 'holes.' (Figure 2.2) The only apparently unresolved problem is that a positron within an EM field would keep on emitting photons. Dirac equation had many remarkable solutions and predictions, such as the instant electron-positron annihilation, whenever they meet, that will be hypothesized in Dirac's and others' cosmological theories to set a (still debated) process of particles-antiparticles annihilation at the very beginning, yet letting unexplained the surplus of matter that could give rise to our universe. Dirac also found out that the famous Einstein equation $E = mc^2$ (relating energy, mass and momentum in special relativity) had a negative solution, namely $E= - mc^2$. Then "Dirac hypothesized that what we think of as the 'vacuum' is actually the state in which *all* the negative-energy states are filled, and none of the positive-energy states." (cf. Wikipedia, Dirac_sea)

In other words, we can make a link with the ISST posing two global states of the particles—either CSR-focus or QST-focus. In Center-Syg-Rhythm focus, it is connected and interacting two-way with the

CSR hyperdimension and the cosmic ISS of the origin. In the Quantum-Space-Time focus, the particle is connected and interacting two-way with the 4D spacetime manifold and quantum scale vacuum. If we have a state-space with two states for each particle, negative or positive, both having the same density of energy but of opposite charge, then there is a possibility that, within a delimited state-space (such as a system), all CSR-focus (negative) energy-states be active at the same time, thus pulling a patch of vacuum into coherent ultra-high vibration. If we translate it at the level of organization of the semantic constellations forming the semantic field of a person, we could envision a state of meditation (a specific constellation 'meditation state') taking the whole mental landscape, or state-space of the moment. ISST postulates that the *sygons*, ejected at the origin from the ISS, are the virtual particles of syg-energy that constitute all syg fields. In this 'meditation state' all the sygons of this constellation are in a fusion with the Self and the Whole, the cosmic ISS. Thus in a CSR-focus, all sygons of this meditation constellation (or altered state) would be *in a CSR, negative energy, state* in sync with the Center-Syg-Rhythm hyperdimension. The person is, at that moment, experiencing a high samadhi state, or the state of consciousness precisely called "the void state" in Taoism and Zen Buddhism.

In 1967, Andrei Sakharov (the father of the USSR bomb who later played a pivotal role as an activist for peace) conceived that the universe began with hydrogen, and before that was the vacuum (devoid of real particles and atoms). And it meant that *the laws of physics had to be different at the moment of creation from the laws in our 4D universe*. If the energy of the vacuum was so gigantic, how is it that we, surrounded by it, were not disintegrated by it, like all matter? He understood that this energy is also inside our bodies, *inside all the atoms and particles* making up our bodies. For Sakharov, the matter-particles (fermions) create a polarization effect in the vacuum that would in-form the gravitational field; gravity itself being akin to a Casimir force. Sakharov's equation links gravity (G) to a specific "Planck cut-off frequency" of the zero-point-fluctuation (ZPF) of the vacuum, that his friend and colleague Yakov Zel'dovich had posed. Zel'dovich, furthering Sakharov ideas, had come up with a way to dispel the bothersome infinity via the idea of a "cut-off" or threshold at the Planck length. At that radius, particles and antiparticles would appear, formed out of the void as *subatomic black holes* of opposite charge that would pop up and instantly annihilate with their anti-particles (along Dirac's equation). They would annihilate each other because these infinitesimal black holes would tunnel out of their own event-horizon that had the same Planck wavelength as themselves.

Sakharov, now building on Zel'dovich's cut-off threshold concept, posed that gravity was a force derived from the fluctuations of the ZPF, a force that had a global direction (a Poynting force, as the one showing the direction of an electromagnetic field). The ZPF was the energy that would create the nucleosynthesis of hydrogen and other atoms, thus forming the matter universe.

However, thought Sakharov, since we are in a matter universe, obviously matter had the last word in this original process of annihilation between matter and antimatter—this is called the baryonic (e.g. matter) problem, or 'baryogenesis problem': how did matter succeed in overcoming antimatter? Says John Brandenburg (*Beyond Einstein*, 167): "Even after Zel'dovich first marvelous scheme, the vacuum remained heavier than the cosmos [itself weighs] in every square centimeter. This has been called the 'most embarrassing number in physics.'" Indeed, the article "Vacuum State" on Wikipedia states that "In physical cosmology, the energy of the cosmological vacuum appears as the cosmological constant. In fact, the energy of a cubic centimeter of empty space has been calculated figuratively to be one trillionth of an erg." (Quoting Sean Carroll, from Caltech, 2006). Of course, we have to add to this 'weight' or density (mass or radiation) per cm^3 of the whole visible matter universe, the density of dark matter (about five times heavier than the visible matter), but still the problem persists. Thus, according to Sakharov, the hydrogen was first formed by drawing energy from the vacuum, and gravity was the ripples of the EM fields pervading the ZPF.

The next leap was made by Steven Hawking, in 1970, when he hypothesized black holes at a subatomic scale. These "micro black holes" were "pure gravity" and spacetime curvature, and they would decay into gamma rays (EM high-energy waves). In other words, the energy of pure gravity of a black hole (virtual massless gravitons), via the process called Hawking Decay, would be transformed into pure thermal electromagnetic energy.

The process implied that a quantum plasma formed at the surface of these subatomic black holes—a cloud of particles and antiparticles that, at Planck length, would tunnel out while the black hole itself disappeared. At that time, Hawking viewed the plasma-filled surface as high entropy, and its temperature as gravity. In recent models, the event-horizon of a black hole is viewed as containing information (cf. Brian Greene, Susskind). Gerard 't Hooft, in collaboration with his thesis professor Martinus Veltman at Utrecht University, was then able to find the theory and equation accounting for the decay of the pseudo-particles *instantons* into thermal EM fields—his thesis was on the neutral pions decaying into photons, a decay process that had been duly observed but was still forbidden by formal arguments. The instantons are deeply linked to topology and the fabric of spacetime, in the sense that, according to David Hilbert and Einstein, spacetime has in itself energy and will store it.

For 't Hooft and Martinus, *the instanton is a self-duality* (and idem for its anti-particle) *that expresses a connection.* It has been used for example to calculate how a real particle could tunnel through a potential barrier, such as a region of potential energy higher than its own energy. It uses Hilbert's Action Principle positing that particles and nature always follow the least action or shortest path through spacetime (Hilbert had been the eminent senior mathematician who helped Einstein with the maths of General relativity.)

> In the individual ISS, as in the cosmic one, each step is a frequency determined by the 'bow' or quarter circle. These bow-frequencies are 'sygons,' virtual or pre strings/particles that vibrate as networks within each particle's ISS (that is, in its 5th curled-up dimension). The two states of the particle, either in CSR-focus or in QST-focus, are the self-duality (of the instanton posed by 't Hooft), and it also expresses a "connection" insofar as it bears the syg-information of the network system. This connection links the particle or system either to the 4D manifold or to the CSR manifold.

Sarfatti: the vacuum as phase transition

Sarfatti develops a "unified vacuum energy interpretation" of both dark energy and dark matter, starting from the concept of a Dirac sea (of negative energy) containing the antiparticles of the 4D particles (of our world), and separated from this 4D manifold by a surface he calls "a foamy Fermi surface" that is "a neutral ionized plasma." The vacuum is both sides of the surface and forms the surface itself (as a two-dimensional boundary) and it is the scale or 'locus' of phase transitions from the flat quantum vacuum toward the curved space of 4D (relativity theory) world. (Sarfatti, 2006, 150-2)

Jack Sarfatti posits that the positron (the antiparticle inside Dirac sea) forms a "bound particle-antiparticle pair" with an electron (in 4D manifold). More interesting, and fitting my own ISS model, the positron would pop out of the Fermi surface and leave a hole into it, that nevertheless maintains its charge information, so that the electron would be attracted back to it and would bind with it, thus forming the now stable bound pair. Sarfatti explains that *the stable 'bound particle-antiparticle pair' bears similarities with a (Wheeler-type) wormhole with two mouths,* one of which, the electron (as a closed loop or vortex mouth) would get integrated in the "Vacuum Coherence Field" forming both matter and causing inflation, while the other mouth would remain in the dark energy Dirac sea. The vacuum as the locus for "phase transition" consists of the negative dark energy (with negative pressure) at subquantum scale and inside the Dirac sea, basically piercing through the Fermi surface as a Psi wavefunction containing an 'enormous number of these virtual pairs' in a given (Bose-Einstein condensate) state. The process would create first a thin layer of dark matter (with positive pressure) containing the 4D particles mouths, then the 'macro-quantum superfluid' that is the vacuum, containing the antiparticles mouths. Thus are explained the interaction of the 'exotic vacuum' that consists of dark matter and dark energy; says Sarfatti (151-2) "The exotic vacuum dark energy and dark matter are simply the amplitude wiggles and ripples of this same virtual pair quantum wave packet."

Puthoff, Dicke, and Haisch: Research on the ZPF

Harold Puthoff and Robert Dicke, then Bernard Haisch, have been forefront researchers on the Zero Point Fluctuations and the Zero Point Energy that fills this field, meaning the limitless source of energy we could harness from it, as well as the possibility of interstellar flight. Puthoff and Dicke elaborated a theory, called Polarizable Vacuum (or

PV) to explain gravity through the interaction of matter with the ZPF—via an electromagnetic exchange of forces with the vacuum. This was an alternative to Relativity Theory's views that gravity is due to the curvature of space. Bernard Haisch developed a theory with Alfonso Rueda to explain inertia through the force the ZPF would put up against accelerating objects.

In the classical version of the ZPF, all particles interact with a background that is random (the vacuum is filled with a random zero-point electromagnetic radiation, homogenous and isotropic).

According to Hal Puthoff, there should exist an "underlying model" describing the real nature of gravity and inertia, and Puthoff showed that they were equivalent. His rationale is that whether in Newtonian physics (where gravity has a scalar form) or in General Relativity (where it has a tensor form), only the actual behavior of this force, its phenomenology so to speak, is described, and not at all the source of its behavior. This situation, according to Puthoff (*Physics Review A*, 1989), "invites attempts at derivation from a more fundamental set of underlying assumptions." In other words, both Newtonian physics and Einstein's Relativistic physics are in need of a more fundamental theory—a deep theory. The momentum driving the research on the ZPF is that scientists deem it a formidable source of limitless amounts of energy, that could be tapped freely from anywhere; not only could it bring a cheap solution to an energy-hungry planet, but it would also take us beyond the devastating ecological impact of using fossil fuels. Nikola Tesla, for one, had understood this possibility. Let me quote a 1935 article called *A Machine to End War*, (while not endorsing Tesla's idea that a weapon of mass destruction would ever end war); the article was published in the journal *Liberty*. The editor notes: "Nikola Tesla, now in his seventy-eighth year, has been called the father of radio, television, power transmission, the induction motor, and the robot, and the discoverer of the cosmic ray. Recently *Tesla has announced a heretofore unknown source of energy present everywhere in unlimited amounts*."[3]

Puthoff furthers Sakharov's model and proposes an electromagnetic basis for gravity, that makes it unnecessary to treat gravity as a separate force or field in a unified theory attempt. Puthoff builds on the assumption that "matter, in the form of *charged point-particles (or 'partons')*, interacts with the ZPF of the vacuum electromagnetic field."

In his *Polarizable Vacuum* model, he proposes that electrons are constantly interacting with the ZPF, as well as planets orbiting a star, and the quanta of energy they receive from the ZPF compensate for the energy they lose while orbiting, and thus the vacuum acts as a stabilizing force. In the section below on Faster Than Light (FTL) energy, we'll see how the ZPF is now considered as a negatively charged medium (disturbingly similar to the ancient ether or aether) that, for some theorists, is totally distinct from the EM fields it interacts with. It is, in reality, the source of an immense radiation pressure, that can be tapped for producing so-called 'free' energy, antigravity and propulsion effect. An important discovery, in the seventies, was that of Paul Davies and William Unruh, who calculated that an accelerated motion in the ZPF would introduce a local symmetry-breaking in the vacuum, that would disturb the more isotropic (uniform) matter-density of the vacuum (in contrast to a uniform motion that would not disturb it). In his 1998 (JSE) paper, Puthoff gives several examples of research (mostly using the Casimir effect) having shown an energy release from the ZPF, that could be translated into another form of energy such as heat. He states: "There is experimental evidence that vacuum fluctuations can be altered by technological means. This leads to the corollary that, in principle, gravitational and inertial masses can also be altered." This opens the way for antigravity spaceships and machines. And concerning the possibility of using wormholes for space travel: "In principle, the possibility exists provided one has access to Casimir-like, *negative-energy-density quantum vacuum states.*"

Dark energy and quintessence

Let's remember that dark energy represents nearly three quarters of the energy of the universe the nature of which we don't have the slightest idea! Physicists working on this domain hypothesize that this dark energy is issued from the void itself (and ultimately, the pre-space).

There are actually two major sets of proven facts that now have to be integrated and explained within any theory whether at the cosmological or the quantum vacuum scale.

- The first set is the discovery of *dark energy, dark matter,* the inflation phase of the post Planck-scale universe, and the acceleration of this expansion (1998).
- The second set includes the discovery of *gigantic clusters of galaxies* and of structures, that we can obviously trace to the initial clustering and anisotropies already existing at the origin, and whose 'afterglow' configuration is now apparent in the Cosmic microwave background (CMB). Astronomers have recently detected huge walls of galaxies stretching over 10 billions light-years (2011-2013); a mysterious *dark flow* having the stupendous velocity of 1 million mile an hour residing toward Hydra-Centaurus (2008) and immensely more rapid than its background; a *Great Attractor* in the direction of the constellation Leo, toward which our local group of galaxies is speeding at the velocity of 1 million mile an hour. (More on this soon.)

The discovery, in 1998, while observing supernovae with the Hubble Space Telescope, of the acceleration of the expansion of the universe led cosmologists to the obvious inference that it had to be produced by a totally unknown gigantic force, that has been called *dark energy*. And now we have to adjust to the fact that this dark energy makes up about 69% of all energy in the universe (as goes the latest count from the Planck team), and our good old matter only about 5%!

What was discovered with greater and greater precision gave us the famous conic picture of the expanding universe since the threshold of Planck, in which we observe in fact different rates of expansion since the origin (see Figure 2.3). Thus, there was an enormous expansion during the Big Bang—the inflation phase in which the universe's size increased 10^{50} times, and up to the decoupling of photons (during the second minute after the Big Bang) that created the CMB afterglow image (+ 370,000 years). Then a steep decrease of the rate of expansion (that's where the large angle cone of the beginning takes the shape of a near-cylinder on Figure 2.3). Followed a very long period, up to 7.5 billion years after it, in which the rate of expansion stabilized. Then, about 6.3 billion years ago, a shift happened to a higher rate of expansion—in short, the expansion accelerated (based on the latest, March 2013, estimate by the Planck team, that puts the age of our universe at 13.798 years old). The most important point to remember (since the numbers are constantly updated in this period) is that this rate of acceleration of expansion underwent several shifts and transitions.

What scientists are looking for with 'dark energy' is an energy that has the opposite property of gravity—it has to be repulsive instead of attractive so as to account for galaxies moving away from each other while the universe expands; this energy has to be in such an amount as to overwhelm the gravity attraction of all dark matter and normal matter combined, and it has to make up 69% of the total energy of the universe.

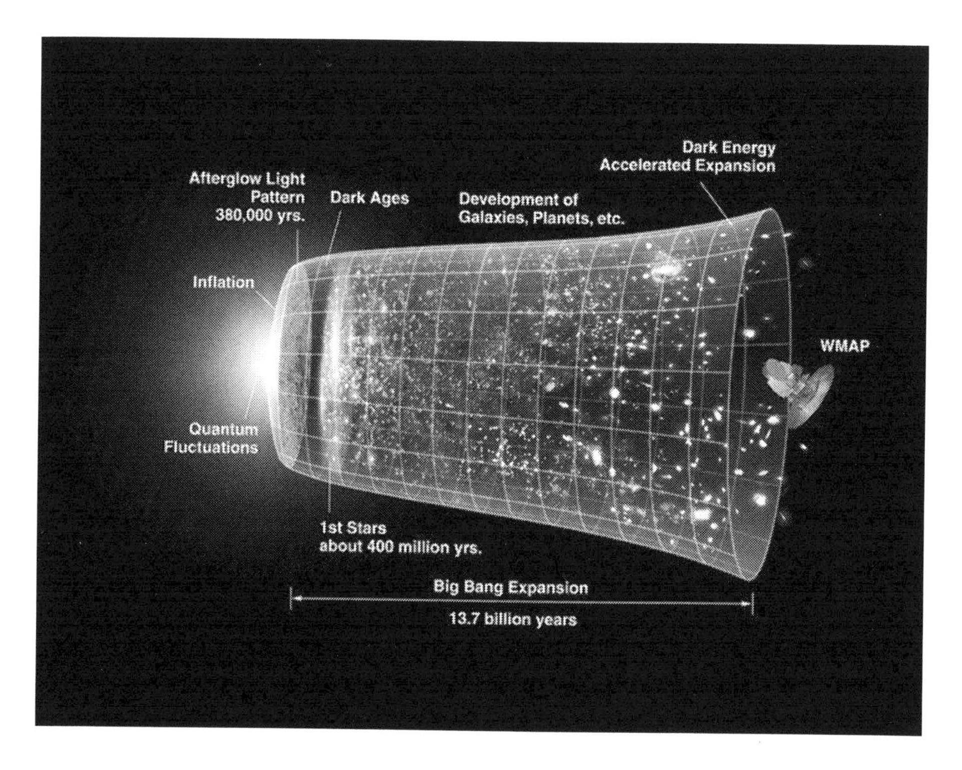

Fig. 2.3. Timeline of the Big Bang and the Universe's curved cone.
Credit: NASA/WMAP Science Team. (Public domain.)

An article on the NASA website, called *Dark energy, dark matter*[4] gives us the actual candidate explanations. Three are listed plus one possibility.

- Dark energy would be an energy inherent to empty space, as proposed by Einstein in his second theory of gravity that includes the cosmological constant. Einstein had first proposed that space keeps coming into being; as it does so, the hypothesized 'energy of space' would be accrued, thus accelerating the expansion of the Universe. However, this explanation clashes head on with the very aim Einstein had while introducing the cosmological constant, which was to render the universe fixated, infinite, and eternal, without a beginning such as the Big Bang. If you remember, Alexander Friedmann was the scientist to take out this cosmological constant and redo Einstein calculations without it, and that's how he proved the universe necessarily had had a beginning. Another point is that—even with discarding again the cosmological constant—the increase of the energy of space is, in this framework, a linear process and wouldn't explain why, billion years after the Big Bang, the expansion got again into an acceleration—a fact proven since the 1998 observation of the supernovae.
- Quantum theory of matter: In this theory, it has been proposed that empty space is in fact full of virtual particles that keep appearing and disappearing. These particles would then translate into a huge amount of energy—in fact so huge that it is 120 digits off the mark to explain dark energy. Says the NASA article: "It's hard to get an answer that bad. So the mystery continues."
- Quintessence: The third listed explanation is "a new kind of dynamical energy fluid or field, something that fills all of space," and has been called quintessence. "If quintessence is the answer, we still don't know what it is like, what it interacts with, or why it exists. So the mystery continues."
- Then the article states that there is a certain "possibility" that Einstein's theory of gravity has to be discarded and that we have to figure a radically novel one. However, in my view, this can hardly be considered a "possibility" in itself, since any solution to the problem would entail a revisiting at best of our actual, Einsteinian, understanding of gravity.

All in all, when we consider the article's three candidate explanations, we end up with only one without major shortcoming: that of 'a new kind of dynamical energy fluid or field' or quintessence.

The interesting term in this description is "dynamical," because we know from our discovery of the anisotropies of the CMB (that is, its irregularities and 'wrinkles') that only with a non-uniform field can we explain the clustering of gases and dust that gave birth to the galaxies. A uniform or isotropic field, such as an EM field, would have local-only properties: it would be uniformly distributed in space, and its energy would decrease as a function of the square of the distance. In contrast, a dynamical field, self-organized as a complex dynamical system with attractors, could have nonlocal properties; it could also be modeled as a chaotic system or turbulent fluid.

> Since the energy field thus evoked with quintessence is of an unknown nature, nothing precludes the syg-energy—that I pose as pervading the universe—to be a candidate. Syg-energy is a dual dynamical force, both energy (having effects on matter) and consciousness (or active information), and it is fueling a cosmic syg-field. It is a force behaving as a nonlocal connective dynamics and creating via the sygons (and their specific meaningful or semantic links) the particular and innumerable semantic fields as dynamical and self-organized networks. And in ISST, the semantic dimension becomes the CSR hyperdimension pervading the universe, yet accessed in each particle and system via a 5th dimension, curled-up and opening at the Planck length.

As for David Bohm, he recognized in this dark energy the 'implicate order' itself, that he associated with 'active information' and consciousness in the universe. Thus, another candidate would of course be the Quantum Potential proposed by Bohm, defined as a field of form and "active information" acting as a nonlocal "guiding" force.

> However, while Bohm stated that "meaning is the bridge between mind and matter," he has not developed the interaction and transition from a unified "implicate order" to the complex self-organized systems in our physical universe. Similarly, how 'meaning' as 'active information' would guide the build-up of such systems has not been modeled. In contrast, SFT+ ISST do model the self-organization of systems via syg-fields and the HD Syg-metadim.

To dig deeper into the subject, Bohm's unified active information force would need to be itself plural and specified (anisotropic and not one-block) in order to in-form highly autonomous conscious systems such as minds; and Bohm's implicate order theory does not provide an answer as to the inherent complexity of the information field allowing complex systems to self-organize. Indeed, meaning in action in individual consciousnesses is highly personalized, innovative and has free will. And this capacity of the human beings to have free will and to create new meaning, novel behaviors, and works of intelligence and art, cannot be attributed, transferred and subsumed to a uniform self-conscious force, omniscient and sending pre-determined information to all beings and systems in the explicit order (the physical world as we know it).

> Meaning is definitely a bridge between the whole and the systems, between mind and matter, I agree, but it has to be (theoretically) assessed as a two-way inter-influence that explains also the crucial input and influence from the autonomous minds (and complex dynamical systems) toward the more global information field (the planetary semantic field, or the cosmic ISS). A solution to this problem (without implying that it's the only one possible) is the layer of semantic fields and eco-fields that I propose in Semantic Fields Theory, that is, the dynamical self-organization of semantic (meaning-creating) systems, via nonlocal connective dynamics and fueled by syg-energy.

About three quarters of the energy of the universe the nature of which we don't have the slightest idea! The most potent hypothesis is that dark energy is deeply linked to, or derived from, the void itself (and ultimately, the pre-space).

Says Teodorani: "The *energy* that mysteriously springs from the void is the manifestation of a latent world, hyperdimensional and atemporal, *where dwells the consciousness of the universe*." (*Bohm*, 6.1; my emphasis.) As I mentioned it, Jung has clearly pointed the 'trans-spatial' and 'trans-temporal' qualities of the psyche (and thus of consciousness). He stated with Pauli that psyche and matter/energy were—in the 'deep reality'—so blended as to form a unique 'stuff,' and that the scientific understanding of this novel domain would usher a radical and global reformulation of our concepts of spacetime and matter, and similarly of our concepts of psyche and consciousness. The domains that were, in their understanding, belonging to this "psychoid reality" or blended substrate, endowed with both mind-like and matter-like properties, were *numbers* (as archetypes and matter-ordering), and *synchronicities*, that is, acausal and instantaneous meaningful links, expressing the dimension of the psyche and the Self.

And similarly, for Bohm meaning as a process or 'active information' was supposed to drive the Quantum Potential, itself acting as the Pilot Wave guiding and orienting the quantum processes (and thus counteracting quantum indeterminacy). Active information is also the foundation of Jack Sarfatti's universal information field.

> In fact, in SFT meaning as a process means a negentropic mind-force in action creating a meaningful world for a specific subject or observer—an 'I' who instantiates a 1st person's perspective. But as far as a whole group, culture, or humanity at large are concerned, all individuals' mind-force in action is co-creating a plural and complex culture and a planetary patchwork; all minds are co-creating the planetary semantic field of which each of us is an intelligent and relatively autonomous facet.

Enigmas in cosmology

Enigmatic 'Great Walls' of galaxies

Valérie de Lapparent, Margaret Geller, and John Huchra (1989, *Science 246*, 897) have highlighted the filament structure of the universe that gives it a spongy texture—clearly visible in the ongoing mapping of the Sloan Digital Sky Survey, or SDSS, who lent its first results in 2003, and in the earlier 2dF Survey. (See Figure 2.4) They analyze walls of galaxies that stretch up to 10 billions light years, and are interspersed by huge low-density regions that can extend over a hundred twenty Mpc (megaparsecs). There is clearly a texture to the universe, with protostellar clouds and galaxy clusters packed in one region of space, while the interstellar space is nearly devoid of matter, to a ratio of one part for 10^{-40}.

The largest cosmic structure observed to date is the ***Hercules-Corona Borealis Great Wall***. It was discovered in November 2013 and stretches (in the direction of these constellations) over 10 billion light-years in length (3 Gpc, gigaparsecs)—that's about 10.7% of the diameter of the observable universe, itself estimated to be about 92 billion light-years—and 7.2 billion light-years in width. It was observed through several types of cameras mapping gamma-ray bursts.

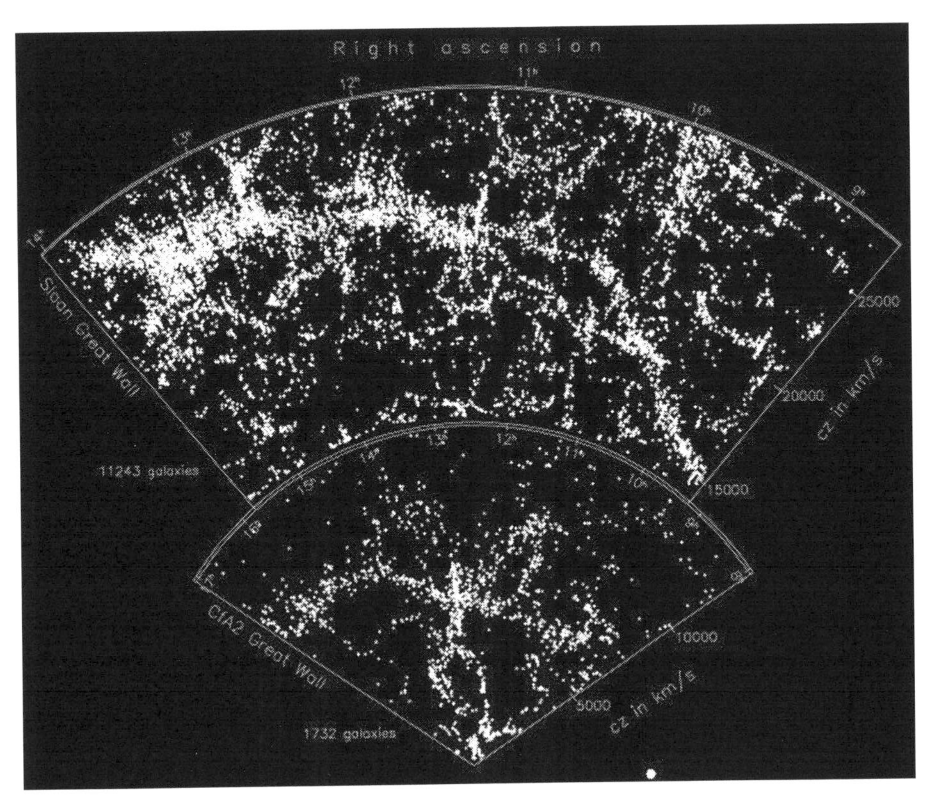

Fig. 2.4. Great walls and superclusters of galaxies within 1 billion light-years. A rendering of the 2dF Galaxy Redshift Survey data (and CfA2 catalog) as well as the SDSS data. Our galaxy, the Milky Way, is part of the Virgo cluster (at bottom center). The famous CfA2 Great Wall and the SLOAN Great Wall are clearly visible. Shown here is only a small circle of 5 Mpc/h size, the typical clustering length according to the standard analysis. Created by Richard Powell, Creative Commons 2.5 licensing. (www.atlasoftheuniverse.com)

The fascinating point made by Istvan Horvath and the Hungarian-North American team of discoverers was that they estimated the probability of such a clustering structure to be extremely low (0.00055% chance) and that rendered an explanation through quantum fluctuations "almost impossible." The Wikipedia article (Hercules–Corona_Borealis_Great_Wall) concludes about explaining it by random distribution: "So the structure's existence under the accepted cosmological models was not only doubtful, but very impossible."

> This is of course of an immense interest for the ISS model that poses cosmic syg-energy at work at the beginning of our universe-bubble—that is, a collective cosmic intelligence in a process of permanent co-creation of the very semantic and matter-friendly environment in which life and consciousness can thrive. This syg-energy metadim (consciousness) that ISST postulates, is operating conjointly with the two other metadimensions Rhythm-Rotation (frequencies and hypertime) and Center-Circle (extension as hyperspace, and also forming the systems' identity).

The next great wall, called the *Sloan Great Wall* extends only over 1.37 billion light-years; it was discovered through the *2dF Galaxy Redshift Survey* and the *Sloan Digital Sky Survey* or SDDS. In its data release of January 8, 2011, the SDDS imaging camera, operating since 2000, had covered only 35% of the sky. Among the largest cosmic structures that have been observed, are the Great Walls, the Large Quasars Groups (or LQGs), the superclusters and the galaxy filaments. The second greatest structure to date is thus called the Huge-LQG, comprising 73 quasars, and extending over 4 billion light-years (discovered in January 2013, its measurement has still to be confirmed). In a light-year, the light (photons) covers a distance of approximately 9.46 trillion kilometers, and so it's hard to imagine structures of such an immense size.

Furthermore, a notable periodicity has been discovered in the redshift of galaxies, called redshift periodicity or quantization, by William Tifft studying the Coma cluster from 1973 to the early 1990s, and corroborated by several astrophysicists in the same decade. (See Beardsley, 1992.[5])

Here is the stupendous definition of quantized redshift: "Redshift quantization is the hypothesis that the redshifts of cosmologically distant objects (in particular galaxies) tend to cluster *around multiples of some particular value*."[7]

In 1987 "V. S. Troitskii argued that light speed had originally been about 10^{10} times faster than now." An astronomer, Barry Setterfield, studied the data on the redshift quantization, and pondered: "The universe could not be expanding if the red shift measurements were quantized. Expansion would not occur in fits and starts. So what did the red shift mean?"[6] The debate was heated and Setterfield was barred from publishing his findings. If there was indeed a redshift quantization, then it meant that the speed of light had been enormously higher at the origin, as Troitskii had proposed. Then several mainstream physicists argued and published that the light speed was hugely higher at the Big Bang scale (such as Joao Magueijo of the Imperial College in London, John Barrow of Cambridge, Andy Albrecht of the University of California at Davis, John Moffat of the University of Toronto).

Bruce Guthrie and William Napier state (from their 1997 study of 250 galaxies): “The redshift distribution has been found to be strongly quantized in the galactocentric frame of reference. (...) The formal confidence levels associated with these results are extremely high.” “Two galactocentric periodicities have so far been detected, ~ 71·5km s–1 in the Virgo cluster, and ~37·5km s–1 for all other spiral galaxies within ~ 2600km s–1.”

All this led some astrophysicists such as Tom Van Flandern and Tifft to hypothesize a *wave nature* to the interstellar void, with a higher density of stellar clouds on the force lines. Van Flandern, in his META theory, then dismissed the possibility of a Big Bang (sticking to a steady-state universe), and put out a theory of gravity in which G derives from “a flux of invisible ‘ultra-mundane corpuscles’ impinging on all objects from all directions at superluminal speeds,” these particles offering limitless free energy. He deduced from observations that *gravity (the graviton) is about twenty billion times faster than light.*

> Of course, the detection of a periodicity at multiple values, that, in my view, reveals sets of wave fronts at a cosmic scale (possibly similar to the spiral arms of some galaxies) could be of vast import for the ISS in-forming the structure of the pre-spacetime and then influencing the organization of spacetime. And as for the much faster-than-light speed of the sygons torsion waves in the hyperdimension and at the origin, this is a tenet of this ISS theory.

Puzzling Dark Flow and Great Attractor: We are hurled in space toward Hydra-Centaurus at a million mile an hour!

In the seventies, George Smoot, Jon Aymon and their colleagues at the Lawrence Berkeley Laboratory, found that not only our galaxy, the Milky Way, rotates on itself as all other spiral galaxies, but that it is hurled in space toward the direction of Hydra-Centaurus. The normal rotation of our galaxy becomes perceptible against the cosmic background radiation (CMB), making the latter appear slightly warmer (blueshifted) in the direction of movement, the constellation Leo, and slightly cooler (redshifted) in the opposite direction of recession, in its tail (the constellation Aquarius). This was discovered by a twenty-two-year-old researcher, Vera Rubin. But (with a much more sensitive equipment) Smoot's team detected a nearly opposite direction of movement, at the incredible speed of 600 kilometers per second, that is, exceeding *a million mile an hour* (2.16 million kilometers an hour) instead of the expected tenth of this velocity. Says Smoot in *Wrinkles in Time* (138-40): "The only way our results could be correct was if there were a *gigantic celestial mass out there*—unseen and unsuspected—that was dragging the Milky Way toward it under the influence of gravity." Furthermore, our whole Local Group (about a dozen galaxies) were also showing the same movement. And such a massive gravitational source contradicted Hubble's Law positing the uniform distribution of matter (clusters and galaxies) in the universe.

Smoot presented this finding in April 1977 at the American Physical Society meeting. And such a gigantic mass meant, for George Smoot, that "the cosmic seeds from which they grew must have been present in the very early universe." And these seeds should be linked to the anisotropies (the "wrinkles") found by George Smoot himself, John Mather, and Mike Hauser in 1992.

Could Smoot's "gigantic celestial mass" pulling us be the mysterious gigantic supercluster of galaxies, that acts as a *massive attractor* for us and other clusters, and was found at a distance of 150 million light-years from us? It resides toward Hydra-Centaurus and comprises several tens of thousands of galaxies (possibly part of the Dark Flow). Indeed, more detailed observations were made in 2008, by Alexander Kashlinsky (physicist at NASA's Goddard Spaceflight Center). He found a stream of galaxy clusters that had a *much higher velocity* in one precise direction, than the CMB, and called it Dark Flow (or velocity field). This stream has precise signatures, e.g. the clusters composing it are among the brightest at X-ray wavelengths, and they emit more hot gas, thus modifying the distribution of the CMB photons compared to the rest of the WMAP research map. This flow is moving along a line from our solar system *toward the constellations Centaurus and Hydra,* with a stupendous velocity of one million mile an hour. This observation has been corroborated in a later 2010 article published by the same researchers in *The Astrophysical Journal Letters* (March 20, 2010). In this second study, the team had doubled the number of galaxy clusters that they checked (now up to 1400) and also their distance. The Dark Flow starts to be visible at 0.8 billion light-years away, and has been traced up to a distance of 2.5 billion light-years (755 megaparsecs). (See Figure 2.5, and wiki/Dark_flow)

Dr. Emily Baldwin (of *Astronomy Now,* in an article posted on March 12, 2010)[8] explains, inserting a quote from the discoverer: "Kashlinsky comments that the dark flow may probe the primeval structure of spacetime on scales well beyond the present-day horizon. 'The standard cosmological model accounts very convincingly for how the present observed properties of our Universe came about as a result of inflationary expansion of our bubble; *the dark flow probes physics prior to inflation*. The two models are complementary to each other.'"

Fig. 2.5. The Dark Flow. Credit: NASA/Goddard/A. Kashlinsky, et al.
From the largest and lowest drop-shape to the highest one:
(soft grey-blue): Clusters from 0.8–1.2 billion light-years away (250 to 370 megaparsecs)
(whitish green): Clusters from 1.2–1.7 billion light-years away (370 to 540 megaparsecs)
(whitish yellow): Clusters from 1.3–2.1 billion light-years away (380 to 650 megaparsecs)
(top ellipse-dark red): Clusters from 1.3–2.5 billion light-years away (380 to 755 megaparsecs)

The NASA/Goddard release states: "The dark flow is controversial because the distribution of matter in the observed universe cannot account for it. Its existence suggests that *some structure beyond the visible universe—outside our 'horizon'—is pulling on matter in our vicinity.*"[9] (My emphasis).

Kashlinsky gives us the precision that the dark flow's structure couldn't arise from gravitational instability, or within our universe horizon, therefore, it had to be created by or via a "structure outside of this horizon." Says he: "The space-time on sufficiently large scales had to be inhomogeneous prior to inflation of the Universe and *that overall structure should have been preserved on sufficiently large scales.*"

In this light, let's ponder the Infinite Spiral Staircase topology and how it creates a dynamical structure driving the expansion of space in the phase prior to inflation (precisely prior to the Planck scale): this dynamical self-organization and expansion of the ISS in-forms space, creating curved and curled-up spires, that have a definite torus shape in thickness. As it unfolds, it displays an increasing circle circumference (or cone), as well as an expanding angle in the direction of its expansion and flow. This is quite in accordance with the shape and width of the dark flow!

Furthermore, considering that the ISS would create the Vacuum and Higgs field, by a jet of bows-frequencies—called ***sygons***—in the direction of the flow (the Poynting drift) of the Spiral Staircase, the ISST postulates that:

(1) A first jet of sygons was emitted (from the bows) at a velocity immensely higher than light before the Higgs field densified via (a) the enlarging wavelengths of the sygons, (b) their increasing interference between them. This, in turn, created the Higgs field and vacuum as a surface of bubbling vortices-particles with a large central bulge (where the interferences would be more dense).

(2) This means that the first sygons (highest energy and frequency) to be emitted at the very beginning of the unfolding of the Spiral staircase and thus before the Higgs field started densifying, were passing the Planck scale unimpeded at enormous speed and formed an advanced wavefront—creating in their wake the *bulk* (the quasi-spatial extended stuff) of the hyperdimension but also the region that will become our spacetime. This sygon wave front should be perceptible through a specific relic background radiation, that

would derive from the sygon characteristics: an immensely faster-than-light, sub-Planckian, non-EM field, with much greater than gamma rays frequencies—with the frequencies of the CSR-HD.

Beauty in theories and mathematics

Another theme we have to ponder is the existence of beauty and our own sense of beauty. While reading works on cosmology and physics, I've found too numerous citations (beyond the statements by the authors themselves) about the importance of the beauty of an equation or a theory to keep track of them. Physicist and Nobel laureate Steven Weinberg's *Dreams of a Final Theory* features a whole chapter titled "Beautiful Theories," and Michio Kaku (1994) has a subtitle asking "Is Beauty Necessary?" in which he does conclude that yes, beauty expresses a balanced theory.

Behold these examples: Renown physicist and Nobel laureate Richard Feynman "You can recognize truth by its beauty and simplicity."

Nobel laureate Paul Dirac: "It is more important to have beauty in one's equations than to have them fit experiment."

A very sweet statement was made by one of the supergravity founders, Peter van Nieuwenhuizen. He exclaimed: "It is the most beautiful gauge theory known, so beautiful, in fact, that Nature should be aware of it!"

Mathematician Henri Poincaré: "If Nature were not beautiful it wouldn't be worth knowing, and if Nature were not worth knowing, life wouldn't be worth living."

And what about non-physicists thinkers, steeped in astronomy, such as the well known author of detective fiction Edgar Allan Poe, who was able to predict a sound solution to the Olbers' paradox eighty years in advance. The paradox is about the mostly dark sky we observe at night: if the universe was infinite and static, then there would be an infinite number of stars and their light would fill any point of the sky, making it bright at night. In his 1848 work on astronomy, *Eureka*, Poe states that the reason there exists great patches of black sky, devoid of star light, is that "the distance of the invisible background [is] so immense that no ray from it has yet been able to reach us at all." And what was his inner gauge to deeply know how correct was his idea? It "is far too beautiful not to possess Truth as its essentiality."

Says Brian Greene: "In string theory many aspects of nature (...) are found to arise from essential and tangible aspects of the geometry of the universe. If string theory is correct, the microscopic fabric of our universe is a richly intertwined multidimensional labyrinth within which the strings of the universe endlessly twist and vibrate, rhythmically beating out the laws of the cosmos." (*The Elegant Universe*, 18)

> I love this description of a beautiful cosmos, yet I'm always sad when I remark, again and again, that, in physics, absolutely no room is given to any phenomenon apart from the strict laws of the four forces: no room for creativity, for sensing and feeling, for thinking and consciousness at large, for collective intelligence. And yet, physicists are in awe in front of mathematical or topological "beauty." Will ONE physicist ever reflect on what their own sense of such harmony, symmetry and beauty really means? Will one realize that without a consciousness in the universe to contemplate it, the nature that they know to be so economical and always taking the shortest path would NOT have created beauty and even less so a *sense of* beauty? I'm not even sure Nature would have created intelligence and self-reflective consciousness because nature works very fine without any theories about its own workings, *UNLESS the cosmos was itself conscious and filled with consciousness of any unimaginable order.*

The resonant music of the cosmos

Perseus' Black Hole emitting a single note as a music through the cosmos

For the first time, intergalactic sound waves have been detected by the Chandra X-Ray orbiting telescope of NASA, emitted by the central *supermassive black hole of the Perseus cluster* (this cluster comprising thousands of galaxies), and located 250 million light years from Earth. The astronomers Andrew Fabian and Steve Allen, from the Institute of Astronomy in Cambridge, England, had made a previous observation in 2002, that showed two enormous bubble-shaped cavities, about

50,000 light-years wide and filled with magnetic fields and high-energy particles (Fig. 2.6, left photo). Both cavities are sources of intense radio waves, and they form an eight-shape figure around the black hole. The jets of material emitted by the black hole create both the cavities and the sound waves while pushing aside the hot gas. The sound waves were last observed spreading out from the cavities; they have provided the heating mechanism that had eluded astronomers beforehand; they calculated that the combined energy from 100 million supernovae was what had been necessary to form the cavities. The X-ray photo showed that the gas filling the cluster was full of clearly concentric ripples forming like the outer rims of many petals around the black hole (Fig. 2.6, right). Fabian has calculated that these ripples are sound waves that, after being emitted, have already traveled a few hundred thousands light-years in all directions out of the cluster's central black hole. The sound wave emitted is only one note, a B-flat, at 57 octaves below middle C; that is, 1015 times below the lowest sound audible to us. The NASA's Goddard Flight Center online news (November 17, 2003) says about it: "In terms of frequency (the time it takes a single sound wave to pass by), the lowest sounds a person can hear is 1/20th of a second. The Perseus black hole's sound waves have a frequency of 10 million years!"[10]

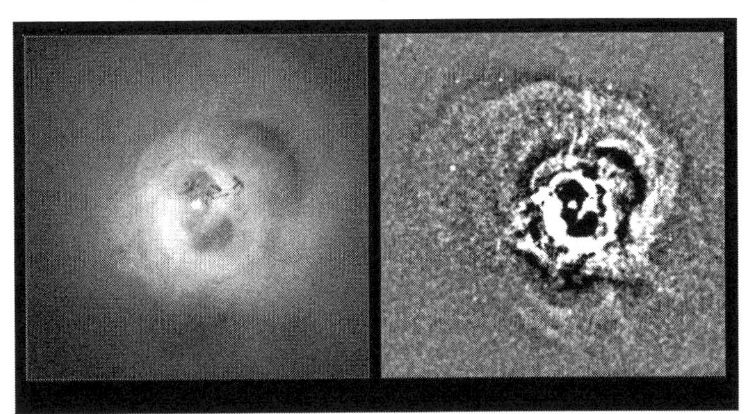

Fig. 2.6. Sound waves generated in Perseus cluster. Courtesy NASA/CXC/IoA/A. Fabian et al.

This unique note is estimated to have remained quasi constant for about 2.5 billion years. Sound waves are in effect pressure waves that travel through the void. In our solar system, the interplanetary space is filled with thin plasma and these pressure waves travel at the typical sound speed of the solar wind (a few hundred kilometers per second), but their frequency are not in the audible spectrum. Says Steve Allen: "These sound waves may be the key in figuring out how galaxy clusters, the largest structures in the universe, grow."

Now that we have seen a clear-cut observation of a cosmic-scale wave with a unique and stable B-flat frequency managing to travel 2.5 billion light years toward Earth, and yet to keep its frequency intact, let's see the new theory of Global Scaling, that explains such phenomena and shows us that we live in a cosmos full of active information carried at enormous speed via resonance—the ISST's metadim Rhythm.

Resonances throughout the universe: Global Scaling

The Global Scaling Theory is sorting out and modeling the scale invariance in physical systems (from quantum processes to cosmological structures), whether in frequency resonances or in form (isomorphic structures). In 1982–84, Hartmut Müller started to elaborate the theory, on the basis of the scaling—the logarithmic scale-invariance—that he discovered in the distributions of masses both at the subatomic level (particles, nuclei, and atoms) and at the cosmological level (asteroids, moons, planets, and stars). Beyond the masses, the scaling expressed also specific properties in these bodies' orbits and sizes. As the 2008 collective online article called *Global Scaling Theory,* by the Global Scaling Research Institute, puts it: "Scaling is a basic quality of fractal structures and processes. The Global Scaling Theory explains why structures and processes of nature are fractal and the cause of logarithmic scale-invariance."[11]

Already in 1748, Euler tried to solve the mathematical conundrum of the oscillations of an elastic and massless *string of pearls*, on which D'Alembert and Bernoulli also faltered, and which was finally solved by Lagrange. And in 1795 Karl Gauss sorted out that the distribution of prime numbers followed a logarithmic scaling invariance.

The theory has its roots in Richard Feynman 1967-69 discovery of scale-invariance in hadrons collisions, developed in his article "Very High-Energy Collisions of Hadrons." Then, still in physics, the work of J.D. Bjorken, and also Simon Shnoll in the fluctuations of nuclear decay rates. During the early eighties, scale-invariant logarithms of development were found in biology (e.g. in morphogenesis and ontogenesis), as well as in the neuro-physiology of perception. This is why we are able to distinguish easily a given note at different octaves. This is true not only for our hearing sense but also for the smell, touch, and sight senses. For example, says the article: "The retina records only the logarithm, not the number of impinging photons. That is why we can see not only in sunlight but also at night." And similarly, our perception of color differences is "logarithmically calibrated" in relation to the wavelengths of light that we see as colors. In the late nineteenth century, Ernst Weber and Gustav Fechner came up with the *Weber-Fechner Law:* "The strength of a sensory impression is proportional to the logarithm of (the) strength of the stimulus." All our senses work in this manner; says the article: *"The logarithmic scale-invariant perception of the world is a consequence of the logarithmic, scale-invariant, construction of the world."* More and more fractal objects and processes were discovered in the natural world. At the end of the 19th century, the work of Cantor and other mathematicians on fractal structures (e.g. the problem of *a line of grains-of-sand*), demonstrated that space coordinates cannot describe adequately the scale-invariant fractals, and it led Hausdorff to postulate in 1919 the fractal dimension D.

But let's see the basics of the dynamics of vibrating strings, that creates the sound harmonics.

As we know, a vibrating string will have a natural frequency depending on its length (setting a full wavelength); then harmonics will form at integer divisions of the wavelength or half-wavelength on the string. The shorter the wavelength, the higher the frequency. In natural oscillations, both the products of frequency-wavelength and frequency-amplitude are conserved. *"Scaling arises very simply—as a consequence of natural oscillation processes. Natural* oscillations are oscillations of matter that already exist at very low energy levels. Therefore they lose (very little) energy, and likewise fulfill the conservation law of energy." If space is finite in the direction of the wave-propagation or penetration, a *standing wave* will form, when the half-wavelength is equal to an integer part of the size of the medium. The frequencies of harmonics will be dependent on the divisions of the first wavelength only by integers, so that the two end-nodes of the wave (where the waves creates an X, at zero height) are always at the two boundaries of the string. This standing wave's rising frequencies can be mathematically expressed as a logarithm. Thus the frequency on a string of definite length could thus rise to infinity, thereby creating what is called 'the ultraviolet catastrophe,' a sophisticated way of saying 'Here we go again, face to face with infinity!' Of course there's no catastrophe in reality while the harmonics rise toward the infinity! Only in the minds of physicists who hate the tendency of the universe—and of their own equations—to abruptly open again and again on infinity; and in order to stop it, they have to constantly figure smart 'renormalizations' that is, methods to input corrections issued from real observations data.

A very interesting effect of these standing waves (which could be not only sound or liquid waves, but also a matter-particle frequency, or else, planets' frequencies in a solar system, etc.) is that they modify the distribution of matter density in periodic, spectral, configurations. In other words, the effect, as the article calls it, is a "logarithmic periodic structural change," and this change of density modifies also the kinetic energy. Just visualize a horizontal half-wavelength (an eye shape), at the maximum amplitude (center), insert a vertical lozenge touching the maximum height and trough of the wave (like a cat's pupil). Now create and superpose the harmonic pattern of three half-wavelengths within the first one (each of the 3 eyes' wavelengths is a third of the first one, and their amplitude (height) is lower). As the amplitude is lower, you will get a fractal shape of the lozenge at this lower scale (perfectly isomorphic to the first one and invariant in form, but at a smaller scale). Now visualize the whole pattern from afar: the nearest to the nodes (the X crossings), matter/structure has been compressed and concentrated and thus the density of matter is higher. The global dynamics, according to the article, is that "oscillation-troughs displace matter that then concentrates in the oscillation-nodes. In this way, a logarithmic, *fractal distribution of matter density* arises in the natural oscillating medium. (...) This change from compression to decompression causes a *logarithmic, periodic structural change* in the oscillating medium; and areas of compression and decompression arise in a logarithmic, fractal pattern."

In 1975, Benoit Mandelbrot started to create his magnificent fractal designs and to apply the fractal dimension to the analysis of many processes and natural phenomena, like the growth of leaves on a tree, coastlines, or crystal formation. And in 1987, Hartmut Müller developed the *Proton Resonance Model* (based on fractal scaling) and used it at first to model and optimize technological systems. Then he discovered that some properties of many natural and cosmological systems are precisely set on the periodic recurrence of some of the proton's frequencies. For example, Müller calculated that in the human body and brain, the main rhythmic processes and spectra of frequencies—such as breathing, heartbeat, neural processing, micro-arterial blood pumping, optical sensor scan, voice, and hearing, all of them were falling on nodes of the proton's resonance frequency spectrum. Similarly, the distribution of the masses of the planets in our solar system, as well as their respective distance from the sun, all this is clearly highly related to the nodes of the proton resonance at an immensely distant octave.

The vacuum as a spectrum of resonances

Now, the stupendous discovery of Müller is the texture and organization of the vacuum. We have seen that the quantum vacuum is actually modeled as the ZPF, in which the 'fluctuations' are postulated to be entirely random, highly energetic, with transient virtual particles constantly popping up and then reabsorbed, and resembling a foam, a highly disordered and entropic turbulence, or else a 'shredder' of any matter lower than Planck length. Yet, the vacuum was described at first as the ground state of matter, that is, its lowest energy state; and in relativity theory (that at first dumped the ether concept), the void was the curved texture of spacetime itself, filled with EM fields. But now arises a new vision: in the framework of Müller's Global Scaling theory, the vacuum is a *spectrum of resonances*, and behaves like an oscillator: "In the vacuum, only natural oscillations are possible. The Planck Formula: $\Delta E = h \Delta f$ [in which E is the energy, h the Planck constant, Δ (delta) is the ratio of change, and f is the frequency] "gives the impression that the energy of *the natural oscillation* of a vacuum-oscillator *is frequency-dependent and quantized*. Consequently, energy can only be absorbed or emitted in determined portions."

Moreover, as the universe is finite, any wave of harmonics in the cosmos, such as our earth's resonance or else the proton resonance frequency, on reaching the edge of our cosmological horizon, would start building up static waves that still would preserve their initial frequency-information.

Now, this is much nearer to the viewpoint of relativity that saw the void—the substance itself of spacetime—as filled with crisscrossing EM fields, and yet, the concept also bridges the gap with quantum field theory by postulating that these oscillations are quantized (that is, discrete and implying particles).

Metadim Rhythm and the resonant vacuum

This is where the scale-invariant organization of natural processes interests me to the utmost regarding the ISS theory and especially the metadim Rhythm, the pre-time. Said Tesla: *"Every movement in nature must be rhythmical.* (...) It is borne out in everything we perceive—in the movement of a planet, in the surging and ebbing of the tide, in the reverberations of the air, the swinging of a pendulum, the oscillations of an electric current, and in the infinitely varied phenomena of organic life. Does not the whole of human life attest to it? *Birth, growth, old age, and death of an individual, family, race, or nation, what is it all but a rhythm?"* (in "The Problem of Increasing Human Energy.")[12]

Now, if we integrate the central tenet of Global Scaling Theory, then all matter-systems in the universe have a specific frequency generating a static wave of resonance, that maintains its fine-tuned spectrum of frequencies in the void. As stated in the article: "A logarithmic-periodic structural change can be observed in all scales of measurement of the universe – from atoms to galaxies."

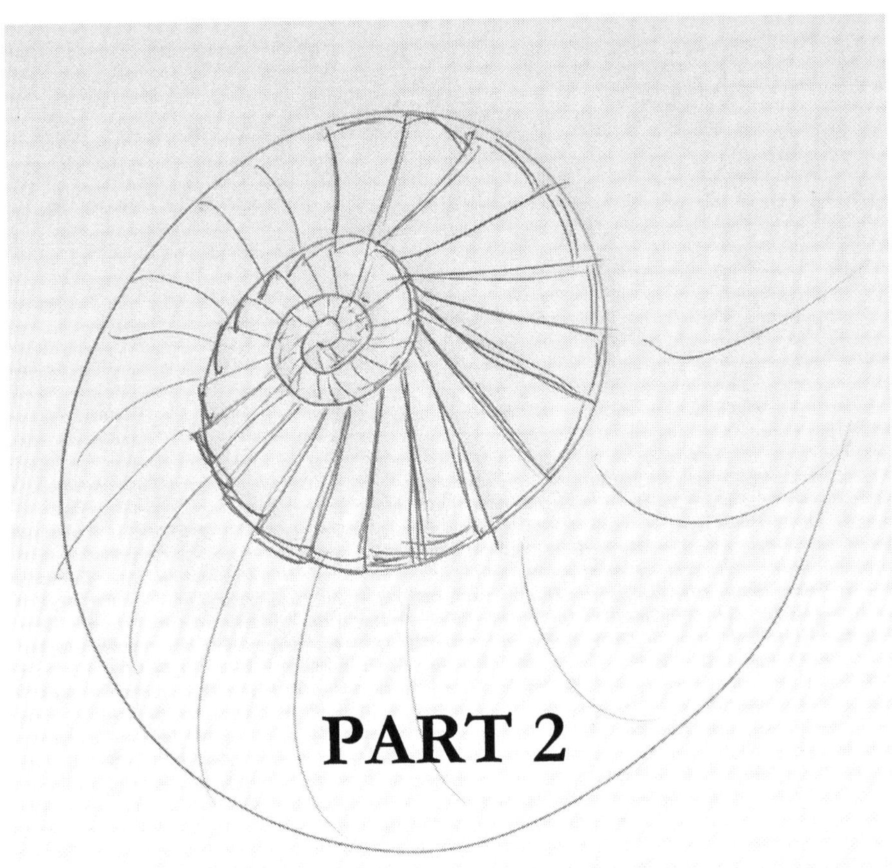

PART 2

THE INFINITE SPIRAL STAIRCASE AT THE ORIGIN

3
PAULI, JUNG, AND THE DEEP REALITY

"Physics must expand to explain mental experiences." Noam Chomsky.
"We need a revolution in physics on the scale of quantum theory and relativity before we can understand mind." Roger Penrose.

Jung's and Pauli's work on synchronicity

Psychologist Carl Jung and physicist Wolfgang Pauli, both giants in their field of research, entertained a collaborative work spanning twenty-seven years and mostly focused on the ground-breaking phenomenon of synchronicity which, for them, was revealing a whole new dimension of reality that neither physics nor psychology had ever tackled. Synchronicity is the occurrence of a strong and meaningful coincidence between a person and an event in their environment. In a paper on the subject, I argued that to distinguish it from a fortuitous coincidence, the event must produce a strong psychological impact on the person living it.[13] As I came to understand it: "Synchronicity shows the capacity of the Self (when the person is engaged in a process of spiritual evolution, or individuation) to succeed in organizing physical reality and events according to its own semantic energy (higher spiritual values, goals, and orientations)."

And indeed, a portent synchronicity associated with a strong PK (psychokinesis on objects) while writing on the subject for my book *The Sacred Network* in Brazil, was nothing less than a striking demonstration of the validity of this postulate! In this part of the article text, I was developing the idea that no actual physics framework (whether Quantum or relativistic) could explain PK—that is, the capacity of a mind to influence matter in an intentional way, for example in healing (bioPK), which has been experimentally proven. I was typing and editing an extract of the previous article, stating that QM could definitely account for nonlocal, beyond spacetime, phenomena such as the entanglement of particles; however, it couldn't account for the meaningful aspect of psi, for example the intention to heal a friend by meditating on her at a distance. To the opposite, I typed "*psi would be responding* to a deep and meaningful psychological process." At the very moment when I finished typing *responding,* following the article's text I had written a month and a half earlier, a strange psi phenomenon occurred: my chime suddenly rang out strongly. "I look up: the chime's weight—a vertical moon crescent in wood of 4.5 inches—turns on itself at least 10 times, extremely rapidly, like a gyroscope." Then it will turn the other way around to release the tension built up in the thread supporting it from the ceiling. The point is, I was involved in a very creative and highly important and meaningful task while pondering these ideas (intense semantic energy), and my Self—using psi—*responded* to my thought!

In contrast to QM, Semantic Fields Theory could explain such influence of mind over matter at a distance, because I had posed parameters of syg energy that were independent of time and space, such as recurrence, intensity, and *semantic proximity* (meaningful links and psychological bonding).

The whole research of Jung and Pauli's on synchronicity was geared toward highlighting how much mind and matter are deeply enmeshed and merged, to the point of forming a layer of "deep reality," a psyche-matter medium in the universe. The synchronicities would be springing from, and expressing, this underlying connective lattice. Pauli and Jung went on sorting out the domains in which this merging occurred (we'll see Pauli's examples, such as the spin complementarity).

Jung's certitude, based on his clinical experience and the science of the Ancients, was that psyche (mind) and matter were one and the same in this deep reality called "The One world" (*Unus Mundus* in Latin) or the "world's soul" (Anima Mundi):

Says Jung in his autobiography (Memories, Dreams, Reflections): "We have all the reasons to suppose that there must be only one world, in which psyche and matter are one and the same, and in which we establish distinctions for the sole purpose of knowing." And in his letter dated March 7, 1953, to Pauli, Jung states "I tried in this latest book [*Synchronicity*] to open a path leading to *the* ***'soul' of matter***, by assuming that *'what is' is endowed with meaning."* 'What is' is of course referring to the whole stuff of reality, including matter. (My emphasis).

Pauli had a very clear dream (the 'Deep Reality' dream) in which his unconscious (the dreaming Self) is making a distinction between on the one hand Quantum Physics, and on the other hand a "deeper reality." On one half of a board, only a title: QM. On the other side, titled 'Deeper Reality,' was a grid of crisscrossing diagonals crossed by an undulating line. The dream was thus hinting at a level of reality deeper than quantum fields and interactions—which, in my view, can only be a distinct manifold or dimension in the universe, possibly a hyperdimension, that would underlie the quantum manifold.

Quantum events are perfectly random and indeterminate according to Niels Bohr's Copenhagen 'orthodox' interpretation, to which Pauli subscribed. This indeterminacy, as I underlined above, is unable to explain or ground the dimension of consciousness—mental capacities such as intention, creativity, decision, free will, and of course psi (nonlocal communication and intended remote influence). And neither can indeterminacy explain beauty and meaning.

The alternative interpretation of QM, that of the *hidden variables,* as we saw, was proposed by Einstein, who stated that even if quantum events appeared random, some hidden factors must be at play that could be unraveled in the future. The theory called *Pilot Wave* was first developed by de Broglie and then by Bohm. With such hidden variables, Einstein intended to reinstall the concept of an ordered and deterministic universe: "God doesn't play dice," did he claim (without any scientific proof, mind you).

Thus posed, in my not so humble standpoint, our situation is dire hopeless: on the one hand we have absolute determinism and/or a creator God, knowing and controlling all future events, thus making our intelligence and creativity, and even our consciousness, wholly superfluous.

On the other hand, we have pure randomness that makes particles of the chair on which I'm sitting wandering hazardously on the other side of the universe. And as the laws of probability have it, there's a non-null probability that all particles and atoms constituting my chair will be at the same instant somewhere else than where I'm sitting. But the crucial problem, of course, is this one: the fact that it goes the same way for all the atoms of my brain! (to which, as we may forecast it, Murphy's laws will add their hellish grain of salt.)

Either pure indeterminacy OR pure determinism?

There must be something lacking!

What's lacking is a third, totally different, principle and level of organization: self-organization in complex systems. That's where chaos theory and systems theory come to the rescue. Murray Gell-Mann has postulated the principle of increasing complexity, with a tendency for all complex open systems to accrue their inner complexity. Numerous researchers have posited that complex systems (such as a forest or a mind) do not evolve in a deterministic and linear way, but that, at certain thresholds, their whole organization and behavior is suddenly transformed—that is, their evolution is nonlinear and dynamic.

Furthermore, in such complex systems, there's randomness at a microlevel and global patterns of organization at a macrolevel. The best example is the tornado: a spiraling wind in which specks of dust are randomly distributed. The global pattern (the spiraling form) is held by the 'attractor' of the system. But a sudden force intruding may break the whole pattern, e.g., the small tornado gets in the path of a huge arson fire; its spiraling pattern is disrupted by a stronger force: huger flames shooting up straight.

As of 2014, with John Bush, chaos theory has been applied to the understanding of quantum systems, via a fascinating novel research on superfluids, that of the 'walking droplets' that were evidenced in 2005 by French physicist Yves Couder to exhibit macroscopic behaviors similar to those of quantum-scale waves-particles, and unexpectedly, they were specifically in accordance with the Pilot Wave theory of de Broglie. (In-depth analysis of the domain in chapter 6.)

Physicists, up to now, seemed to all agree that the evidence of instantaneous interactions between particles (their *entanglement*) not only proved nonlocality but also the pure indeterminacy of quantum events (as proposed by the orthodox QM interpretation of Bohr, Heisenberg, Pauli, etc.). The proofs have only been accumulating since the indisputable results of Alain Aspect in 1982-84, using Bell's theorem protocol. Thus were supposedly disproven Einstein's and Bohm's hidden variables supporting a deterministic universe.

Does it ring a bell? Isn't it similar to the ping pong game played at one point between the proponents of light as waves (interference patterns) and those of light as particles (quanta and photons)? During the long debate—from Huygens opposing Newton in 1678, to Einstein solving the photoelectric effect by light quanta in 1905—both schools could cite successful experiments proving clearly that their theory was supported by facts, primarily that of Young's 1801 famous double-slit experiment demonstrating wave-interference patterns (and still to our day spurting out unsolved paradoxical results). And yet, after reigning over physics for a long while, this certitude was shred to pieces when Einstein solved the riddle of the photoelectric effect by positing a definitive particle measure, that of Planck's quanta of energy, released or incremented through particle exchanges: his famous 'light quanta,' later called photons.

Of course, a leap was needed to reach the only possible solution to the strange nature of light: that both schools were correct because reality is paradoxical. Not only light, said De Broglie in 1927, is both and simultaneously wave and particle, but all matter particles (such as electrons) are associated with a wave. However, the crucial point, in my view, is that *this leap happened in logic, not in physics per se*.

So are we confronted here with the same type of problem? Namely that we are in need of a new logical leap because the nature of reality is immensely more complex and shaded than our current scientific logic can grasp?

The concept of entanglement of particles is derived from Pauli's equation of spin, that states that the spins of a pair of particles (issued from a common source) must always be complementary; thus, if one rotates to the right, the other rotates to the left, so that their sum always equals zero. Isn't that the perfect example of an absolutely binding global order at the level of the global system that the two particles form? And yet there's an indeterminacy at the level of the particles, for the global law doesn't bind any of the particles to be in such and such state, and this is why we can interfere in the system by changing the spin of particle A and watch how particle B will instantaneously change its spin (the EPR paradox experiments).

Pauli moved a step further in terms of logic when he proposed that the law of complementary spins is of a synchronistic type. Now, *synchronicity* as I said, was defined by Jung as an *acausal* process. In synchronicity, no force/event A precedes and causes the behavior of force/event B. Rather, the forces/events interact on each other in an instantaneous way, expressing a connection that is not bound by space or time (since the change occurs quicker than the speed of light through space, as proven by an EPR experiment that involved bouncing one of the particles on the moon surface). Synchronicity operates in another dimension of reality, says Jung, one that is deeply endowed with meaning (semantics) and is related to the 'soul of matter,' that is, the fused mind-matter (hyper)dimension of reality.

The main theorist of the hidden variables school, physicist David Bohm, also raised 'meaning' (in his *Implicate Order* theory) as the key to all exchanges at the quantum level, that of the 'dance of particles' and to the creation of all events in the world (the explicate order). But he didn't mean 'meaning' as an inert data, but the process of creating meaning, what he called 'active information' carried by pilot waves. So what can be such 'active information' if not an *energy* endowed with meaning, given that energy is classically defined by its action (or 'work') on matter-systems, and here the pilot wave determines the path of the particle and therefore it does create/modify events in the world?

In Semantic Fields Theory (SFT) likewise, the semantic fields are by definition semantic energy—therefore consciousness in process—and they operate as self-organized networks. The main dynamics of SFT is that of spontaneous network-connections with similar or resonant semantic fields, and this is also the basic process involved in thinking, relating, and sensing (i.e. attributing meaning).

As we'll explore it, the SFT-ISST model seems to fit in very nicely with the specifications posed by all three researchers, Pauli, Bohm, and Jung. And yet, it sets for itself a new framework, one that is neither deterministic not indeterministic. It opens the new ground for what Teodorani calls the 'physics of synchronicity.'

Pauli's law of spin and work on his dreams

With his University friend Werner Heisenberg, who initialized Quantum Mechanics, Pauli is definitely one of the fathers and main brains behind the rise of quantum physics in the early twentieth century. At only 18 years old, Wolfgang Pauli had grasped the depth of Relativity theory to such a degree that an article he wrote on the subject got him the praise of Einstein himself. He passed his doctorate at the university of Munich, with, as thesis director, the eminent mathematician Arnold Sommerfeld, a recognized light of theoretical physics and quantum theory. (Sommerfeld became the thesis director of several other Nobel winners, beyond Pauli, such as Werner Heisenberg and Hans Bethe.) Born in 1900, Pauli formulated his major discovery—the Exclusion Principle—in 1925. At Munich University, Pauli and Heisenberg used to discuss and work together, exchanging ideas during long strolls. This year 1925 the two brilliant friends must have experienced an extraordinary state of creative breakthrough since both had a stroke of genius: Heisenberg posed the bases of QM a few months before Pauli came up with the Exclusion Principle for which he received the Nobel prize in 1945. Pauli's Exclusion Principle states that an electron can't occupy the same orbit (i.e. have the same atomic state) as another one, unless they are paired and their spins are in the opposite direction. With this principle, Pauli proposed a fourth parameter to describe particles: the spin. He posited that only two electrons can share the same orbital, at the condition that their spins be opposite (the spin of such paired particles can only take the values +1/2 or -1/2, expressing the pairing of a left-handed spin with a right-handed spin). This is the principle that has been tested and corroborated in the famous EPR paradox. (Since then, we have added spins of 1/3 and 2/3 in quarks.) Pauli's pairing of particles with opposite and half-integer spins instantiates one of the two great principles governing particles, that of *anti-symmetry,* which applies to all *fermions* particles (or matter particles, such as electrons, protons, neutrons, muons, neutrinos and quarks). To the contrary, the *bosons* particles (or energy particles), with integer spins, comprising the photons and mesons, follow a symmetry principle.

The principle of anti-symmetry (with the fermions) sets up the stage for the exquisite variety of atoms, and thus their whole classification through their atomic number—Mendeleev's table of elements that we all know from school. The Exclusion Principle that governs anti-symmetry is thus forcing each electron to be on a different orbit and a different energy state (unless they are a pair with opposite spins). Dmitri Mendeleev, in 1869, discovered that the chemical elements (metals and gases) that present the same atomic number (or number of protons), or else that are in the same column if arranged by their periodicity, exhibit similar chemical properties.

The symmetry principle, to the opposite, sets up the stage for the capacity of blending, of being organized in a 'coherent state,' such as, for example, the laser beam, the polarization, the supra-conductive state. Thus, the early universe is thought to be in perfect symmetry within a sort of unified medium—the field or ocean of Higgs—in which the four forces would be somehow blended. The step by step emergence of the four forces with their associated particles are major phases of symmetry-breaking.

But let's dive into this fascinating topic of the Higgs bosons who, at the very beginning of the universe, just after the Planck length, were like a glue forming the field of Higgs. In July 2012, the CERN in Geneva disclosed it had detected a new particle in the mass region of the predicted Higgs boson, produced by collisions within the Large Hadrons Collider (See Figure 3.1). During an experiment using a CMS detector, a collision between two ultra-high-energy protons produced a Higgs boson which decayed into two jets of hadrons (at 11 am) and two electrons whose paths appear as lines (at 5 am). Even if the discovery, practically certain, requires more experimental data, it already produced new insights. The most important one is that mass is not an intrinsic property of particles, but rather a dynamical effect of their interaction with the field of Higgs in which they move.

At the origin of the universe, the Higgs bosons appear at 10^{-10} second, that is, at ten billionth of the first second (some physicists propose from 10^{-12}s to 10^{-10}s), and the temperature is one million billion degrees. It is on cooling down that the field of Higgs undergoes some change in its properties and becomes now able to interact with particles (notably quarks). The more the virtual particles interact with this field which makes them slow down, and the greater their mass; and vice-versa, if the particles hardly feel the Higgs field, their velocity is greater and they get a lighter mass.

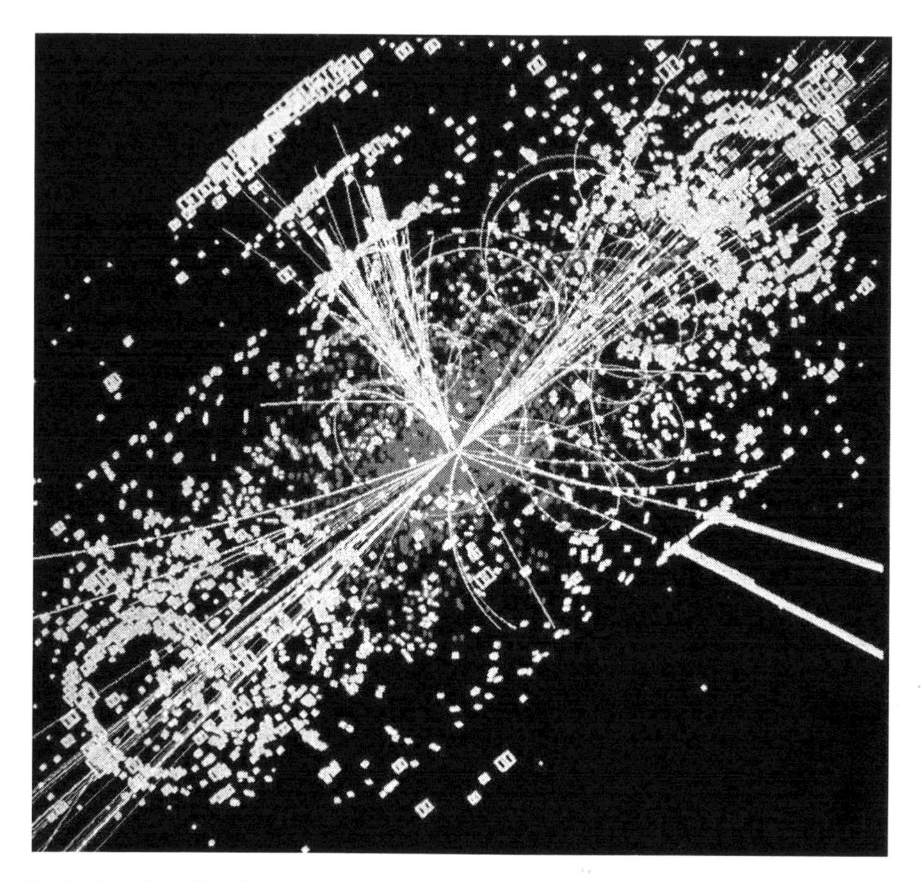

Fig. 3.1 Detection of the Higgs boson: Simulation of a collision during a CMS detector experiment in the LHC at CERN. (Credit Lucas Taylor. Public domain.) A collision between two ultra-high-energy protons produced a Higgs boson which decayed into two jets of hadrons (center toward 11 am) and two electrons whose paths appear as thin lines (at 5 am). The energy these particles deposit is shown in blue.

The interesting point is that mass isn't any real 'stuff' anymore but rather a parameter of particles similar to charge—some sort of 'gravitational charge.' This leads to another discovery that the primeval particles are like tiny waves or ripples on the surface of the Higgs field, the Higgs bosons being the smallest ones in the Higgs field.

The Higgs boson was predicted in 1964 independently by Peter Higgs, François Englert and Robert Brout. According to Peter Higgs, the field of primeval bosons (the actual photons) is nowadays absolutely everywhere, from space, to galaxies, to inside our tables, and even inside our brains and bodies (the Higgs boson has itself a mass around 123-126 GeV). So that while the atoms (and their properties) are derived from their number of electrons in different orbitals—and losing an electron from the exterior, or valence, orbital means a shift to another element—, the particles' mass is derived from their early interaction with the Higgs field of the primeval universe.

The shifting from one atomic element to another was something Pauli gave a lot of thinking to, as it was exemplified clearly in the *Beta decay* of radioactivity, which he was the one to clarify. The radioactive nuclei of atoms are instable, and they transform themselves into other elements via a complex process: *a neutron gives rise to a couple proton-electron, plus one antineutrino (of electron source).* Both the electron and its antineutrino are ejected out of the nucleus. What is fascinating in this process is that *the new atom is a different element*, and it can be itself either radioactive or stable. This is what's happening for example when carbon-14 decays into nitrogen-14. The atomic number of carbon is 6, whereas that of Nitrogen is 7, making them neighbors in the Table of periodic elements. During this Beta decay, the atomic number (the number of protons) increases by 1 unit (from 6 to 7) while the mass number (number of protons + neutrons) stays identical at 14.

It is while posing this dynamics that Pauli predicted in 1931 the existence of the neutrino, having no charge, and thought to have no mass (although new evidence suggests some neutrinos at least have a tiny mass). The 'neutrino' was thus named and integrated by Enrico Fermi in 1934 within his theory of the radioactive decay, and it was experimentally evidenced starting in 1959 for the electron neutrino.

It is interesting to note, as the astrophysicist Massimo Teodorani showed in his book *Synchronicity*, that Pauli considered both the exclusion dynamics (the shift in orbit being concomitant with a shift in the energy level) and the Beta decay (the transformation of an atom/element into another one with the creation of one particle by another one) as processes totally unrelated to causality. Indeed, Pauli thought they were revealing an acausal dynamics, that is, the *synchronicity* he and Carl Jung were studying. "Both radioactivity Beta, with the emission of an antineutrino, and synchronicity are linked in a manner that has not been understood yet." (cited by Teodorani, in *Synchronicité*, p.63.)

Pauli was physics professor at the ETH of Zurich from 1928 to 1958 (the Swiss city where Jung lived and had his psychoanalysis practice) with an interval from 1940 to 1946 in which he was professor of theoretical physics at the Institute for Advanced Studies (IAS) in Princeton. At 28 years old, following his divorce, problems in his emotional life brought Pauli to seek the help of Carl Jung, who referred him to his assistant Erna Rosenbaum for an analysis; this is how a research collaboration and a friendship was to develop over the years between Jung and Pauli, giving rise to the fascinating correspondence between the two scientists. In a few months of analysis, Pauli had already produced about a thousand dreams of exceptional quality that Carl Jung used to exemplify and expand his keen analysis of the transmutation processes of the psyche, as guided by the Self, the subject of the personal unconscious. Indeed, Pauli's dreams put into play remarkable symbols of the alchemical process that were a delight for Jung to interpret in his books (the Pauli dream materials are Jung's main source for the dreams he analyses).

Prior to his Jungian analysis, Pauli's psychic and emotional life had been rather chaotic (his first marriage, with a dancer, lasted less than a year!) and mostly oppressed and silenced by his too focused and demanding intellect. Yet, as soon as he discovered the fecund Analytical Psychology of Carl Jung, he plunged with extreme ease into the complex dynamics of the psyche and soon became a skilled interpreter of his own luxuriant archetypal dreams. The truly amazing thing about Pauli is that the moment he understood how to pay attention to what his own unconscious had to say (through deciphering his own dreams), he started using the sophisticated information he was getting to prod new advances in his scientific research. And he did so to fulfill two goals.

Pauli's first goal was to get in a dialogue with his own Self, via his *anima* (the archetype of the feminine in the psyche, and also the incarnation of the *Sophia*, or wisdom). He thus triggered the onset of what Jung calls the *individuation process*—the progressive harmonization between one's ego (the I of our waking state, absorbed by its conscious relationships, goals, and actions in the everyday world) and one's Self (our higher consciousness, the I of our personal unconscious, in sync with the collective unconscious). The concept of a personal Self (soul or spirit) is present in a wide variety of cultures under as many names (e.g. the Hindu *Atma*, the Yaqui Indians *Nagual* self, the Christian *soul*, the Greek *nous*); and similarly the pattern of 'initiation' exists in most cultures on earth, from shamanism to Eastern religions or philosophies. This initiation process is described as a dialogue and synergy between the ego and the higher self, leading to their progressive blending and harmonization until the human being becomes a unified ego-Self. As Jung has shown, the human psyche is dual: each person acting out a male persona has an inner and unconscious feminine counterpart called the *anima*; and each person acting out a female persona has an inner and unconscious masculine counterpart called the *animus*. The anima/animus is like the voice of our Self, and our Self is in constant communication with the collective unconscious, the planetary semantic field. In the first stages of the individuation process, the anima/animus is the seminal partner and the channel for receiving insightful and precognitive information that prods us to develop the full mental and spiritual potentials of the Self.

Pauli's second goal was to further his scientific research and specifically in quantum mechanics, and we will see several examples of his stupendous ability to extract information from his Self and his unconscious, via dreams or real-life synchronicities. Thus Pauli's dreams were not solely symbolic of inner psychic processes, as most people's dreams. His dreams presented him with precise scientific information and were literally guiding his physics research. This is why I will, further in this chapter, use the dreams materials of Pauli to move further along the lines of research he had envisioned.

It seems to me that Pauli was soon able to launch a superior blending of his left-brain analytical capacities with his right-brain intuitive and holistic capacities—what I believe is our next stage of mind evolution on earth. He started reading eagerly the antique philosophers and eminent alchemists pointed by Jung in his books. As evidenced by his letters to Jung, he quickly achieved a deep understanding of psychic processes, a comprehension all the more crucial for him that he was the siege of repeated psi phenomena of a spontaneous and non-intentional nature, namely large-scale PK, even at a distance. For example, during some of his visits at the Institute for Advanced Studies in Princeton, all the computers would crash simultaneously. This is how the 'Pauli Effect' came to be acknowledged by the computer and physics scientists working at the laboratory. One day, all computers crashed, and everybody was expecting Pauli to pass the door any minute, but this didn't happen. What transpired afterwards was that, at this very time, Pauli had been commuting at the town's train station, a few miles distant.

As we have seen, Pauli kept on making prominent discoveries in quantum physics, such as the existence of the neutrino in 1930, and the statistics of the spin in 1940. With Pauli, we have an exceptional unconscious Self able to think and compute using mathematical and physical concepts and equations, and moreover able to peer into a new dimension of physics, one that the keen and cognizant conscious intelligence of Pauli recognized as the science of the future. (SFT indeed posits that the unconscious Self has the capacity to reflect and to think, using a nonlocal connective dynamics within the semantic field, yet branching into the brain's neuronal networks.)

Then a 'Great Work of Wisdom' started forming at the horizon of the two scientists and scholars, Pauli and Jung, for they engaged together in prolonging Jung's discovery of the phenomenon of synchronicity, a process instantiating an acausal principle and putting into play the Self as being trans-personal and operating beyond space and time. Pauli's main objective was to find the foundation and the language of a novel psycho-physical science.

Pauli's visionary scientific dreams

The Contracting and Rotating Space dream

The dream of the *Contracting and Rotating Space* begins with Pauli's anima (generally appearing as 'the exotic woman') *starting to oscillate.* This is, of course, a quite baffling superposition of material physics properties and soul qualities, since a mind or soul is rarely viewed as 'oscillating'. Then, *through its oscillation, the anima is influencing the very geometry and dynamics of space—which starts contracting and rotating.*

Chaos theory is the physics of turbulence phenomena—such as whirlpools, hurricanes, boiling water, etc. It stresses that when phenomena are nonlinear, they may change totally their global organization (called global order). For example: you put a pan of water on the gas. At first, the more heat generated by the same flow of lighted gas (you don't change the intensity), the hotter the water becomes (this is a *linear* process). But suddenly, at a certain threshold (around 100°C depending on the altitude), the water changes its internal organization and boils: Benard cells start forming (the rotating vertical vortexes in the water). This is a nonlinear process, triggering a 'bifurcation' (a change in the behavior or 'trajectory' of the system) and the emergence of a novel global organization. Chaos theory (which can explain physical as well as psychosocial phenomena) emphasizes that before undergoing a drastic change of its global organization (or of its trajectory), a system goes through a phase of instability marked by oscillations—a phase in which its parameters are constantly shifting from one state to another one. This oscillatory phase precedes the modification of the *attractor* of the system which drives its dynamic organization (such as increasing or decreasing in size, or a shift in the type of attractor). In Pauli's dream, the anima, porte-parole of the Self (the subject of the individual's whole consciousness or *semantic field*) is going through such an oscillatory phase. This announces a shift of the attractor governing the behavior and formal organization of the system. The system being transformed in the dream is space; and its transformation is (1) a contraction, and (2) the emergence of a rotating movement: The multidimensional anima (or Self) is thus mysteriously coupled with a spatial system that starts contracting and rotating.

There exists a well-known physics system in which space is both contracted and rotating: the Black Hole. The supergravity both bends space into a more and more dense and thus contracted sphere; meanwhile a quasi-flat spiraling *accretion disk* is formed, a sort of spiral plunging into a sink (i.e. a funnel shape). (Figure 3.2)

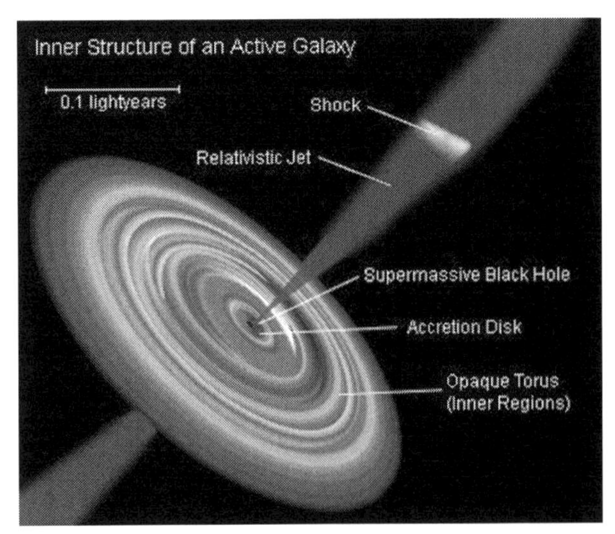

Fig. 3.2. Supermassive black hole at the center of a galaxy, emitting two opposite jets of plasma orthogonal to the accretion disk. (Image: Credit Mrbrak, Wikimedia Commons, GNU Free Doc. License.)

The spiraling disk (i.e. rotating) attracts all matter that happens to be on its outside (outside the 'events horizon') and sucks it into the black hole's singularity in which supergravity will bend, distort, and finally crush all matter (be it a galaxy) into near nothingness. Only a *x-ray radiation*[14] will escape and be shot out of the black hole, as two thin opposite jets perpendicular to the accretion disk. The *Hawking radiation* is called *black hole evaporation* because it is a black-body radiation that reduces the mass and the energy of the black hole.

Due to the fact that all matter is sucked into the singularity, we can detect stellar black holes only indirectly, through the distortion of the gravitational field of a galaxy going past it. The galaxy's apparent form shows a *lensing effect*, that is, a gravitational bending of its emitted light and radiation and a curving around the black hole. (Figure 3.3)

Fig. 3.3. Gravitational lensing effect and Einstein's ring as part of a nearby galaxy is sucked into a black hole. (Wikimedia Commons, GNU Free Documentation License.)

In a black hole, huge amounts of gas can be seen by orbiting telescopes to rotate around the event-horizon, then to spiral inward toward the center of the stellar black hole and being sucked into it, while the gas emits massive radiation and its temperature rises steeply.

When all the gas surrounding it has been sucked into the singularity, the black hole becomes a quasar, whose enormous and striking bright light is brighter than the galaxy it is in.

One Wikipedia article called *Black Hole* offers a stupendous video simulation of the gravitational lensing distorting the image of a galaxy passing behind a companion black hole: a secondary image of part of the galaxy is created, curved in a ring (called *Einstein's ring*) surrounding the black hole, and the luminosity of this part of the galaxy appears neatly amplified to a very distant observer.[14] In 2002, astronomers R. Schödel and his associates detected that our galaxy, the Milky Way, has at its center such a supermassive black hole, with a mass about 4 million *solar masses*, near the *Sagittarius A** region of space. It was confirmed in 2008.[15]

We know now that there are *supermassive black holes* at the center of every galaxy, some emitting jets of plasma at near the speed of light (i.e. relativistic speed) escaping along its poles, as we saw in Figure 3.2 (p. 102). The black hole in *Centaurus A*, weighting about 55 million solar masses, is active and ejects jets (in the X-rays and radio wavelengths) that take the shape of spiraling *lobes* and the usual long jets. (See Figure 3.4.)

As for *micro-black holes* at the subatomic scale, the search is on at CERN since 2008 for the detection of the Hawking radiation (in the form of gamma-ray flashes) stemming out of them like a specter, part of this radiation being sucked back into them.[16]

Let's return to Pauli's *Contracting and Rotating Space* dream. Why the anima, or the Self, would be linked to singularities of spacetime is a quite baffling question. However, let's memorize that Pauli's whole consciousness (the conscious and the unconscious) knew that the psyche and mind were linked and coupled with both the quantum dimension and the Einsteinian-Newtonian physical matter. About the latter, let's remember that he was himself the subject of large scale PK, called *Pauli effect*, that triggered on his passage the disruption of all machines in the IAS lab, and he was not only highly interested in this synchronistic effect on matter but also amused by it. Thus Pauli's conscious intention was to find why a synchronistic PK could exist and how it was working; and this intention was relayed to his unconscious Self—who then expressed its own answer in the dream.

Fig. 3.4. Centaurus A galaxy's central black hole emitting lobes and jets. Composite of 3 images from LABOCA, Chandra X-ray, and WFI. (Credit: ESO public outreach, 2009; Wikimedia Commons.)

Thus, this dream of the contracting and rotating space presents to Pauli's conscious mind (on awakening) that a person's Self is able to have such an influence over space and matter as to create a black hole (one that doesn't need to be at an astronomical scale, but would rather exist as an anomaly at the quantum scale or else in an extra-dimension of consciousness). Let's keep in mind that the process of influence of the psyche over matter (and thus over space and time) is still cryptic, but has been proven by numerous experimentations.

We cannot not remark that Pauli's unconscious was, through the dream, describing precisely the geometry and gravity parameters of a black hole; yet, the Self (or anima) was at its center, playing the role of a 'point attractor' at the center of the black hole—the singularity itself. His unconscious was thus (beyond looking for a psychophysical science) foreseeing the scientific model of the black hole, hypothesized by astronomer and mathematician Pierre-Simon Laplace at the very beginning of the nineteenth century (the same Laplace who assumed a clockwork universe), then by Karl Schwarzschild in 1916 on the basis of the just released General Relativity. However, the physics of the black hole (the theorems of singularity) were developed only in 1970 by Hawking and Penrose (thus some three decades after Pauli's dream). The phenomenon (as we saw) will be proven by the astronomical observation of supermassive black holes at the center of galaxies, at the turn of the millennia.

Massimo Teodorani, while commenting on Pauli's dream about *The contracting and rotating space* sees in it: "a novel concept of space subjected to rotation, an absolutely novel concept of theoretical physics; maybe... [the intuition of] the basis of the physics of synchronicity." (Teodorani, 2010) My understanding of what he means by 'physics of synchronicity' is a physics of acausality and nonlocality, an entanglement between mind and matter in which meaning and consciousness play an organizing and guiding role, maybe that of a 'control variable.' Thus Teodorani, with his keen intuition, foresees that the "novel physics of synchronicity" must somehow have something to do with rotation. An astounding and noteworthy statement from the part of an astrophysicist!

Pauli's Circulation of light dream

Another of Pauli's dreams, that of *The circulation of light*, may shed more light on this spatial contraction and rotation process. Let's remark though, that the fact we are able to apply successfully a dynamical law (or pattern of behavior) to a specific system or process doesn't exhaust this law and its meaning, nor restrain its domains of application to the sole one being described. This particular dream is related in a letter from Pauli to Jung dated October 28, 1946.

The dreamer reads a book about the Inquisition trials against the cosmological heliocentric theories of Galileo and Giordano Bruno, and on Kepler's conception of the ***Trinity****. A blond man [the expression of Pauli's psyche involved in science] says: "A complaint is lodged [filed]* *against **men whose women have objectified their rotation**." Then the scene of the dream changes and the dreamer is himself in a court room and being accused. He calls on his wife for help.*

*Then the dream shows again the first decor but with his wife present. The blond man says: "The judges have no idea what is a rotation and that's why they can't understand men. But you, you know very well what is a rotation!" "Of course, I answer, **the cycle—or the circulation of light—this is part of the primeval or first causes**."*

The dreamer then wishes goodnight to his wife, and on remembering the accused scientists, weeps. The blond man adds "Take hold now of the first key. (Pauli & Jung, 2014) (The text in brackets is mine, so as the translation).

> *[Synchronicity, 12/30/12: I take hold of the dictionary (Harrap's Shorter, 2072 pages) to check the translation for 'Causes premières' (first causes') and, opening it by chance, my eyes fall first thing on the word 'circulation.']*

The last sentence of the dream "Take hold now of the first key," means that Pauli has found a key. The dream predicts that he'll be opening soon a new gate in his self-exploration process and thus he has achieved a huge leap in his communion with his Self/anima (his *individuation* process).

Now, if we interpret the term "the women who have objectified their rotation," logics demands that what is (or has been) objectified in the process—had perforce to be non-objectified beforehand. That is, either the rotation was not set into motion, or else it already existed but in such a subtle way as to be in the immaterial realm, as opposed to the objective, material realm. We have indeed numerous references in Eastern religions, and Eastern medical and philosophical lore, to a 'spiritual energy' (called Kundalini, Chi, or Ki) linked to the energy centers or chakras within a "subtle energy body." *Chakra* is a Sanskrit term signifying 'wheel' or 'lotus' because all 7 (or 8 in some systems) chakras are energy vortices that, on being activated, rotate with extremely rapid rotation.

Each Chakra is represented by a mandala, a lotus flower with a given number of petals, colors, etc.

In the Eastern yoga systems (Hindu, Buddhist, Taoist, Zen, Jainist, etc.), the awakening of consciousness is synchronous with the activation of the chakras and with the Kundalini energy ascending along the vertebral column, where it will activate each chakra one after the other. Now, the Kundalini energy, once awakened, is considered to be the Shakti, the Great Goddess herself.

The yogi (whether a man or a woman) undergoing this (inner) initiatic transformation—the awakening of his/her consciousness to higher and transcendent states—is called a 'Shakta,' that is, the complementary aspect of the Shakti, and the process will lead to the 'conjunction' or 'sacred marriage' between Shakta and Shakti at the head chakra—the epitome of the awakening process in Eastern lore.

Now, if we follow the Jungian framework (which Pauli had adopted to trigger and study his own individuation or awakening process) then the 'woman of a man' is his feminine counterpart, inner, and unconscious, that is, the anima, the feminine facet of the Self totality.

The inner woman (the 'spouse' of the dream) who has 'activated' (or 'objectified') her 'rotation' process is thus the inner Shakti (the divine feminine energy of the Self) who has triggered the chakras' rotation, the turning of the 'wheels' of the energy centers. Indeed the *circulation of light* is a well-known yogic expression referring to the circulation or flow of light unimpeded and harmonious in several or in all chakras. The kundalini energy emits large luminous auras around the awaken and rotating chakras—the activation of the lotus at the top of the head emitting the magnificent auras around the heads of Buddhas, divine beings, and wise men and women that are represented on the paintings, murals, and thangkas. (See Figure 3.5)

The dream thus means, at the personal level of the dreamer Pauli, that his time of hardship (the trial time) is over, and that his own process of awakening and of activating his inner Shakti-energy has started ("you know well..."). If accused, he will be able to make reference to the inner process as an objectified or empirical proof of the existence of an inner path of knowledge that receives information, guidance, and wisdom directly via the connection with a divine energy and source—the Atman or Self, or else an archetypal being.

Fig. 3.5. Aura around the head of Avalokiteshvara. Kano Motonobu, *White-robed Kannon, Bodhisattva of Compassion,* first half of the 16th century. Japan, late Muromachi period. Hanging scroll. Ink, color and gold on silk. Wikimedia, Public domain. http://en.wikipedia.org/wiki/Buddhist_art_in_Japan

Plotinus and the First Causes

Now, the dream has also a wider and 'cosmic' interpretation in terms of throwing light on an ontological and cosmological framework. This interpretation is clearly hinted at in the concluding remark of the 'I' of the dream: *"Of course, the cycle—or the circulation of light—this is part of the primeval or 'first' causes."* Now the first causes were the primeval beings or principles that gave rise to the whole process of creation/evolution of beings and the universe. In monotheistic religions, the First Cause is of course unique, God himself. In Plato it is both singular as a deity 'The Grand Architect' and plural, the great "Ideas" that resemble the archetypes of Jung. In the (badly named) 'Neoplatonic' system of Plotinus (third century CE), the Unity, the One "without form," is the first cosmic entity (or *entelechy*) of a Triad of Light:

1. the One,
2. the Supreme Intelligence,
3. the Universal soul.

These three cosmic entities are endowed with consciousness and intelligence of the highest order, yet they are not personalized. Each cosmic realm has a counterpart in the individual psyche:

1. the One has for counterpart in the individual: the state of contemplation of the One—the fusion with cosmic consciousness.
2. the Supreme Intelligence, that of the intelligence of Being itself, the Self and the cognitive level of intuition.
3. the Universal soul, that of the individual soul (level of reason).

In Plotinus' system, the universe had at its core a Supreme Consciousness—"the One"—who acted as a Supreme Light, whose "irradiation" was activating and *animating* the process of an intelligent birthing of beings. In Latin, the meaning of *anima* is soul; and thus 'animating' means literally 'giving their anima or soul to' the progressive emanation of being(s). First the birthing of "intelligence" (in Plotinus: gifted with intuition) and then of "soul" (in Plotinus: gifted with discursive reasoning). It is the level of intelligence that corresponds to Being in Itself (*dasein*) and thus to the Self of Jung. Plotinus attributes to this level of Intelligence the intuitive and holistic thinking—the Self being able to re-connect with the dimension of the

One and thus to get immersed in transcendental fusion states or unity states with cosmic consciousness.

Funnily (compared to our excessively high valuing of reason since more than three centuries), Plotinus states clearly that reason (the third level only) is not complete in itself, and that "(it) will access intelligence when (the soul) contemplates (the higher realm of intelligence)." In other words, reasoning will not be complete or intelligent and 'enlightened' unless it is merged with the Self's intuitive processes. I can't be in greater agreement! The way I often formulate it is to predict that human beings will make a leap in intelligence when they develop a seamless merging of their left brain with their right brain processes (that is, respectively, logic and analysis, with intuition and system's thinking).

Thus Plotinus' framework points to a 'circulation of light' among the 'three' primeval causes or principles, the first cause being the One—the Cosmic consciousness that Porphyry, his first disciple who was the editor of the *Enneads*, calls "the God who has no form." Porphyry relates in his *Life of Plotinus* that, four times at least, Plotinus had the 'vision' of the One while in contemplation of the first principle.

Thus in Plotinus' framework, the *cosmic* 'circulation of light' goes down from (1) the One, to (2) cosmic intelligence, to (3) the 'soul of the world'—the *anima mundi* of the alchemists and the *collective unconscious* of Jung. Plotinus views the *individual* 'circulation of light' as a reverse process: an ascension back to embracing the reality of the One—from the reasoning level to intuition and then to contemplation (the states of Unity or of fusion). Plotinus thus gives us a key for understanding the circulation of light as a double flow-dynamics (upward and downward, thus turning a wheel) happening both at the individual and at the cosmic scales. What is prodigious in Plotinus' *Enneads* is the presentiment of a universe as a hologram, conscious and alive, in which the One is existing in any 'part'—that is, as we would say now: in any particle or system. Plotinus wrote in *Ennead* 2.3: "If we consider the universe in its wholeness, we can see that it constitutes a *vast organism*, that all beings are parts of this Whole and, through the *sympathy* that is uniting them together, it creates a unique harmony." Thus the harmony of the One is infusing and organizing the universe.[17] And also "This universe (...) has in itself a *soul* (psyche), who pervades all its parts, meaning that she dwells in all beings who are *parts of the universe*." (*Ennead 4*, 613) Such concepts are astonishingly resonant with the basic tenets of holographic universe models and systems theory!

The ring symbol and the Golden Spiral

The path of initiation or awakening has sometimes been called a 'path of reintegration,' that is, a reconnecting (from the low level of an ego too anchored and too involved with material reality) with the levels of intuition and then with cosmic consciousness (the semantic dimension, the collective unconscious). But regarding the question we are now addressing—the cosmological organization of reality—we must not take the standpoint of the individual consciousness but consider instead the cosmic consciousness and the birth of the universe (from the X-point up to Planck scale). We have to interpret Pauli's contracting space dream as referring also exactly and literally to the organization of the universe from (1) a point/wholeness (the X-point) of pure consciousness (The One), to a cascade of emanations unfolding into (2) intuitive intelligence, then (3) logical thinking. These three semantic dynamics are to be understood as the semantic dimension in-process—the Self layers of the universe, before we reach the layers of matter. It is, in fact, reflecting the cosmological birth of our universe-bubble—as a holistic consciousness field, unfolding into a hyperdimension (pure energy) and then a spacetime manifold (energy and matter particles).

Another dream of Pauli—that of *The Ring of the i*—puts into play his soul (anima) teaching him to play piano; then she hands him a ring saying that it is "the ring of the i." We know that 'i' is the symbol for imaginary numbers in mathematics. As, in the dream, the imaginary numbers are thus supposed to form a ring, a circle, we can infer that Pauli's unconscious superposes to a suite of imaginary numbers the form of a circle and a rotating or spiraling progression (to form this *Ring of the i*). But if we look at natural numbers, the 'sequence of Fibonacci' embeds exactly the progression of a perfect spiral, called the Golden Spiral because it is based on the golden or *Phi ratio*, a spiral that unfolds out of enmeshed golden rectangles. (See Figure 3.6.) So that we could right away conceive of the exact same spiral progression (based on Phi), in imaginary numbers, that is, negative numbers.

A Golden Rectangle and its Sacred Middle Part

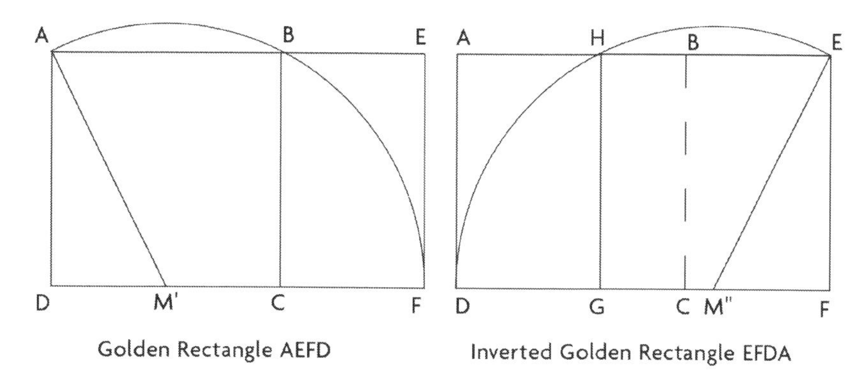

1. Draw a square (3 x 3) ABCD
2. Put your compass needle at point M' (halfway along DC) and draw the circle passing by apexes A and B. You get point F on the prolonged basis of the rectangle.
3. Draw the perpendicular EF to the basis: AEFD is a golden rectangle.
4. Do the same on the opposite side, drawing the square EFGH, then the circle of radius M"E. EFDA is a golden rectangle.

   AEFD (or EFDA) is a golden rectangle: its length AE divided by its width EF equals 1.618 (phi).

   HBCG is the sacred middle part.

   BEFC, the rectangle (on end) added to the square, is itself a golden rectangle.

Embedded Golden Rectangles and Golden Spiral

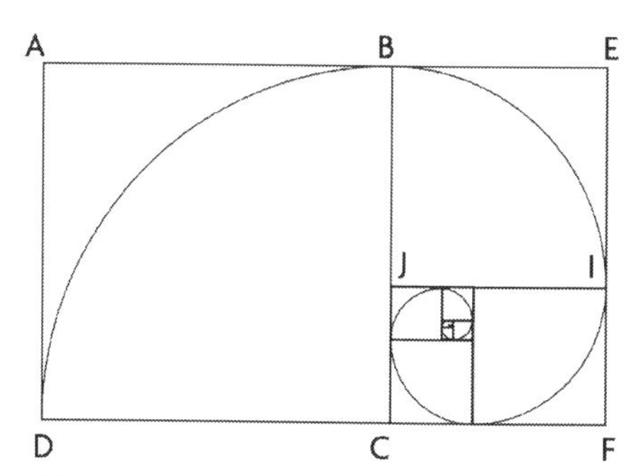

1. In each golden rectangle (on end), extract the square.
2. Draw an arc having as a radius the side of the square and you will get a golden spiral.

Fig. 3.6. The Golden Spiral, based on the Phi ratio, unfolding out of enmeshed golden rectangles.

4
THE ISS: TWO ENMESHED AND INVERTED SPIRALS

"You can recognize truth by its beauty and simplicity. When you get it right, it is obvious that it is right (...) because usually what happens is that more comes out than goes in."
Richard Feynman, Nobel Prize in Physics.

Another type of dynamics is implied by Pauli's *Circulation of light* dream: it is the *rotation of the primeval light itself before the inflation phase.* (Of course, we take 'light' as meaning energy, specifically a dynamic consciousness, and not as rays of not-yet-existing photons.) And we'll see that it will reveal, and be in agreement with, stupendous features of the cosmic ISS. But let's see first the actual stand of physics on the pre-BigBang universe.

Spiral rotation of the primeval universe

It has been mathematically posed that before the sudden expansion (or 'inflation') phase, there had to be a *spiraling motion* within the infinitely curved point-universe, due to the gigantic forces of gravity crunching the tiny infinitesimal sphere (as it has been pointed out by the Bogdanov). Therefore, already at the phase of the X-point without dimension (the 'point-that-is-a-sphere') this sphere is already in a rotating motion (which indeed reinforces the curvature). This rotation is a spiral that keeps enlarging the sphere. This is why we get the light cone: it is the X-point-sphere whose radius keeps increasing and, by a spiraling movement, gives birth to an ever larger sphere and then a quasi-flat disk.

At this very beginning, from the origin to the threshold of Planck, the two dynamic and simultaneously geometrical (topological) principles of the emerging universe are thus:

- the *Phi ratio* steering the enlargement of the radius of the spiral, from the X-point—according to the Phi (ϕ) ratio = 1.6180.
- the *number Pi,* that triggers the creation of the quarters of circle from increasing radii, and therefore of the outward spiral (according to the Pi (π) ratio = 3.1416). (See again Fig. 3.6, p. 113)

Since we don't have space or time yet, scientists use the term *pre-space*. The ISS pre-space, then, is organized in an endogen manner (purely in and of itself, indeed self-organized) by two proportions—that of the radius to circumference (or Pi) and that of the enlarging spiral (or Phi). The X-point of origin itself, to which physicists attribute the mathematical property of infinity, corresponds to the metaphysical and ontological dimension of the One-who-is-the-Whole: the One-infinite. Indeed, if one thing is logically self-evident, it is that neither nothingness, nor a uniform (thus too fixated) stuff, nor absolute disorder (total lack of organization), can lead in any possible way to a finely organized cosmos that nevertheless has enough disorder and looseness to keep on self-organizing and evolving. This remark is also perfectly valid in terms of ontology, that is, considering the basic beingness of all-that-is.

Here are the crucial facts to consider: (1) we observe with our minds, thus with consciousness in-process, (2) astronomical and physical laws regulate an organized cosmos and their mathematical and physical cosmic values are so finely tuned that any slight change of any one of the two dozens cosmological variables would not lead to the universe as we know it. Yet (3) this universe exhibits self-organized systems, emergence, a dimension of randomness and indeterminacy, and the minds' capacity for autonomous decision and free will.

So that we can pose right away some absolute prerequisites for such a self-organized cosmos. These are:

1. the universe before Planck scale must have both self-organizing properties, and finely tuned values,
2. at the point-scale, even if in unity and harmony, the universe cannot be uniform and without extremely fine variations.
3. to the opposite, it has to display a very structured, yet exquisitely complex and dynamical, self-actualizing, configuration.

The second point is what astrophysicists had predicted and have now corroborated regarding the primeval storm of photons that gave rise to the *cosmic microwave background* (CMB) radiation (or *relic radiation*): without infinitesimal variations of density, temperature, and frequency, the birth of galactic clouds couldn't have been possible; and therefore, neither would have emerged the universe with its innumerable galaxies as we now observe it. The latest CMB map, effected via the satellite PLANCK sent in May 2009, and its analysis of the anisotropies (tiny fluctuations) were released in 2013. Jean-Michel Lamarre, the visionary inventor of the PLANCK probe, explained that "because PLANCK has been conceived for measuring the polarization of the radiation and to introduce the least distortion possible in this measurement, we can proceed to measure the remnants of the gravitational waves produced during the inflation phase of the Big Bang."[18] The Wikipedia article[19] on *Cosmic microwave background radiation* states about the release on March 21, 2013 by PLANCK's team of its latest analysis of the CMB map: "The map suggests the universe is slightly older than thought" (it is now estimated to be 13.798 billion years old). "According to the map, subtle fluctuations in temperature were imprinted on the deep sky when the cosmos was about 370,000 years old. The imprint reflects *ripples that arose as early, in the existence of the universe, as the first nonillionth of a second (10^{-30} s)*. Those are the ripples that gave rise to the present vast cosmic web of galaxy clusters and dark matter."

The third point is derived from the experiments on the Higgs boson conducted with the LHC (Large Hadron Collider) at CERN, Geneva. According to the 1913 results, the Higgs field was, at its emergence, in a high-energy and 'metastable' state, meaning that it was highly improbable and far from a low-energy and stable ground state.

Metadim Center-Circle: The point-sphere of the origin

Let's resume what physicists have inferred or posited about the universe before Planck scale (and long before the Big Bang):

- (In a Relativity framework) the point of origin is a singularity; before space, time, and matter existed, the universe had an immense density, crushed by the supergravity into a ball;
- (According to the LHC 2013 results) it was in a high-energy, non-random and meta-stable state (this inferred from the later Higgs field configuration)
- the radius of the sphere, and thus the size of the primeval universe, have kept increasing;
- before the universe size reached Planck length, of the four forces, only gravity existed. Particles and matter couldn't exist before Planck's threshold.
- toward the X-point, energy tends to the infinite;
- the enlarging sphere was a spiral.

ISST's hypothesis is that the enlarging sphere was formed by a dynamical self-created spiral, specifically one steered by the Phi ratio.

Now, we have to ponder more on this spiral, which seems to imply (just as the concept of a progressively increasing radius) a progressive increment—somehow arithmetical and smooth. But let's consider in depth the golden spiral in its geometrical form. The archetype of the spiral, and the best way to design it perfectly, is by using a matrix of Golden Rectangles, embedding the Golden Ratio Phi $\phi = 1.618$. The Fibonacci sequence nearly embeds the logarithm of Phi.

Look again at Figure 3.6 (page 113), but this time we'll imagine drawing the spiral from a small size (JIFC) to a larger size, in order to mimic the increase in size of the universe, from the X-point toward Planck scale. Each time, take the golden rectangle's longest side (JI) to draw a square (JIEB), and draw an arc (BI) with this length (from apex J). Thus, each time a square is added to the largest to-date golden rectangle—and it creates an even larger Golden Rectangle. (This means that in order to draw the spiral, one has to put the compass needle, in the added square, at the apex the nearest to the center of the spiral, open the compass to the length of a side of the square and draw the arc that connects the two apexes nearest to this center.) To this larger golden rectangle, take the length of the longest side of the whole structure to date (BC) to add another square (BCDA), draw the new arc (BD) from center C, etc.

What it all means is that the Golden Spiral is not a progressive, arithmetical, augmentation of the radius from its center, but that it grows by the increment of a quarter of a circle, whose radius, divided by the radius of the previous quarter of circle, gives no less than the Phi ratio. And the more the spiral grows (the bigger the integer numbers in the Fibonacci sequence) and the more a radius, divided by the one before, will approach the number Phi, that is, 1.6180.

This traditional and antique way to draw a perfect spiral dates back to the Greek mathematician Pythagoras, probably earlier to Egypt and Sumer; and we know the Fibonacci sequence was studied in India in 200 BCE by the mathematician Pingala. It enables us to discover this new property of the golden spiral, in which each rectangle that has been drawn (by adding a square) is a golden rectangle. But now, each radius of an added square is incrementing the previous radius by the ratio of approximately 1.618. In our drawing (on Figure 3.6), the two larger squares are respectively 3.5 cm and 2.19 cm, and 3.5 divided by 2.19 is 1.6—and the greater the squares would be, the more the result of this operation would tend to approach 1.6180.

So what does this tell us? No less than the expansion of the universe—even before Planck length and the inflation—is not at all a linear process: its radius constantly *jumps* from one value to another one, and in doing so, it constantly keeps the ratio of Phi between one state of the system and the next (and of course between this state and the previous one). In other words, the budding universe's structure is formal—a field of form—whose process of enlargement is not linear and smooth (as in the arithmetical increment of one unit to the next unit 1, 2, 3, etc.) but to the contrary happening by sudden leaps, yet harmoniously driven by a constant *ratio* or *proportion*—that is, a number totally *unrelated to the spatial scale of the system*. In brief, the underlying structure is 'discrete' as opposed to uniform, just as the quantum world is discrete, based on quanta, that is, wave-packets translated in discrete packets of energy, all multiples of the quantum of energy, *h*, the Planck constant (6.626×10^{-34} Joules second).

Therefore, this geometry—or more precisely, this topological matrix of a universe—is fine-tuned to harmonics of the Phi ratio, shifting to greater orders of form-resonance, in dimensions that do not need any preexistent space and time. We can say that Phi, being the logarithm of the leaps in radius size, is the ratio of formal evolution. However, the evolution of the universe is not only formal: it is also *dynamical, rhythmical, and meaning-laden*. In terms of dynamics, it implies a fantastic energy to launch the regular iteration of each new leap and thus we have a sort of dynamical topology.

At this point, we can pose that we have a topologically evolving universe, driven by a logarithm, and making constant leaps. We definitely have a growing evolution, a constant shift of scale, and thus we have to call it a *dynamics*; however, this dynamics (for now) is only of a topological order.

But here is where a second surprise is in store for us: let's draw from each arc the full circle. As we can now infer, all the values of these circles (diameter, circumference, surface, etc.) will show the *same Phi ratio* when compared with these values in adjacent circles. This Phi ratio is so deeply ingrained that it appears magical, in the sense that it is self-replicating or reiterating itself *ad infinitum*. It just works as a non-dimensional constant, that is, a constant number that can be applied to different measurements using different matrixes and at other scales—just as the Planck constant shows the age of the primeval universe as a fraction: 10^{-43} of the first second, and also its frequency 10^{43} in Hertz.

- But let's consider again the antique drawing of a single golden rectangle. (See again Figure 3.6, the top part.) The two squares retrieved from each side delimit the sacred middle part of the Golden Rectangle. Now, each square contains a virtual circle (of course any square can house a perfect circle touching the midpoint of each side). But if we look at the whole rectangle, then the figure it contains is an ellipse. Thus, a golden rectangle can be formalized as an ellipse—the sacred middle part being most of the elongated part of the two half circles. This ellipse is perfectly round on its outside small sides.
- So that basically we have two sets of endlessly growing circles (along the Phi logarithm). The first one is the set of virtual circles drawn by prolonging each arc of a quarter of a circle. The full circle's circumference would be driven by $2\pi r$, i.e. the radius multiplied by 2

times Pi. Thus from the center we get the new radius, which gives us the new circumference, all this extended in quasi-space. This is why 'the point that creates the circle' or Metadim Center-Circle is a fundamental process. The second set contains all virtual circles comprised within all squares—and they themselves show a spiral. The radii of both sets are incremented by Phi.

Spiral patterns and Pi in crop circles

Now, astonishingly, this infinite spiral staircase of the origin and the Pi and Phi numbers appear very often in some patterns of crop circles.

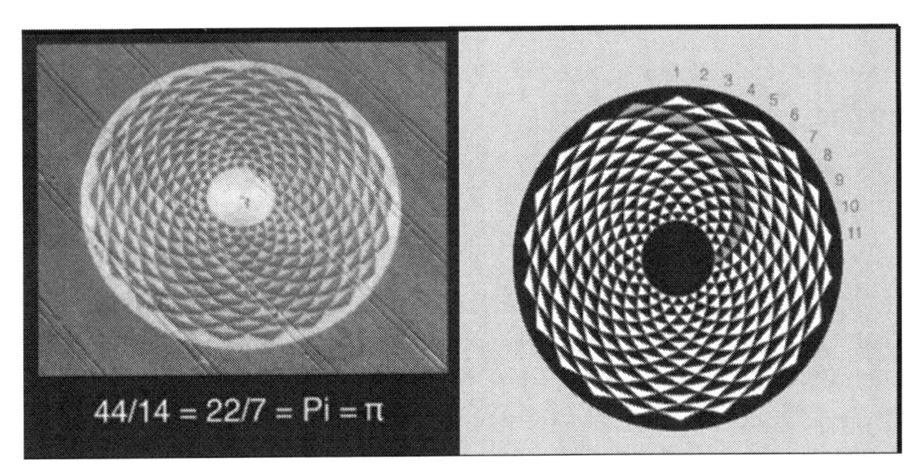

Fig. 4.1. Spiral patterns and Pi in crop circles. *Picked Hill* crop circle, UK, 8/13/2000, Credit: Michael Glickman. http://www.michaelglickmanoncropcircles.com/blog/why-pi/

Contemplate for a while the spiral patterns based on Pi (π) within the *Picked Hill* crop circle that appeared in England on August 13, 2000 (Fig. 4.1.), and the embedding of Pi in the *Barbury Castle* crop circle (June 1, 2008), and see their astounding decoding by expert Michael Glickman (*et al.*) in his article "Why Pi ?" posted on his blog[20] on June 28, 2012 (see an extract on my own blog.)[21] In the Picked Hill crop circle above, 44 clockwise spirals make patterns of interferences with 44 counterclockwise spirals to create 14 concentric circles. The ratio of 44 to 14 gives the number Pi. However, remember that a golden spiral is based on the proportion Phi, so that Phi and Pi are enmeshed to create a magnificent and dynamic form.

Now, the same pattern of embedded and inverted spirals with 89 one way and 55 the other way (two following numbers in the Fibonacci sequence) gives the sunflower's pattern of florets *and* Phi, according to mathematician Helmut Vogel. (More on this soon).

The topological dynamics of Phi and Pi enmeshed

This golden spiral dynamics tells us something more about the growth of our budding universe. It tells us:

1. that the point-circle (or Center-Circle) is a root-process or arch dimension of the universe (this is why the term metadimension, or in short *metadim* is appropriate); I will call it metadim Center-Circle;
2. that all progression goes by leaps (the smoothness of the spiral as a geometrical form is only apparent; its true structure is discrete);
3. this will prefigure, starting at Planck scale, the quanta (wave-packets) as discrete packets of energy, i.e. the discrete nature of the subatomic world, and
4. it also foreshadows the abrupt shifts in orbits of the particles (such as electrons), accompanying leaps in their energy-state—as described by Pauli's Exclusion Principle. And in this respect, we can only note how near is the Phi ratio of the orbital leaps of the spiral (ϕ = 1.618), to the Planck length of 1.616×10^{-33} centimeters. How strange!

But we have not yet reached the second great surprise I was announcing, and here it is: Each set of virtual circles has a different initial condition—one set with the smallest arc (¼ circle), and the other one with the smallest virtual circle in the smallest square. However each circle in these two primeval sets is instantiating the topological law of Pi (π): that is, the proportion between the radius and the circumference, thus 3,1416.

In effect, we have two proportions, two numbers, that are the engine of the self-organization of the point-sphere universe; the snapshot circle driven by Pi, and the expansion dynamics driven by Phi—that is, the Phi (ϕ) ratio of 1.618, and Pi (π) equals to 3.1416.

Surely, the proportion between these two root-numbers (or arch-numbers), cannot be overlooked, given that they are working conjointly in the budding universe: This proportion is 3.1416/1.618 = 1.94. Or else: 1.618/3.1416=0.52. Nothing special at first sight, but let's keep them in mind.

As we have seen, the Phi ratio instantiates the dynamics of topological growth, in a universe still deprived of space-time-causality.

In this system, the virtual circles prolonging the arcs show all the possible states of the system—that is, they form the infinite virtual suite of these states—what we call in mathematics and chaos theory the 'state-space' of the system.

However, this system will shift to new global orders at precise thresholds: First, at instant 10^{-43} second, at the very moment one of these virtual circles reaches, as a diameter, the threshold of the Planck length of 1.616×10^{-33} centimeters, space, time, quanta, and thus particles and matter come into being. At instant 10^{-36} second, the sudden inflation phase (the Big Bang) makes the sphere increase its size 10^{50} times. An immense time later, at 10^{2}s (1 minute and 40 seconds), the photons decouple, setting the primeval explosion of light. This is the very tempest and field of photons which we can now, 13.79 billion years later, observe and measure, through its relic imprint lingering at 370 thousands years after the Big Bang, extremely cooled down to 2.7 degrees Fahrenheit. And with this fantastic radiation of photons, the spacetime of our universe was suddenly hugely expanded.

Thus, the ordering principles of the pre-Big Bang universe (precisely before Planck scale) had at least two parameters (or rather dynamical forces) steering and driving the topological growth: Phi and Pi.

The cycle or circulation as a first cause

So let's get back to Pauli's dream in which he (as the dreamer) said: "*The cycle, or circulation of light, is part of the first causes*." First thing, this statement is equating the cycle (measured by a number of cycles per second or hour) to a 'circulation of light.' Since light is made of photons, and photons have no mass but have energy, then we are talking about a 'rotating energy,' having a measurable frequency. We know that at the Planck scale, the universe-bubble has a frequency of about 10^{43} hertz (it vibrates ten million of billion of billion of billion of billion times in a single second). And given that just before reaching Planck's length, it had a spiraling growth, then it means it was rotating the same number of times in a second—at that precise virtual circle (or exact snapshot of the spiral) which had, as a diameter, 10^{-33}cm.

Philosophically speaking, the first causes are the first or primary ordering principles that governed the emergence of both matter (the universe), consciousness, and human intelligence. As we saw, in a religious paradigm, the first cause (singular) is God the creator—who created all that is, including Man.

If we adopt a scientific perspective we have to admit as a proven fact that the universe was, in its original phase, a point with no dimensions, before time and space existed. But we certainly cannot say that it was a creation or created; because the term 'creation' is loaded with the assumption that before the instant of creation, there was nothing, apart from a 'creator' (with the flimsy exception of the philosophical concept of a *creation ex nihilo*, i.e. out of nothing). And the concept of 'creator' is itself a complex and plural constellation of meaning; its central concept, implied, is that *the* creator was an intelligence—unfathomable, yet gifted with intention: the act of creation is an act of will.

Furthermore, the other concepts that are implied or contained in the idea of a Creator are:

1. before the universe and 'Man' were created, only God existed.

And (for most cosmologies):

2. God the Creator (singular, male) has created only one universe and one humanity.

Now, we have to admit that a factor (or cause) which has been predominant at the very beginning of the universe, must logically linger as an 'ordering principle' at all phases of development of the universe. There is no problem with the fact that some laws and dynamics may emerge at a given threshold or time during the evolution of the universe, but the 'first causes' or more exactly the 'first organizing principles' have to remain active all along (even if in a mysterious and badly understood way). Thus, for example, the spiral is governing the form taken by numerous systems—from the galaxies, to hurricanes, to turbulence patterns in liquids, to biological life, to flora, to minerals, and, the most important of all, to the DNA and even the basic movement of particles.

Concerning the *Circulation of light* dream, we may infer that, in Pauli's unconscious, the rotation of light—creating a cycle—is one of the primary ordering principles of the universe. And indeed, the spiraling growth from the X-point up to the Planck sphere is such a rotation, whose frequency (or number of cycles) was getting only lower and lower down to Planck's frequency (10^{-43} hertz/second). The nearer the spiral is to the X-point, the higher the frequency—at which point it is tending toward an infinite value. So that, yes, given that physicists have hypothesized a spiral motion before the Big Bang, we can definitely take the rotation (the cycle, the circulation) as one of the initial dynamical and organizing forces (or first causes) at work in the primeval universe.

As for the term 'light,' of course we would be at a loss to speak about photons before Planck scale—where only after that threshold the first bosons particles appear, together with space and time. However the concept of 'light' is much wider than that of photons, and even more remote from the specific photons having the range of frequencies of the visible light. For example, the photons of the now relic radiation which were at first in the high gamma-ray frequency have now reached the invisible microwave frequencies. And also, there's the recent discovery of biophotons—or photons emitted within biological tissues by many biological processes—by Fritz Popp, whose breakthrough research is now leading to more and more experimental data and proofs. (See Popp, 2002)

The coupled phenomena of rotation and rhythm

With the rotation (the spiraling movement) comes the associated phenomenon of rhythm. The virtual circles of the Golden Spiral, and their unfolding in huge numbers, create rotational rhythms or cycles, that is, an immense range of frequencies, operating as vibrations or pulsations. This frequency (calculated by physicists to be at 10^{43} hertz at Planck scale) is of course derived from the number of cycles *per second* at that scale that now includes the time parameter of the 4D world.

However, the point-scale universe was spiraling and growing for an immense span of imaginary *time* before reaching Planck's scale. More precisely, a specific pulse happens each time there's a shift to a larger scale of the spiral—each time the radius of a quarter of a circle is accrued by its multiplication by Phi. So that the frequency is not that of a virtual complete movement along one single turn or 'spire' of the spiral, but a pulse at each quarter of the spire—this being only approximately so, because the center of the spiral is displacing itself constantly, as it is, on a spiral itself. Let's call the ¼ circumference, such a quarter of circle, a *bow*. This bow is defined by its frequency—it is a vibrating curved string; thus *it is also a virtual particle: a sygon*, which has the frequency of its bow. (A bow has the frequency of the segment linking its two ends; e.g. for the bow or arc BI, the straight line linking the two apexes BI of the square, that is the diagonal of each added square. When attached to the cosmic spiral, it is a bow-frequency—a specific frequency in the cosmic DNA. And when emitted as a torsion wave from this bow, with this same frequency and at superluminous speed, it is a sygon, a virtual particle.)

What we get, then, is something rather astonishing: before Planck's scale, and moving out of the near-infinite smallness of the X-point, the first pulsations of the primeval universe are ever decreasing frequencies, punctuated by the Phi ratio. And the Phi ratio, as we know it, is expressed, as numbers, by the Fibonacci sequence. In consequence, this sequence shows all possible states of the system (it is its state-space).

The Fibonacci sequence

Why? As the budding universe's diameter (and thus the wavelength of each quarter of circle's bow) as well as its ISS spiral get smaller and smaller the nearer to the X-point, and furthermore as the X-point is tending to the infinite (as scientists have stated it), the highest frequencies are the nearest to the X-point. Frequency, as we know, is inversely proportional to the wavelength λ (lambda). (For example, for an EM wave moving through the vacuum, the frequency f is equal to the speed of light c divided by the wavelength λ.) Thus, the shorter the bow, the shorter its fundamental wavelength, and the higher the bow's frequency.

Now the sequence of Fibonacci, as the mathematicians know well, is an infinite suite. You can always, starting with number 1, add this number to itself 1+1=2, then calculate the sum of the last number found and of the preceding one, and repeat the process ad infinitum. You will get: 1, 1, 2, 3, 5, 8, 13,... The unfathomable, unexplainable mathematics, is that if you take any number of the sequence and divide it by the one preceding it, you will approach the Phi (ϕ) ratio 1.6180.[22] And the larger the numbers, the nearest the result of the division will be to the first four decimals of the Phi number. This is why the Fibonacci sequence is infinite (just like the sequence of the natural numbers and that of prime numbers). As it is, with computers, it has been calculated to an immense number of digits.

The Italian mathematician Leonardo Fibonacci posed the sequence in his book *Liber Abaci*, in 1202, based on the work of earlier Indian mathematicians (the earliest one we know of being Pingala around 200 BCE) and its ancestral use in Indian prosody.

The Fibonacci sequence has been mathematically found to drive the organization of numerous biological organisms and growth processes, such as the arrangement of leaves on a stem, the branching in trees, the uncurling of ferns, the fruit sprouts in a pineapple, etc. For example, the patterns of florets in a sunflower was shown by mathematician Helmut Vogel (in 1979) to follow the ratio 89/55 of the Fibonacci sequence, with one set of spirals left handed and the other one right-handed (see again Fig. 4.1, p. 120). Similarly, the Yellow Chamomile instantiates 21/13 numbers of enmeshed spirals.

So that, if you imagine that you're moving from Planck's length backward in time toward the X-point, you get an ever higher frequency, derived from the ever smaller radius. The number 1 is the first number in the Fibonacci sequence, and as the frequency moves up the scale toward the X-point, then *obviously we have to put the number '1' exactly at the Planck scale.* Thus the Fibonacci numbers (positive for the frequency) get bigger from the Planck angular frequency (= 1) toward the X-point where they tend to infinity. In contrast, these numbers grow in the negative with decreasing radii and wavelengths, from Planck radius (= -1, in imaginary numbers) and tends to the infinitely small at the X-point of origin.

As this pre-time dimension of Rhythm encompasses the future Planck-scale dimension of time, Rhythm reveals itself as a meta-dimension of time (*meta* in Greek meaning more global).

Given that it is nonsensical and inappropriate to speak about frequencies measured in Hertz (meter per second) in a dimension where no time (in hours and seconds) exists yet, a handy solution may be to use a new parameter of cyclical time. As the problem is similar with the ISS pre-space, in which meters don't mean anything anymore, I propose to use new units for the ISS hyperdimension: the Bindu (meaning the mathematical 'point' in Sanskrit). And we may also use the term 'bindu-scale universe' (rather than point-scale) for the period from the origin up to Planck length.

I propose the following equivalences for the Bindu-scale universe:

- Planck frequency = 1 Br (one BinduRhythm unit)
- Planck length = -1 Bc (minus one BinduCenter unit)
    - 1 Br (or 1 BinduRhythm), that is, 1 Bindu unit of the metadim Rhythm) = 1 cycle of the Planck Sphere = Planck angular frequency (in Hertz or s−1 units in the SI system) = 1.85×10^{43} s−1 (the time for light to move along the Planck length of 1.616252×10^{-33} centimeters).
    - i1 Bc (or -1 BinduCenter unit, that is, -1 Bindu unit of the metadim Center-Circle) = radius of the Planck sphere = 1.616×10^{-33} centimeters. This is due to the fact that the primordial universe at Planck scale is a Planck sphere whose size is Planck length.

Let's note that the constant of the speed of light C doesn't apply in the ISS hyperdimension, in which the frequency (or BinduRhythm, Br), as well as the angular momentum, will trespass the light speed toward an infinite velocity and frequency.

So that we have now a second metadimension of the pre-space—Rhythm—and more precisely the rhythmical evolution from the highest frequencies near the X-point of origin, getting progressively lower, down to frequency 1 BinduRhythm at the Planck scale. And the evolution of this frequency is that each bow's frequency instantiates a new number in the Fibonacci sequence; so that the rhythm (just as the spiral and its cycles) is driven by the Phi ratio.

Rhythm—the metadim of time.

Rhythm, indeed, has to be viewed as a dimension of time preceding the Planck-Einsteinian time. Just as we had a pre-space preceding Planck-Einsteinian space, we now have a pre-time.

But now we realize abruptly that we have thrown our anchor at the shore of a very stunning island—because the hyperdimensions of Rhythm/pretime and Center-Circle/pre-space are nowhere near what we are accustomed to in our physical universe. To simplify our model (and to be able to visualize it easily), let's imagine for now that the center of the spiral stays at the same 'place' (we know it is moving itself on an 'internal' spiral, but we will address this added complexity shortly) and thus that the quarters of circles are on concentric orbits (same center).

Now, in terms of metadim Rhythm, we can visualize only the beginning of the Fibonacci spiral starting with a frequency of 1Br (one BinduRhythm unit), which we posed equals to the Planck frequency. The Bindu frequency gets positively incremented (together with the number of Bindu rotational cycles) in pace with the size of the Bindu universe shrinking with each pulse toward the X-point.

In order to visualize the principles of this novel dimension—metadim Rhythm—let the X-point be the center, and let the Planck circle be the periphery. At the Bindu scale, the frequency of the bows is only increasing by leaps, but it is bound to integers of the Fibonacci sequence (that is, using only integer numbers, positive for the frequency). In contrast, the radii decrease (from -1Bc) in imaginary numbers (negative). Thus the frequency moves up in Br: 1, 2, 3, 5, 8...

Metadim Rhythm as a space-like temporal field

It is impossible to draw the ISS at the exact scale, because the shrinking/enlarging of the spiral is so wide that only a few circles can be drawn in a normal page format. Or else we have to draw on a much larger surface and then reduce it. What this means is that the enlarging of each spire by 1.6 represents indeed huge incremental leaps and pulses. So, in order to understand better the dynamics and principles, we have to draw the ISS without sticking to the exact scale, so that we have a great number of bows. (See Figure 4.2)

Imagine that each bow gives out a specific pulse—a one note that is function of the length of the string on the bow, and because the frequency rises, this note rises at each bow toward the trebles. Basically, the bows are thus tones-frequencies, and as we know that string theory equates particles with vibrating strings (as frequencies), then the bows are in effect pre-strings, and they are emitting the sygons torsion waves. Try now to visualize that the notes follow each other so closely that, basically, *you hear the melody, but each note rises from a different place on the ISS*—as if each bow was its own speaker. What do we have then?

In the Bindu-scale universe, the cosmic ISS is *a melody, a complex rhythm, spread in space*. Rhythm—as set by frequency pulses—is the new dimension of time (in BinduRhythm units).

Fig. 4.2. The Infinite Spiral Staircase and its bow-frequencies (Phi ratio not preserved). (Artwork: Chris H. Hardy)

So that *the hyperdimension of time—the hypertime—is spread in virtual space*. The Bindu-scale universe has a space-like time, that is, its hypertime, Rhythm, is a 'time field' in which Rhythm/time has discrete values within the field. This spatialized time is exactly what Minkowski and Relativity theory had predicted would exist outside the light cone—in the 'elsewhere' region of the light cone. And we can consider it as a ***temporal field*** but neither a uniform one, nor an abstract mathematical one. Rather, with the ISS, we have a time-field (physical) that is itself discrete and nonlinear—as if each bow had its own virtual field, together with its very unique bow-frequency and sygon.

The sygon (virtual faster-than-light or tachyonic particle) that each bow will eject out of the spiral staircase will of course have the exact same frequency as its bow matrix. In terms of archetypal reality, this concept is also very well grounded. For example, this 'musical hypertime' is probably why, from the depth of the syg dimension, mathematicians are so notably excellent musicians! (and I hope they will back me up on that!) Moreover, the Bindu time is cyclical, and many a spiritual and philosophical lore speak about such a cyclical time at the mythical origin.

Don't we get into a transcendent mind state?

Basically the topology of this *ISS field* is that of a *colimacon* or helicoidal staircase (pronounce it *colimasson)*. This type of staircase is turning and rising in a spiral on its central axis; but it can also be a square spiral or an entwined double-staircase, such as the square one designed by Leonardo Da Vinci (Figure 4.3a above). Da Vinci is rumored to have conceived the most astonishing double-helix staircase, that of Chambord Castle, while residing in his Clos Lucé manor in Amboise, where he spent the last years of his life invited by the king Francois I.

Fig. 4.3. Helicoidal staircases or colimacons. 4.3a. (Left page) Design of entwined square double-staircase, by Leonardo Da Vinci. (Credit Institut de France, Photo C.H. Hardy). 4.3b. (Above) In the Vatican Museum, designed by Giuseppe Momo. (Public domain). 4.3c. (p. 132, top) Double-helix staircase of Blois, France. (Credit: thestars-themoon.tumblr.com).

In Chambord Castle, two helicoidal staircases are entwined so that people ascending one never meet the people descending the other one. Among the most magnificent colimacons are that of the Vatican museum ((Fig. 4.3b), and that of Blois (Fig. 4.3c), an innovation prefiguring the stunning double-helix staircase in Chambord Castle.

But the ISS (unlike most colimacons) is not only turning and rising around its central axis, but it is also accruing its width. It is a spiral that has some depth, meaning that it has some volume, as if it was a flattened helicoidal stone staircase, probably resembling the nearly-flat disk of our galaxy, with its central bulge. However, an added feature is that the ISS central axis is not a straight axis. Remember, when drawing the golden spiral, that you have put the compass needle at the new apex of each new square. Thus the centers of the quarters of circle are displaced in a spiral, thereby forming an inner and smaller spiral inside the larger one created by the bows. (See Figure 4.4, right.) This is plainly visible on a colimacon such as the one inside the Lighthouse Phare de la Baleine in Ile de Ré in France (4.4, left).

In brief: Given the spiral of golden rectangles and the constant shift of the center of the quarters of circle, the ISS is rotating on itself, like a light beam and the axis is itself rotating along a golden spiral away from the original point and extending in size. We thus get a small spiral inside a larger one, both isomorphic since they are both on the Phi ratio, yet with different initial conditions.

The ISS field is thus the global topological field: the colimacon with steps as depth, whose central axis is itself moving in a spiral. And on this colimacon, the arc (of ¼ circle) forming the curved outside edge of each step is the bow giving its note through the 'depth' of the step.

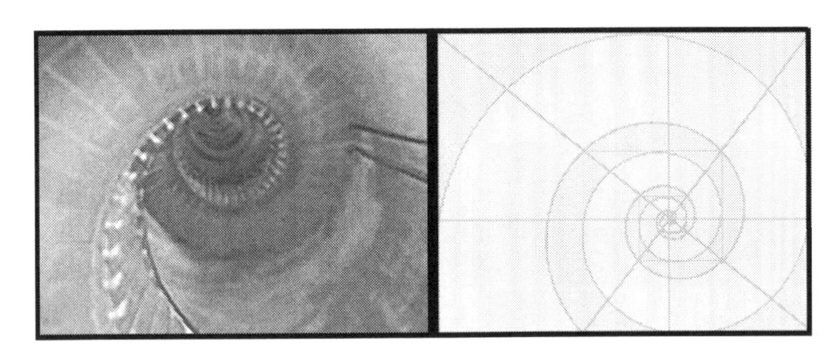

Fig. 4.4. The inner spiral of a colimacon. (Left) Lighthouse Phare de la Baleine in Ile de Ré, France. (Wikipedia Commons.) (Right) Inner spiral created by the centers of the bows in the golden spiral (in a suite of embedded golden rectangles). Credit: CHH.

As I see it, this is also a reason why the universe is not perfectly flat but has a very slight curvature—with Ω_k (omega-k) roughly −0.0027)—because in the absence of depth and curvature the strings-particles in spacetime would not create patterns of interferences in the quantum field, and thus dynamics, accretion of matter in rotating disks, coalescence of atoms and gas into larger, denser, molecular systems... all the way to cosmological systems. And also without the Center-Circle hyperdimension setting both a curvature and a system's identity, the holistic, harmonious whole would not exist as an entity, a collective consciousness entity—such as Gaia for the Earth.

Indeed, the topology of the ISS, the spiral, is expressed by many natural shapes and forms—such as certain seashells, flowers (the sunflower, the spiraling of rose petals), turbulences (such as hurricanes). Extremely astonishing is its resemblance to the DNA double helix, with its pairs of bases set perpendicularly like a ladder; and the fact that numerous particles in a Cloud Chamber display a spiraling motion. (See Figure 4.5)

Fig. 4.5. Spiraling particles in a Cloud Chamber. Credit: *www.universe-cluster.de/*

The discrete quantum property of deep reality is dampened in our Newtonian material world: the adjacent molecules and cells (for example in a textured leaf or a bamboo stick) seem to us to be linked tightly together in the same continuum of space, and to form one solid stuff. Just as a marble table appears solid, but we know that there is a large space between the electrons and each and every nucleus, just like there is a huge 'void' existing between our sun and the planets in our solar system.

Thus, what was discrete and discontinuous but interconnected in the deep reality becomes, in our natural physical reality, continuous and coherent—but only apparently so, as we know well.

In the ISS field, the field in itself is a larger whole which binds together the discrete notes, and creates a melody with them.

The ISS field: two inverse enmeshed spirals

The ISS is a system comprising two enmeshed spirals. The larger external spiral bears the bows (and their array of frequencies); the smaller internal spiral connects all the 'centers' of all the quarters of spires. These two spirals are embedded—entangled—one in the other.

Intuitively, we cannot escape the insight that each spiral is turning in an inverse direction. We thus get two entangled and opposite spins—just that they are cone-shaped spiraling rotations. Indeed, if we put the X-point of origin at the tip of the light cone (as physicists do), then we obtain a complex curved cone in which a small spiral is unfolding inside a larger one. (See Figure 4.6) The centers move along the small spiral. However, the cone is not circular; given the Phi ratio, the spiral, and thus the cone, are elliptical; however, the circularity exists also (as we saw) via the virtual circles embedding each discrete bow of a quarter spire.

In the ISS hyperdimension, the Phi and Pi numbers are launching the dynamics of a colossal increase in size of the spiral from the origin onward, nearly twofold at each spire (precisely 1.6 times bigger at each discrete leap). The Bindu-scale universe's density of energy was the more titanic the closer from the X-point.

This could explain why, with less energy and lower frequencies, during the inflation phase (from instant 10^{-36} second to about 10^{-32} second), the primeval universe increased its size 10^{50} times.

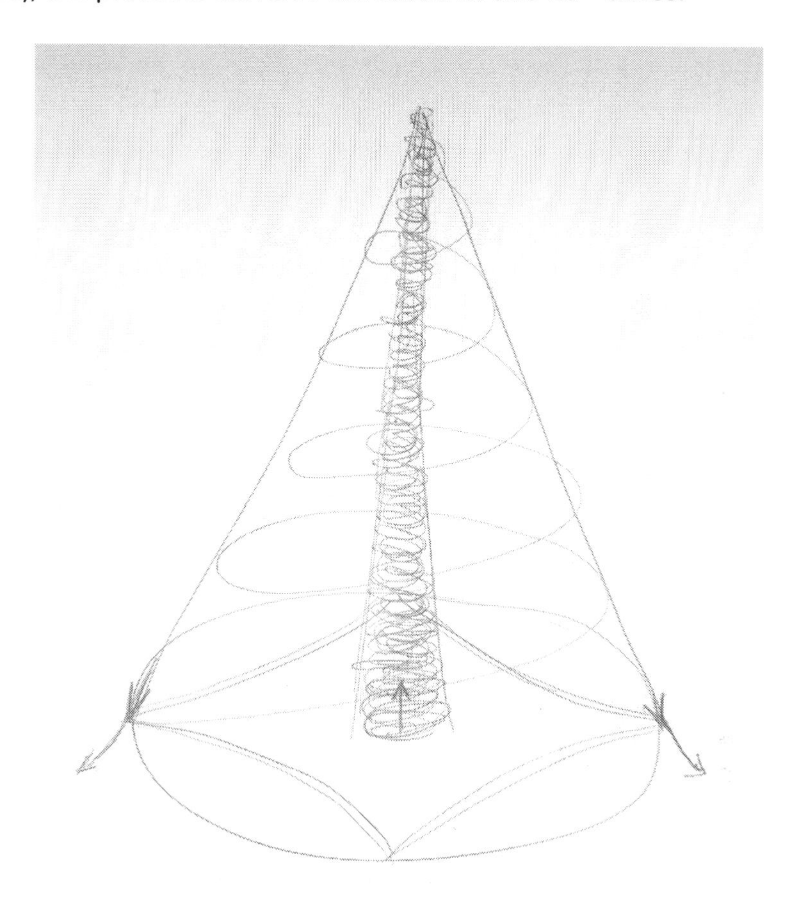

Figure 4.6. The ISS curved cone, spiraling out of the X-point and the inner spiral with its inverted rotation and opposite direction, toward the X-point. (Artwork: Chris H. Hardy)

Let's dig deeper into the two inverse spiraling motions—one with positive spin and the other with negative spin.

The larger external spiral, right-handed (clockwise) unfolds toward a larger and larger circle, all the way to the Planck scale, and thereafter in our spacetime universe. (Its direction of flow is forward toward our time.)

The internal spiral, left-handed (anticlockwise), is moving in decreasing scale toward the X-point. (Its direction of flow is backward toward the X-point.)

We thus obtain a constant **two-way synergy** between the X-point and the present time of the unfolding universe in the light cone (13.79 after the Big Bang)—or more schematically, between the ISS hyperdimension and the present state of the universe (i.e. the material and semantic world we inhabit). This creates a two-way and instant exchange of active information that of course happens in a nonlocal, synchronistic, way.

This two-way synergy is of paramount philosophical and ontological consequence for us, because it is grounding our freedom of choice and the existence of probable futures and thus open paths, for Earth, humanity, and the universe as well (as opposed to a predetermined evolution of the universe).

We have pondered this crucial topic in chapter 1, stating that negentropic and self-organizing systems (such as minds, complex natural systems…) as we observe them in reality, can only exist in a mixture of order (constraints, causal laws), turbulence, apparent disorder, and fuzziness (chaotic attractors, nonlinear dynamics), and indeterminacy (randomness). As we know from neural nets and from self-organized systems, only a feedback (a retroactive information and process) can allow systems to self-organize, to adapt to their environment, and to adjust themselves or find the optimal behavior. If we had only an active information and causal processes going one-way from the origin to the actual state of the universe, this universe would be totally deterministic and constrained. In such a universe, consciousness and minds would be superfluous.

Just like consciousness has to exist as a metadimension even at the instant zero of our universe-bubble, this two-way synergy is absolutely fundamental, and essential, at any phase of the universe's evolution; it has to exist from the very beginning, at the scale of the cosmic ISS of the origin; and therefore we ought to sort it out and ground it in the dynamics of the pre-Planckian universe.

If we pose this two-way and constant dynamics both within the cosmic ISS, and in between the ISS and the present universe, then **the cosmic ISS can be understood as a collective Self of cosmic**

proportion. All information about the paths taken by any system (whether willingly chosen or relatively determined) will be reaching the cosmic ISS in a synchronistic way. A multidimensional snapshot of the state of the world, and of all the events and systems it contains, will be registered (at each instant of our world's 3+1d existence) in the bows of the infinite spiral staircase—that is, in the Center-Syg-Rhythm or CSR hyperdimension.

And the other way around: any information about past or possible states of the universe (and even about past universe-bubbles), all information about conscious civilizations in the universe, is potentially within our reach via the connection of our inner Self with the collective Self (planetary and/or cosmic semantic field), via also the access each one of us has to the CSR manifold within our own individual Self.

The cosmic ISS is thus the compact hyperdimension of our universe-bubble. It is our cosmic DNA and is imprinted, at each instant, with the information about all the beings and systems and their evolution in our universe.

Parameters of metadim Rhythm and the ISS

Let's dive a little more in this novel **dimension of rhythm**, that we have to view as a meta-dimension (or metadim) since the linear time of Newtonian-Einsteinian physics—represented by a vector, a straight line from past to future, a worldline in Minkowski's cone—is only one possible state of the system, and is thus included in it.

The metadim Rhythm:

1. is **a temporal field**. It extends non-linearly in a space-like configuration, and its main parameter is the array of bow-frequencies as cycles in virtual circles. If we were to observe it in its entirety (as a 4D observer), it would look like a giant spiral in which each cell on the spiral would have a different radius becoming smaller and smaller, and a unique frequency (therefore a slightly different musical note and a slightly different hue). Just like the sunflower, the ISS spiral is tending toward the infinitely small radius at the center of the flower, thus plunging toward a sink at the center of the ISS field—the X-point opening on infinity. Imagine that each tiny hexagon (or floret) of the sunflower's core is a bow and a unique musical tone! (See Figure 4.7)

How strange to find patterns in galaxies that resemble a geometrical spiraling crisscrossing grid! (Figure 4.8)

2. It is a **spiraling rotation of discrete bows**, each with its own frequency and orbit, resembling the steps of an expanding colimacon staircase.

3. The discrete increment of the bow-frequencies follows the **phi ratio**.

4. **No one bow has the same frequency as another one**—according to Pauli's Exclusion Principle. Each bow is on a different quasi-orbit (set by its radius length).

Fig. 4.7. Spiral shapes and crisscrossing arcs in a sunflower. (Public domain.)

5. **The ISS field has a gigantic energy density and strength**, beyond any other force in the universe. As such, no existent force can ever

break the ISS field as a whole, or disrupt it, or disturb it. It would instantly spring back into existence. Yet, networks of bows are constantly reorganized and tuned by any single person's Self.

6. **As a field of form, the ISS field has an enormous resilience of its global wholeness**—this being due to its bonding via the spiral and via its formal Phi ratio.

7. **The ISS' field of information and its state-space are non-finite;** it exhibits tremendous internal diversity, flexibility, and plurality of states.

8. **The two embedded spirals and their inverse rotation create a synergic dynamics**, that is, a two-way synchronistic exchange (as opposed to a one-way deterministic command) between the ISS hyperdimension and the universe unfolding within the light cone.

Fig. 4.8. Spiral shapes and transversal arcs in a galaxy.

9. The unfolding universe's information is contained, yet non-determined, in the ISS field, and simultaneously, **the ISS field is imprinted and in-formed by the paths and states emerging in the universe**. The ISS field contains also all the information about preceding universe bubbles, and its very existence points to, and is grounded into, a collar of such universe-bubbles preceding ours.

10. **Each bow knows the totality of the ISS field**. Each bow—being an exact and unique note + hue + orbit—knows all the bows of the entire sequence, of the totality of the ISS field. As a hologram or a complex dynamical neural net, the information of the whole ISS field, as well as the actual snapshot state of the universe (inscribed in it as activated bow-tones) is contained in each of its sygons or bows.

11. At the very beginning of the unfolding of the ISS: As soon as a new bow is formed, ***a sygon*** **having this same frequency and wrapped in the virtual circle**, is ejected as a torsion wave with immensely faster-than-light velocity, in the direction of flow of the ISS. All sygons (bearing the information of a preceding Universe-bubble) move as a conic beam quicker than, and ahead of, the unfolding spiral itself. Their interferences at Planck length will create the Higgs field that will morph into the vacuum. The first sygons to be launched (the highest frequencies), by deploying as a cone, will create the bulk of the hyperdimension, a larger, encompassing, bubble in which the 4D spacetime bubble (our timeline cone) will take shape.

The bound hyperspace and hypertime

Let's ponder again Massimo Teodorani's comment on Wolfgang Pauli's **Contracting and Rotating Space** dream. Shunting a lengthy interpretation, he gets instead a striking insight that brings him to exclaim in no meek words: "A new concept of space subjected to rotation; an absolutely novel concept of theoretical physics; maybe... [the intuition of] **the basis of the physics of synchronicity.**" (Teodorani 2010; My underlining and translation.)

When this astrophysicist talks about "space subjected to rotation" as a brand new theoretical concept, he doesn't mean a rotating system in space—whether in quantum space (spin), or in Newtonian space (a planet's orbit). He means a novel dimension of **space-as-rotation**.

And indeed, this is exactly what we have already with the ISS field: As in Pauli's dream, we have oscillations and a contraction of space and also a hyperdimension of time: a space-like field of rhythm, that is both rotating and getting more and more contracted and small until it becomes a X-point open on the infinitely small.

Our two novel metadims are indeed hyperdimensions:

- **Metadim Center-Circle** or point-sphere, with its creation of space-boundary (the circle) from the point-center, and thus launching the Bindu space. Metadim Center not only supersedes the classical 3D-space, but encompasses it, making it one of its possible configurations: it is the hyperspace.
- **Metadim Rhythm-Rotation,** with its topological progression of bow-frequencies in a spiral, launching the Bindu time. Metadim Rhythm not only supersedes the Planck-Einsteinian dimension of time, but encompasses it, making it one of its possible configurations: it is the hypertime.

Let's recall how, at the dream's beginning, the anima starts oscillating. What we have seen up to now is that, as modeled in chaos theory (the mathematics of turbulence patterns at all scales), the oscillating phase always precedes a transformation of the global order (or organization). But now we know what organization/order we are talking about: the deeply entwined metadims Rhythm-Rotation and Center-Circle. This fused manifold is our endless infinite colimacon staircase stretching and contracting to infinity. Each bow is both a note and a sygon—a unique frequency in the metadim Rhythm that contains all possible bow-notes that can ever exist.

A lattice of crisscrossing infinities: Phi and Pi

Indeed, the Fibonacci sequence is endless that is, non-finite; it tends toward the infinite. There's always a new number that may follow the preceding one, as goes the ordinary suite of numbers 1, 2, 3, 4... (called natural numbers).

Furthermore, Pi (π = 3.1415926...) is also a non-finite number because, however far after the dot we calculate it (with computers only), it never ends anywhere and it does not show any periodicity. Already in 2007, some 200 billion decimals had been calculated for Pi and no pattern has been detected. Pi has been proven to be not only an irrational number but also a transcendental number (that is, non-algebraic; Lindemann-Weierstrass theorem, 1885).

Some researchers deem Phi ϕ and Pi π to be 'universe-numbers,' meaning that any kind of finite suite of numerals could be found in its decimal unfolding (after the dot), although it is still an objet of debate.

Just like the X-point is a sink into infinity, yet a point without dimension, our bows (the ¼ circumferences) are produced/calculated with Pi: more precisely, each circumference is birthed by a center (a point) and projected on a virtual space by 2πr (with the Phi ratio having each time accrued the value of r, the radius). This is why I call this metadim Center-Circle, because the center/point projects the circle or a sphere in virtual space by the power of Pi. This enclosure, as we will see, creates, with the circle, the boundary of an individual system, its organizational closure. In terms of Systems Theory, it creates the identity and self-organization of the system by a constant back and forth between the center and its boundary—be it a cell, a membrane, or any natural system.

But, if we consider that the number 3.14 is non-finite, then there is an *inherent fuzziness to where exactly the bow will be projected*, depending on how precisely—how many decimals after the 3—nature will execute the partition.

To this we have to add that the 'center' from which will be projected the bow (with a novel radius) is itself a point, and a point is nondimensional, yet infinite, being a sink toward a smaller and smaller scale, from its sphere to its center—in essence, just like the singularity at the origin of the universe is understood.

Is this all? No: the maths of vibrating strings tell us that there exists a non-finite array of possible frequency variations for the same string (called the 'ultraviolet catastrophe' because these frequencies rise toward the extremely high ones at ultraviolet wavelengths). If we assume the bow is vibrating like a string (and bows can be viewed as pre-strings with regard to string theory), then each bow exhibits the properties of standing waves on a string. The fundamental frequency creates harmonics, that is, subsets of itself at ever increasing multiples of the first frequency. And this array of frequencies within the same bow means that each bow has a very original and unique fundamental frequency and nevertheless it is complex and plural, with a state-space of possible internal states.

So that, in the Metadim Center the infinity is already (1) in Pi, and (2) in the center, and (3), as well in the projected bows (due to the infinity of Pi); and in Metadim Rhythm the infinity is (1) in the multiples of phi and (2) in the number of bows or sygons.

Now something strange happens when we calculate, using the Fibonacci sequence, either the suite of radii or the suite of basic frequencies. If we calculate the radii's as wavelengths and derive their attached frequencies, then the frequencies will be fuzzy fields. And inversely, if we calculate the frequencies, and derive the radii's wavelengths, then the radii's wavelengths will be fuzzy fields. This seems a clear mirroring of the Uncertainty Principle (working at the quantum scale) all the way to the CSR hyperdimension, and this mirroring would derive from the fact that the universe is self-consistent (see chapter 10).

Of course, calculating the sygons frequencies could be of much greater importance for us (as opposed to calculating the radii), because, according to ISST, these frequencies should be resonant with the main particles of the Standard Model.

Hyperspace and Hypertime as dynamic

Let's note though that in ISST the metadim of space (Center) and the one of time (Rhythm) are modeled not only as topological configurations, but also as an unfolding dynamics, and both create spacetime with their unfolding. In contrast, the Newtonian space was just an unknown 'box' or, as Greene puts it, a 'theater stage,' in which stellar bodies as masses attract each other (as an inverse function of their distance), but not having any property in itself. The philosophical debate has been going on for ages as to the intrinsic reality of space and time. In contrast, quantum physics and QM evolve on a flat surface, which raises an added problem for their integration.

Most physicists view time as curved, given that it is integrated in the Einsteinian spacetime that has the topological property of being curved (as in Riemann's and Einstein's curved space). Yet Einstein endowed the fabric of spacetime with EM fields and the capacity to store energy.

In ISST, I use the term quantum-spacetime (QST) to refer to the region where dwell all the particles we know of (often called the Standard Model particles), that includes the spacetime of Relativity Theory and the quantum void or vacuum of QM. Basically, the QST domain starts after Planck scale, when particles are allowed. This fusion of the terms reflects the fact that quantum physics (the vacuum and the Zero-Point fluctuations) is more and more integrated with Relativity theory (curved spacetime), and both model the same domain. Outside of the QST domain stand both the subquantum domain and various hypothetical hyperdimensional domains (hyperspace, pre-space, etc.). In this Infinite Spiral Staircase theory, the hyperdimension, Center-Syg-Rhythm or CSR, is complex and triune.

Metadim Center is the dynamics of the center (or node) creating its circle to set the organizational closure of its own system—and in the process it creates its own hyperdimensional space.

Metadim Rhythm triggers also a dynamical process: by oscillating, the bows put the circle/torus in rotation and create waves-particles with specific frequencies: ***the sygons*** that, due to their entanglement with metadim Syg (which we will ponder shortly), are active information. The sygons will be propelled by their initial thrust and energy and will create our whole universe-bubble with its two regions, the HD bulk and the quantum-spacetime or QST.

A new conception of the universe at the origin

Now we get a global conception of the Bindu-scale hyperdimension:

- it is self-organized, self-expanding; it creates its own pre-space, its own pre-time, and it is also its own music.
- it is open on infinity via several basic entities: it is a universe full of holes and intrinsically porous.

  Synchronicity: Geyser of sparks in the candle flame (these candles never do that!)
- the three metadims Center, Syg, and Rhythm, that inform the CSR hyperdimension, are operating conjointly, in an entangled fashion, as a topological manifold and as self-organized dynamical systems.
- moreover, it is a mesh of three crisscrossing metadims, that permits to the unfolding universe-bubble to be fine-tuned in an infinite number of possibilities (or configurations), yet the cosmic ISS has reinforced the best paths and networks-systems that favor the blossoming of life and consciousness.

5
A COSMIC COLLECTIVE MIND

"Every movement in nature must be rhythmical. (...) It is borne out in everything we perceive—in the movement of a planet, in the surging and ebbing of the tide, in the reverberations of the air, the swinging of a pendulum, the oscillations of an electric current, and in the infinitely varied phenomena of organic life. Does not the whole of human life attest to it? Birth, growth, old age, and death of an individual, family, race, or nation, what is it all but a rhythm?" Nikola Tesla. [23]

Now that we have a good grasp of two of the three enmeshed metadims composing the infinite spiral staircase, let's focus on the third one: metadim Syg, the metadim of consciousness in action in the early universe and in the cosmos. And again, Pauli's dreams will reveal a lot about the prime forces of the primordial universe, and the most stunning and challenging piece of information is certainly that the cosmos is conscious, animated by a free and artistic anima.

In the present ISS model, the *anima mundi* or cosmic consciousness is a collective consciousness of cosmic proportion, in the form of a syg hyperdimension pervading the spacetime systems and events at all scales, that is precisely the dimension in which our individual minds operate, with creativity, freedom of choice and exquisite variety and flexibility. But first, let's realize how much the ISS is an archetypal form and how its dynamics infuses life forms as well as matter systems.

The infinite colimacon: an archetypal form

If the triune CSR hyperdimension and the ISS turn out to be a correct description of the pre-space's self-organizing parameters, then we'll have to fathom that the ISS topological configuration—an enlarging spiral + orthogonal rings or tori also increasing in size—is an archetypal form, *the* arch-archetype itself (Figure 5.1, left).

Consequently, logics demands that if indeed this is the case (that the triune ISS describes adequately the baby universe's organization and form), THEN we should get this configuration not only in many natural forms (as it is the case, all the way to the DNA helix), but given that it is also an archetype in the collective psyche, in artworks as well. The DNA's double helix, the foundation of life in all biological organisms, even in the viruses and the early one-cell organisms of the life chain, is of course a profound instance of the spiral archetype; moreover, the pairs of chemical bases constituting it are set perpendicular to the central axis of the double-helix. Many flowers display a spiral or multi-spiral configuration, either in the rows of petals (such as the roses), or in their core (such as the sunflower). And of course lots of shells have not only the enlarging spiral shape but also the orthogonal coils or tori.

In matter systems, we also find the spiral dynamics from the astronomical scale to the microscopic scale: not only galaxies rotate around their supermassive black holes, but many particles show a spiraling motion in a Cloud Chamber, a left-handed or right-handed spin, or else a double spin. Spiral dynamics are very common in the physics of plasma, fluids, and gas turbulences, from the water eddies and whirlpools, to the wind tornados and hurricanes, and microscale vortices.

And in fact, we do find also this complex shape with the orthogonal tori in traditional artworks. Some very sacred and antique silver rings in Northern Mali (a mostly desert land) have the shape of a torus (just like a usual, round in width, ring), but unlike other rings, they are made of coils set perpendicularly to the torus (as if the torus was a round sequence of coils, each very neatly designed and sometimes finely sculptured. The coils, well separated when the ring is not too worn, show the discrete nature of the coils despite the overall shape of a torus or spiral. Some of these rings get larger toward the top of the hand and smaller on the side of the palm, with a very regular increase and decrease of the width of the torus, as the one drawn on Figure 5.1 bottom right. I found a similar shape (but without any modification of the width) in an antique drummer ring from a Bon Po shaman in Tibet. This ring from Mali is an initiation ring, and it is used by shamans in the desert—among other usages—to find underground water. I donned it for years as a pendant, and it had for me the feel of an archetypal shape—and yet I was at a loss to explain what resonances exactly its shape could trigger.

Fig. 5.1. a. Archetypal form of the ISS spiral and its perpendicular set of bows (top left). b. (bottom right) A similar form in a Mali ring. (Artwork Chris H. Hardy)

There exists a shell that presents a similar structure as the Mali ring (Figure 5.2): it has both the large spiral (looking like a torus) and the perpendicular coils or axes, in this case, slim ridges that are perpendicular to each interval segment of the internal wall of the torus (and not pointing to the center of the shell, despite the fan shape of the ridges due to the quite round form of the torus).

Fig. 5.2. A shell with transversal bows within a spiral torus. (Public domain.)

So that, basically, the traditional and sacred Mali rings show a torus in the shape of a spiral, and a suite of tori (the coils) set perpendicularly to the main one, each one in a spiral. The overall design looks like an electro-magnetic coil, with extra-large wire spires around it. But more importantly, these Mali rings are an expression of the archetypal form of the ISS spiral and its perpendicular set of bows.

The Mali ring is not the only artwork that, for me, had an archetypal resonance, albeit unexplained. I've been, over the years, deeply moved by a specific shape in artworks that is found in Asia, namely China, Japan, and India. This form is a flattened torus, that is, two concentric circles with the space in between filled up.

Now, this form has been held sacred since the Neolithic age in China. The bi Jade disks are ancient Chinese artifacts, the most ancient dating from the Liangzhu culture (3400-2250 BCE). All are circular flat disks with a hole in the middle, generally large. The engraved motif is often a field of tiny rounds, as on the Han dynasty bi, in which the tiny rounds are set within a geometrical grid of crisscrossing isosceles triangles. (Fig. 5.3a, left page.) I also found this archetypal shape on a modern sculpture in Delhi (5.3b, below).

However not any two concentric circles will do; they need to have a very specific proportion in order for my unconscious to be moved to the core and buy (if I can) the artwork. If the two concentric circles were to be much nearer in size, or much more apart, the figure wouldn't trigger any deep feeling.

Fig. 5.3. Archetypal flattened torus in artworks. a. Antique Chinese bi jade disk, Han dynasty. Creative Commons/ Yongxinge, GNU. wiki/Bi_(jade); (left page). b. Modern wooden sculpture from India (right). (Photo: CHH)

I could recognize grossly the proportion (from the deep surge) but up to now I was as much at a loss to explain it as with the Mali ring. But now I understand that the right proportion between the two circles is one approaching the Phi ratio (specifically the 144/34, that is, 34 multiplied by 3 times Phi). I'm not the only one to be able to intuitively recognize or draw a figure that will trigger an archetypal resonance—the modern Delhi artist had probably no cue about the phi ratio (even if the Fibonacci sequence was used in India's antique mathematics) and yet his unconscious knew the archetypal shape.

Metadim Syg: The Cosmic Anima playing piano

At this point we have a very interesting resonance with the dream Pauli had about *the ring of the i*. If you remember, a woman (his anima) was teaching him piano; she gives him a ring and says "this is the ring of the i." As we have seen, the "ring of the i" shows a topological configuration of the imaginary numbers along a spiral, a circle, or a torus. As for *pianos*, they feature one specific note for a key; when a key is struck, it puts into vibration a specific string having the frequency of this note. So that we have a striking similarity between the piano's array of notes in the 'ring of the i' dream, and the ISS spiral of bows: each bow is a specific note (frequency) and the 'ring' resembles the rotating dynamical ISS which binds all the bows-notes into a unique system.

My first interpretation of this dream drew the link with the Fibonacci sequence, the only self-evident link between a suite of numbers and rotation. At that earlier point in my development, I had brought up this dream to deepen the interpretation of the other dreams of Pauli that had to do with circulation and rotation. However, what we don't have modeled yet is the specific role of the *anima* being capable of influencing the metadim Rhythm, or to trigger a synchronistic process having a tangible influence on physical systems. Let's focus now on the anima and consciousness—the *metadim Syg*.

In fact, Carl Jung in his book *Synchronicity*, beyond positing that it was an acausal phenomenon and necessarily meaningful for the person, gave three definitions of the phenomenon. All three imply a *subject's mind* who is experiencing a coincidence with a meaningful event that is happening either (1) in the present, or (2) at a distance in space (clairvoyance), or else (3) at a distance in the future (precognition). Thus Jung's basic definition of synchronicity implies a consciousness not only observing the synchronicity but experiencing it in a highly meaningful way, that is, a person interacting with a meaningful world. (This is highlighting what we call in cognitive sciences a *1st person perspective*). Another trait is that synchronicity works nonlocally (beyond space and time), since it allows clairvoyance and precognition.

Despite the fact that Jung had a first-hand knowledge of a fourth type of synchronicity, this one involving a mind triggering an alteration or change in matter systems (the famous Pauli Effect, that is, psychokinesis or PK), he didn't include it in his description of synchronicity. But the deep synchronistic entanglement that Teodorani is speaking about—the one that could exhibit the dynamics of the "physics of synchronicity"—has necessarily to imply the *influence of mind over the organization of matter itself*. In other words, if the mind was just observing but, while so doing, was unable to trigger a change in the observed systems, then there would be nothing added to Newtonian-Cartesian physics based on the absolute separation of matter and mind. All quantum physicists since Heisenberg know better, and the Observer Effect has been a bedrock and a hot topic of QM up to now (we'll ponder it in depth in chapter 6). However, whether in probabilistic QM or in the deterministic Pilot Wave theories, the dynamics by which consciousness could influence matter systems has not been modeled. QM simply posits that what we observe has already been modified by our observation; and the pilot wave (precisely Q, the quantum potential) guiding the path of particles has not been connected to consciousness in any clear way.

The (integral) theory that would link physics to consciousness dynamics and mind dynamics is still lacking on all fronts, whether within physics or within theories grounded in cognitive sciences and psi theories. What we have up to now are general physics principles, such as the entanglement or the collapse of the wavefunction, that *could represent* (or be interpreted as) the way a mind influences quantum and matter systems.

In contrast, Jung and Pauli in the 1950s added a novel dimension to this understanding with the principle of acausal synchronistic phenomena intruding sort of transversally into the 4D spacetime of matter events. And in the present ISS theory, the synchronicity dynamics are precisely how the CSR hyperdimension intrudes into 4D systems and events and modify their organization and paths.

Thus, the physics of synchronicity envisioned by Teodorani implies a consciousness as a 5th dimensional syg energy entity who is observing and influencing, while in deep entanglement with matter-systems via the entwined metadims Rhythm and Center.

A cosmic piano

In the introductory scene of the *Ring of the i* dream, the dream-anima of Pauli was in the process of teaching him piano. Pauli's anima, as a reflection of the cosmic anima, was thus setting the stage ('teaching him') before delivering to him some hard-to-fathom knowledge about the 'deep reality' and its parameters, like the imaginary numbers. This specific context was to act as a priming to orient his understanding and the interpretation of his dream on awakening.

This context was music—and specifically piano. And indeed the wondrous infinitely rising steps of the spiral staircase can easily be seen as a futuristic colimacon piano—each 'step' as a key with a somewhat futuristic and high-tech geometrical perspective (See Figure 5.4). The analogy is perfectly sound, and literally so, since we saw that each step is indeed a note.

Fig. 5.4. ISS bow-frequencies as keys in a futuristic piano. (Artwork: Chris H. Hardy)

Now, in Pauli's *Contracting and Rotating Space* dream, it is the anima (the same one who was playing piano) that will trigger the contraction and rotation of space, thus influencing the metric itself of space. The anima influences space's geometry (its topology will be contracting and shrinking as a spiral piano) and its dynamics (its rotating behavior). In other words, Pauli's anima appears as the cosmic soul of the universe—the cosmic anima, or arch-anima (the first, archetypal, anima of the world). However, this cosmic anima is not an eternal self-similar consciousness (as goes our image of god) but she is entangled with the synchronistic self-evolution of the whole universe. Let's focus now on this 'agency' in Pauli's dreams, who is orchestrating changes in the world by 'playing' and creating a piece of music.

What the dream puts precisely into play is the anima of Pauli starting to oscillate. And before this 'oscillation' of the anima, there's no 'contraction-rotation of space.' Oscillating implies a constantly shifting set of frequencies. We cannot project any limitation, in terms of frequencies, or in terms of complexity of the oscillatory movement, based on this dream-image.

What we do know for sure is that the cosmic anima is oscillating in and of herself. In the ISS, the two enmeshed metadims, Center and Rhythm, are at the very least self-organized and self-conscious, but we don't know how free they are. In contrast, the arch-anima is a soul, a cosmic soul immersed in her own inner music. And while she is able to contemplate her own inner music and the deep fabric of the world (the ring of the i), now she starts vibrating in her being.

And lo! It affects the geometry of space. The staircase, the piano-like spiral, starts folding and getting smaller and smaller, rotating. That is, the arch-anima starts on a journey toward the point of origin, toward a more and more contracted and rapidly rotating pre-space and pre-time. And along the way, sounds and harmonics are activated on the musical spiral staircase.

In a normal piano, it is the strings attached to the piano keys (hidden behind the vertical wooden case) that are put into vibration, thus creating music. In our enmeshed Center-Rhythm metadims, the arch-anima plays the cosmic piano by putting the bows (the sygons or pre-strings) into vibration, in resonance with her own soul vibrations or *soul music* (the oscillations).

> *Synchronicity during the corrections: I open the dictionary by chance and my eyes fall on 'rap' (music)! That happened one minute down the text from 'soul music' that, having an insight, I had just inserted! (lots of semantic energy involved!)*

And this activation/momentum (because the pre-space is described as 'contracting') moves along the staircase toward the origin, toward the infinite point of the birth, the emergence of the universe.

How marvelously fitting that the dreamer's anima (an apparently 'simple' exotic woman, but a persistent guide and awakener) talks to Pauli's conscious mind (who will interpret the dream and get the message)—prodding him to learn to view the world in a new way, not only the maths and physics of energy-states and orbits, but the highly significant inner experience of a soul immersed in the artistic, esthetic, and spiritual experience of one's own being-in-the-world.

We cannot miss here another reference to Plotinus, for whom, as we have seen, the spiritual cosmic realms were conscious entities, and moreover, the realm of the One (realm 1) or the cosmic soul (realm 3) had to be experienced in a transcendent state. This state would take the wise person into not only understanding and realizing, but also tasting, and participating to, the ultimate reality—as a beautiful, blissful, beneficent, and mind-opening experience.

This is also the concept of a blissful exploration of the reality of Self (atman) and cosmic consciousness in the Vedic and Hindu spiritual path: the quality of Ananda (bliss) is concomitant to experiencing the ultimate reality as a field of cosmic consciousness (brahman, the whole)—that is, realizing in oneself the ultimate reality of boundless consciousness.

Now, as she has an infinite piano—even if the 'keys' (the bows) have a specific frequency—the arch anima's inner music (thoughts, intentions, artwork, knowledge and refined concepts, subtle feelings and consciousness states) is in turn evoking the harmonious resonance of whatever ensemble of notes are keyed in and tuned in with them. And from this primeval original music that speaks of the infinite, in a magnificent explosion of meaning the sygons will free themselves from the ISS, and will create their own conscious hyperdimensional bulk.

> *"Et la danse du dieu, libéré des matières lourdes, atteignit aux premières couleurs."*
> *(And the dance of the god, liberated from heavy materials, attained the primeval colors.) (CHH)*[24]

Nonlocal consciousness

Consciousness is such a primary part of reality that without it we would not be aware of ourselves nor able to perceive or think—and even less so to invent or fathom theories and laws about reality.

Given this fact, logic forces us to assume that consciousness is a facet of the fundamental reality. It is not the by-product of an evolution—it was there all along. However, when I posit such an axiom, I do that in purely scientific terms: because consciousness accompanies any cognitive act (of perception, action, feeling, bonding, etc.), it is a fundamental part of the universe, and we have to account for it in any integral theory of the universe. (We'll ponder this in depth in chapter 10.)

In fact, integrating consciousness in our theories is immensely more needed than to reach the unification of the four basic physical forces—which by the way, are just what we knew of until 1998, while we did miss a gigantic one at work in the air we breathe and also inside us—namely dark energy, 68% of our universe's energy and reality—pervading all matter.

Posing this axiom (about 'consciousness as a facet of the fundamental reality') in scientific terms demands that we avoid using concepts bound to ungraspable cultural and personal variations, such as "creation," and "God," especially as referring to a Causal Agency—the causal agency of all. And in this respect, talking about a cosmic anima is very refreshing and doesn't channel or bridle our thinking process. The arch-anima as a 'collective cosmic consciousness' without gender or form is even better.

> *Synchronicity during the corrections, about one minute down the text from "atteignit" (the poem's extract I just added and for which my translation of 'atteignit' (the past tense of 'atteindre') as "reached out to" left a question mark in my mind. I now open the dictionary by chance to look for 'canaliser/to channel' and my eyes fall at first glance on the verb 'atteindre' in French, and I find with 'attained') a better translation! This synchronicity is quite remarkable in the sense that it proves that my unconscious pursued the thinking process by itself (while I kept moving down the text) and it spontaneously offered me a solution. (A bedrock of SFT)*

Already, the aim of the Semantic Fields Theory (SFT) was to give consciousness its due: we wouldn't be here reading or writing, feeling, socializing, or theorizing without it. We have to keep in mind that when science reaches out to build a so-called "Theory of Everything," the best part (and only palatable one!) of "everything" does have to get into the picture, doesn't it? As this didn't happened to be the case, and the TOEs appear by now to be a definite class of theories, I prefer, following Ervin Laszlo, to talk about an "integral theory."

Metadim Syg

Of course, the arch-anima is the third of our metadims, the *metadim Syg* instantiating consciousness-as-energy, that is, the metadim of a universe-as-meaning self-organizing and able to transform itself continuously. While I have spent the last two decades exploring and formalizing this semantic dimension within the human mind and its interactions with other beings and the world, I'm still in the process of fathoming how exactly it operates in the Bindu-scale universe. However, there are some logical and ontological prerequisites I may list for now (and derived from my previous research).

- The syg (or semantic) dimension, that of consciousness-as-energy, pervades the whole universe—all living beings, and all matter-systems, from particles to galaxies. It is extremely similar to our individual semantic field, and comprises all semantic fields in the universe.
- The syg dimension acts as a creative and negentropic force, prodding all innovation, creation, intention, feeling and action.
- It also contains all the memory of the universe. It includes what was and what will be. However, it is not bound and constrained by the imprint of all past states of itself as a whole.
- The syg dimension is at the same time a whole (the One), and an infinite plurality. As a whole it is the cosmic consciousness of our universe-bubble—the collective unconscious of Jung, the Realm of the One of Plotinus, the world of ideas of Plato, the noosphere of Teilhard de Chardin, the brahman and Tao of Eastern religions, and possibly the 'implicate order' of David Bohm. As a plurality, it is made of all the semantic fields of biosystems and eco-systems in the universe—intelligent entities, sentient beings, and self-organized complex systems.
- We all co-create Earth's planetary semantic field. Our personal semantic field (as a Self) blends in with the cosmic consciousness, setting a synergy through which any Self may have an influence on the whole, just as the whole influences its cells. Together, we form the nodes of the giant network of our collective Self, Gaia.

However, the whole (our planetary semantic field, Gaia for the Earth) is much more than the sum of all semantic fields it contains.

- The semantic dimension of our universe includes all planet-scale semantic fields, and thus all intelligent self-reflexive civilizations in the universe, as well as all complex star and cosmic systems. Whether housing an intelligent civilization or nor, they still have a semantic field. Similarly, all complex systems on Earth, however tiny, have their own semantic fields—or more precisely, they are also semantic fields, in the hyperdimension.

Nonlocality of the mind

We know by now—via a large amount of experimental data—that consciousness exhibits nonlocal properties (in other words, it is a process unconstrained by space and time) and that it is capable of modifying the organization of matter, such as in healing or bioPK.

Therefore, we have to account for a collective or cosmic consciousness that is both:

1. deeply enmeshed and entangled in matter (physical and biological, in the infinity small as in the infinity large) and yet

2. non-dependent on matter, space, time, and causality; in other words, the syg dimension is relatively autonomous from them.

'Non-dependent' means that consciousness doesn't necessarily needs to be incorporated in a 'body' or 'physical form' in order to exist and act. A work of art, for example, is a semantic constellation that will survive and go on existing whether or not this artwork is destroyed. Thus, the 'Lost book of Aristotle' or the formerly famed 'Gardens of Babylon' are still alive semantic constellations in our world culture.

It emerges from the research into the deep nature of reality, from eminent thinkers (such as Jung, Pauli, Teodorani) that consciousness is not, however, separated from matter, as in Descartes' old paradigm view. My way to solve this problem (in front of which cognitive scientists have been stalled, just like physicists in front of Planck threshold) is **to view consciousness as *the* organizing and negentropic force of the universe**. Consciousness may operate either through matter-systems, or in and of itself (as non-dependent on matter and spacetime). When it does operate through a physical brain and body, or through a physical system, it is then deeply interwoven with this system; yet it is not confined, limited, or driven, by the physical system it is coupled with or by the 4D universe at any time. Consciousness is like a network of enmeshed and superposed layers of reality, with the head in the sky (immaterial syg connections and exchanges) and the feet in gross matter (such as neurons or the geomagnetic field). Yet all layers are in synergy.

It is also a fact of experience for many researchers that there exists a collective dimension to consciousness. Yet, the theories put forward give only a basis for the 'memory' (or data reservoir) aspect of this collective consciousness or field—such as that of Laszlo, or Sheldrake. Much more difficult is to account for the creative, innovative, intentional, and free thinking aspects of consciousness, as well as for its extreme flexibility, variety, and divergent processes. And most of all, for its capacity for self-reflection (self-reference), anticipation, and the power to transform itself internally.

The thorniest, and at the same time the most exciting question is by which processes, or means, does consciousness interact so deeply with matter that it can, with a simple intention, modify the organization and growth of biological matter? These are facts amply proven by hundreds of bio-PK experiments, but whose *modus operandi* (mode of operation) remains clouded in mystery, just like the nature of gravity, or the entanglement of particles.

Based on my previous research into the nature of consciousness, its cognitive and psi dynamics, I may propose some hypotheses as to the role and presence of consciousness—put into play by the arch-anima—in the primordial universe.

Cosmic DNA

I thus propose to consider consciousness as existing at the birth of the universe (before matter, space, time and causality existed), as a Syg-metadim deeply entangled and enmeshed with the Rhythm and Center metadims, thus forming a 3-thread dynamic structure of enmeshed forces—the CSR manifold or hyperdimension—operating through the Infinite Spiral Staircase.

I'm now going to offer some specific hypotheses about the nature of metadim Syg in the universe. But let me underline first that it is essential for us, in order to grasp a reality that escapes our previous worldviews and blinkers, to be able to view it as a 'collective field of consciousness' (a syg-field at the cosmic scale), thus in line with the way we think about the collective unconscious and how we are part of it, bathing in it and yet capable of influencing it by a striking new artwork, a discovery, and the like. However, it is as essential to be able to view it as a cosmic anima, the arch-anima, a whole that is more than the sum of its parts, a cosmic consciousness with both prescience and memory spanning an immense time, and endowed with all the gifts of intelligence, sensitivity, feelings, intuition and creativity that we ourselves enjoy. And nevertheless, just as our own individuality, *this whole is plural* and, fortunately so, not a one and uniform stuff, because if it was the case it would be unable of inventiveness, reflection, growth, and choice. And yet, we all meet in spirit in this hyperdimension and we are in conversation and in synergy with it and among us.

All these ways of apprehending and understanding the syg hyperdimension are needed for each of us to find our place in it and how to operate in this hyperdimension. It is my deep understanding that we are meant to operate with this syg energy, and to discover and learn how to play these energies as if they were an astounding musical instrument. It is also my deep hunch that we'll discover quite soon the physics of these sygons waves (their frequencies and traces in the universe); however, since they are not only physical but also syg energy and consciousness dynamics, we won't be able to grasp their nature and how to tame these immense energies until we treat them as more than matter. What all this means is that, this time over, the scientific revolution is not just about acquiring a bunch of new laws and their derived technology. What is in store for us is that with our minds and psyches, we penetrate the hyperdimension, and this will make us understand what this new level of reality is. And in order to get there, we need to decisively bypass and leave behind the down-to-earth stand of naive physicalism and its materialistic paradigm, together with its tailing hordes of skeptics who seem not to have even reached the logic of the quantum world yet. Here are some seed-ideas (in flexible concepts):

The ISS is a collective and cosmic syg-field (consciousness as arch-anima). It is co-created by the cosmic DNA's field of information (with weighted paths) in interplay with all consciousnesses in the universe (all syg fields). This cosmic syg-field is open, self-organizing, evolving; it is not constrained by and limited to the ISS itself, because since the origin it has extended itself in the hyperdimension's bulk and exists also as a compact 5th dimension within all systems' syg-fields.

The arch-anima (metadim Syg) is creative, self-organizing, and self-referent, a full-blown consciousness deeply enmeshed and entangled with the metadims Center and Rhythm—thus forming the Center-Syg-Rhythm manifold. Part of her (in all times) is the primordial universe's infinite musical spiral staircase. She is a supreme observer and thinker, and yet she revels in beauty and art.

The Infinite Spiral Staircase (ISS), as *cosmic open DNA*, contains all possible states of the future universe; it is the state-space of our universe-bubble. Yet it comprises also a specific subset of life-forms and matter-systems already memorized and experienced in a previous universe-bubble. This parent universe's matter systems shrank as a cosmic black hole and sublimed itself into a *pure CSR field of information.* As such, it emerged at the X-point as the seed of our own hyperdimension, the cosmic DNA of our universe-bubble, forming our ISS and Big Bang. However, this *past musical chore of a preceding universe* is not in a deterministic or fixed state. The cosmic DNA is like an electronic piano, a synthesizer, and whatever systems (such as sentient life-forms and intelligent civilizations) have been explored in the past, and thus memorized, they constitute only a field of experiences and doesn't impinge on our own creativity as an intelligent species. In accord with its fundamental negentropic nature, the cosmic DNA is open to fluctuations and to novel states, in virtue of its syg energy creating meaning and organization.

Once our ISS is coming to be at the origin, the syg-energy or arch-anima will start improvising an inner music out of her own Self—while the ISS' metadims Center and Rhythm will respond with a music of their own playing with forms and resonances. The music of our specific universe-bubble (to the end of our bubble time) will be an experiment in consciousness, in unmarked and enigmatic territories. Yet, as collective cosmic semantic field, the arch-anima knows of other such experiments—some being her own anterior experiences, and others happening synchronously 'elsewhere' in the syg-dimension of the multiverse. The challenge is for the arch-anima (all of us sentient beings) to make the experiment evolve into a stunning realization in which she will both express and discover new potentials of herself—as consciousness in individual beings and as collective realizations at all scales. The arch-anima weaves new experiences from her collective field, but she also improvises (just like a jazz musician) with her own alive consciousness, her imagination and sensitivity, and with inspiration. Consequently, all of us, self-reflexive consciousnesses and intelligent civilizations in the cosmos, we are part of the arch-anima, we are facets of this giant collective semantic field; whether an individual mind or a planetary civilization, we all co-create the specific music of our universe.

Thus, the Infinite Spiral Staircase is a DNA of a specific kind: it memorizes snapshots of all events and beings in the unfolding universe and influences them using inspiration, intuition, and insight (and not in a deterministic way). However, the ISS is also a dynamical system in an evolving process. Via its own hyperdimensional virtual spacetime, it can access any 3+1D spacetime event or system. It is itself influenced by each and every semantic field existing in the universe, whether individual or collective, thus setting a synergy, a conversation, between on the one hand individual minds and systems, and on the other hand the collective cosmic consciousness. Cosmic consciousness, the arch-anima, is a cosmic semantic field comprising all self-organized and self-referent semantic fields in the universe. All minds and complex systems in the universe steer together the constant co-creation of our semantic universe. Cosmic consciousness is indeed a collective work of art in which each one of us is involved.

The physics of synchronicity is that instant two-way exchange of syg-energy/information between the cosmic ISS and the 4D beings (about all minds, events, and systems in the 4D universe). It means (1) the instant imprinting of the dynamical state of all systems on the bows of the cosmic ISS; and (2) through the syg-energy reaching us from the ISS, we have access to a boundless field of information, and the hyperdimension is always open to us, albeit mostly unconsciously. Through the upward spiral, as creative individuals, we may imprint a new melody on the ISS; through the downward spiral, cosmic information and archetypes are sent back toward our consciousnesses (or semantic fields). The physics of synchronicity at the cosmic scale between us 4D beings and the hyperdimension sets a co-creativity with an immense field of accumulated wisdom.

Forces at work in the CSR hyperdimension

A. Syg energy: the negentropic force in the universe

The universe's most fantastic energy has necessarily to be the force driving the self-organization of systems, the rise in complexity and information, that is, consciousness as a negentropic force. Most of all, syg energy accounts for a world that is meaningful for each and every being living in it at its own scale—whether a human or an ant. This is why several physicists readily link consciousness to the immense energy-field that is dark energy or quintessence.

I call syg energy the consciousness in process and acting at any level of reality, pervading matter and capable of modifying its organization. We are talking here about the creation of meaning as a process and also the organizing and negentropic force in the universe. Syg-energy (the sygons) creates, pervades, and organizes the semantic dimension (or metadim Syg), and can be seen as the dimension of all the Selfs or as the collective Cosmic Self; it is also the arch-anima, and the collective unconscious. The semantic dimension is a collective entity in which each human consciousness is a facet; a facet capable of having an influence on the whole, even if tiny, at the level of the ISS.

At the human level, we all breathe in and out this syg-energy at any moment of our lives; all our thoughts and mental operations, as much as our feelings, sensations, volitions, goals and actions, all are fueled and animated by this syg-energy. The sygons are definitely the prana, the energy of consciousness that one breathes in the air. If you think this is beyond credibility, just think that you have in you and in the air and things around you the very photons from the first tempest, and your bodies and brains and desks and houses are made of the very first atoms that came to existence in the first twenty minutes of the universe after the Big Bang. A great percentage of the atoms of your body are about 14 billions years old: How does that feel? The other part, most of the heavy atoms, are issued from supernovae explosions. The difference, huge I admit, between photons and the Free Sygons that pervade and organize the hyperdimension (both the bulk and the compact one in each particle), is that the sygons are faster-than-light, tachyonic, virtual particles, not bound by spacetime, matter laws, and causality. So that the syg energy is everywhere and interconnected through frequencies of thoughts and formal similarities, via the three

CSR metadims operating in a non-local way. As syg-energy is nonlocal and expresses a totally distinct manifold (from the 4D spacetime), the cosmic flows and the exchange of active information (via sygons) are synchronistic, of the harmony and resonance type, especially in meaning, but also in form and frequency.

Now, the important point is this one: Because each one of us creates and feels and gives meaning to our human and physical environment, because we have free will and intentions and goals that we can strive for, the syg-energy cannot be the sole attribute of a preset super-program that would drive the evolution of the universe as a towering and eternal causal agency—forever causal and not in itself evolving or influenced by what universe and forces it launches.

Syg-energy, consciousness-as-energy, has to be a free-for-all force, each being participating in the continuous co-creation of the world we inhabit. It is with free-for-all syg-energy that we ceaselessly shape and in-form our planet so that its future is but the responsibility of all conscious beings on Earth, in the proportion of their potential action on the planet.

We have to refrain from the culturally-driven tendency to project outside of us the energy of consciousness, whether as towering causal agencies or as deterministic and causal chains, as we tend to interpret all dynamics in the world in a materialistic and causal paradigm. A change in under way, as exemplified by the fascinating new research just published (just in time for me to integrate it in this book) and that demonstrates the self-organizing dynamics of the DNA and RNA, so that we know now that the nucleic acids and DNA precursors can self-assemble to form longer chains of polymers and then more complex life-forms, they just need liquid crystals (condensed water). This study (Fraccia *et al.*, 2015), led by Tommaso Bellini (Univ. of Milan), suggests that self-organizing properties inherent to the precursors of DNA could be the origin of life. Liquid crystals in a bath prod an ordering dynamics in ultra-short segments of DNA, that results in a linear aggregation via "duplex" and "end-to-end attractive interactions." (Fraccia *et al*, 2015) "The spontaneous self assembly of small DNA fragments into stacks of short duplexes greatly favors their binding into longer polymers, thereby providing a pre-RNA route to the RNA world," says Noel Clark, one of the researchers. Already in 2009, a study by Geoff Baldwin *et al.*, (J. of Physical Chemistry B) showed that nucleotide sequences from

DNA strands, immersed in a water bath, could recognize sequences similar to themselves and gather together, ignoring dissimilar sequences. Says the authors "The sequence recognition can reach across more than one nanometer of water." (Fraccia, T.P. *et al. Nature Communications*, 2015; 6: 6424. DOI: 10.1038/ncomms7424)

Similarly, the cosmic DNA is a self-organizing force; it is not a deterministic program or causal agency: it is the cosmic and collective field of information at the origin. It is neither bound by causality nor by linear time; to the contrary, it is operating two-way, in a synergy with the existing beings, through the synchronistic hyperdimension.

Metadim Syg, as a cosmic DNA, is the collective intelligence in action, the negentropic force of the universe; but it is also a field of active and generative information containing:

- all possible events in this universe bubble,
- at the X-point of origin, the musical score of previous universe-bubbles are imprinted on the ISS as already activated 'melodies' on a fraction of its quasi-infinite bows.

This cosmic consciousness will prod the creation of a viable material universe with its numerous forms and fine-tuned physical parameters. Nevertheless, each human being is endowed with the capacity of expressing the same creative powers, qualitatively speaking. As we'll see in chapter 7, each particle has a quasi-replica of the ISS as a compact hyperdimension, and our Self is each one of us' inner hyperdimension. In reality, all humans can accrue their mind-power and syg-energy by numerous techniques of self-development and self-improvement that have been developed by all cultures in the world.

B. Sygons and the frequency domain

The steps of the spiral staircase are bows expressing a specific frequency multiple of the phi ratio—each bow being totally original and unlike any other one, according to Pauli's Exclusion Principle. We shouldn't be astonished that this principle applies also to the pre-space (or hyperspace), given that it is dealing with energy-states and frequencies, as well as spin (rotation) and orbits, and that the bows are exactly frequencies (sygons) within circles and in rotation (torsion waves), as well as energy states. The ISS theory poses that only a small part of all possible bows-frequencies of one universe's ISS can be activated simultaneously. (This could reflect the small part of our individual human DNA coding for proteins, compared to latent DNA, typically called "junk DNA" by the late twentieth century biologists.) Any single bow can be activated and get into oscillations/vibration mode, thus launching a sygon and its torsion wave in the direction of flow of the cosmic spiral. Within the ISS, each bow is a pure frequency of one type only but having several potential states: (1) it is either in virtual mode or activated mode; (2) it has an array of harmonics; and (3) it has a network of links with some other activated bows. In the ISS, the sygons are pre-particles/strings. The high frequency ones, the Free Sygons, will create the bulk of the hyperdimension as tachyonic torsion waves; and the low-frequency sygons will become the strings/particles of the Standard Model after crossing the Higgs field. The frequencies of the bows and sygons before Planck scale, where only the Bindu virtual time exists, are expressed by Metadim Rhythm in Bindu-rhythm units (in Br) and not in Einsteinian time because the 4D spacetime doesn't exist yet. The arch-anima uses ensembles and networks of frequencies to express the complex melodies of systems. Even fine-tuned physical laws of the spacetime domain may thus be 'ciphered' in the virtually infinite bow-system. In this way the ISS contains vast archives of past melodies as well as continuous snapshots of the actual state of innumerable systems in our present 4D universe. Yet, even in a 13.79 billion years old universe like ours, the immense majority of the bows-frequencies are not activated yet, leaving infinite possibilities of evolution open.

C. Expansion and inflation of the point-scale universe. This expansion from the infinitely small X-point and culminating in the inflation phase or Big Bang soon after Planck length, is known to all physicists. In the ISS theory, the expansion is driven by two numbers: Phi and Pi—with $\phi = 1.618$, and $\pi = 3.1416$. It creates a topological space which is in effect a pre-space, thus perfectly in accord with the known facts.

D. Cycles and frequency as pretime. In the formation of the infinite spiral, issued from the void and the infinitely small, the bows are all quarters of circles and thus they instantiate potential circles and cycles of rotation. This pretime is definitely a rotational, cyclical, time, that is, in effect, a pre-time. It sets the launching, by the ISS, of the sygons' torsion waves, each sygon bearing the frequency of its original bow.

E. Rotational dynamics and angular momentum. This rotational dynamics of the spiral is prefiguring, or rather launching the rotational dynamics of matter after Planck scale, from the vacuum to the astronomical systems, such as rotating galaxies and presumably clusters of galaxies, and probably our whole universe-bubble. It also launches the spin of particles and of microscopic systems. The ISS produces a rotational velocity and an angular momentum.

F. The discrete nature of ISS. Given the leap, at each step, produced by the logarithm of 1.618, both in radius length and in frequency-bow, the ISS is thus a discrete reality, prefiguring perfectly the discrete (i.e. non linear and non-continuous) nature of the quantum field starting at Planck scale.

G. The bows and sygons as pre-strings. In superstring and M-theory, the particles have been modeled as strings vibrating (like musical strings) on specific frequencies. The bows and correlated sygons, expressed by a frequency (function of Phi) are definitely pre-strings in the ISS domain, thus giving a smooth origin to the strings-particles of the spacetime domain, as well as a smooth transformation into the matter-energy domain.

In synthesis, the ISS models many facets of the primordial universe:

- A near-infinite syg energy and consciousness at the origin.
- A boundless field of information (potential and actual) inscribed in the Cosmic DNA (the ISS).
- A discrete nature prefiguring the discrete nature of the quantum particles and events.
- An enlarging Point-Sphere, thus an evolving universe.
- A topological ordering principle (a field of form) with Phi and Pi.
- A rotational and torsional dynamics (bi-rotational in effect).
- An immense non-finite frequency domain able to express any possible event and evolution in any number of matter-universes.
- The birth of free sygons, the virtual string-particles that will create the hyperdimension bulk OR become the string-particles in the quantum and 4D world (the quantum-spacetime or QST manifold).

This ISS theory fits so incredibly well the post-Planck scale physics that it seems to be a very likely candidate model for the pre-space.

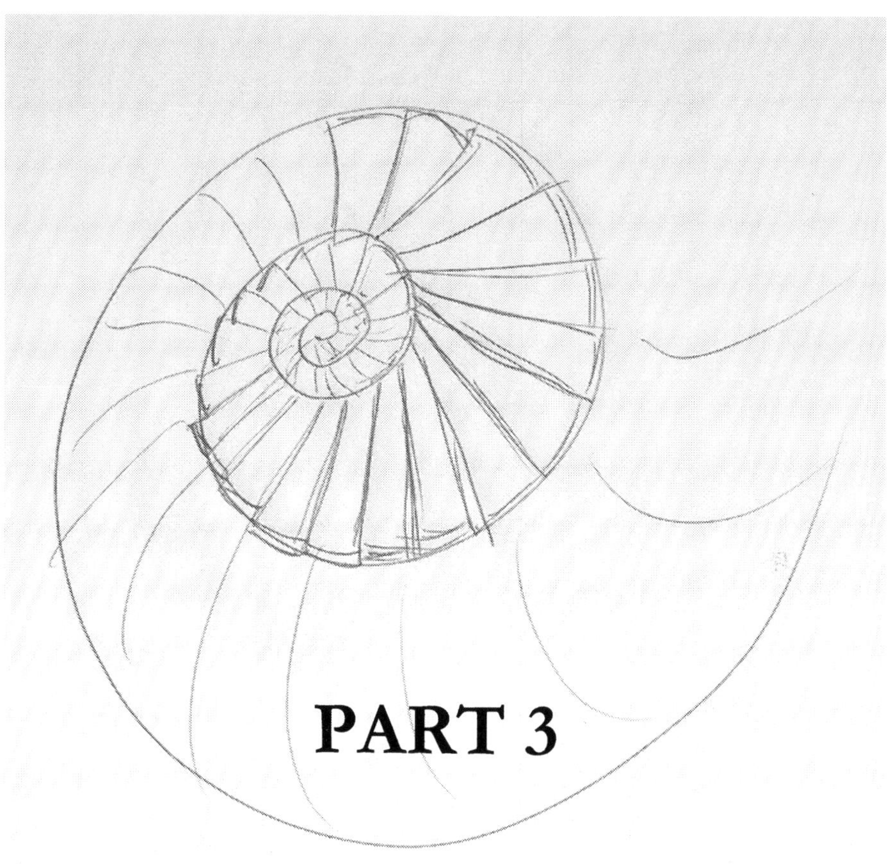

PART 3

NONLOCAL CONSCIOUSNESS IN PHYSICS

6
CONSCIOUSNESS, PSI AND NONLOCALITY IN PHYSICS

"This systematic denial on science's part of personality as a condition of events, this rigorous belief that in its own essential and innermost nature our world is a strictly impersonal world, may, conceivably, as the whirligig of time goes round, prove to be the very defect that our descendants will be most surprised at in our boasted science, the omission that to their eyes will most tend to make it look perspectiveless and short." William James.

"The true physics is that which will, one day, achieve the inclusion of man in his wholeness in a coherent picture of the world." Teilhard de Chardin.

Following the lead of Heisenberg, Schrödinger, and Von Neumann (the Copenhagen school) quantum mechanics had posited in the 1930s that the act of observation was triggering the collapse of the wavefunction and therefore that the reality we observe has already been modified by our observation. Several early QM theorists included consciousness in their models, as being the observer (Wigner, von Neumann) and even as making choices (Stapp). In the 1970s, a number of physicists had put themselves to the task of bridging the gap between mind and matter. Around the time that Fritjof Capra published his famous *The Tao of Physics*, a group of scientists linked to the Lawrence Berkeley Laboratory (Berkeley University), among them Nick Herbert, Fred Alan Wolf, Jack Sarfatti, Saul-Paul Sirag (whom Evan

Harris Walker and Brian Josephson rallied), had launched the *Fundamental Fysiks Group*. Meanwhile, in synergy, physicists at the Stanford Research Institute, such as Harold Puthoff and Russell Targ, not only studied psi experimentally (notably Remote Viewing) but also explored theories based on QM and the entanglement in order to explain psi phenomena. In Princeton University's PEAR lab, with the dean of the Engineering department, Bob Jahn, laboratory manager Brenda Dunne and lead researcher Roger Nelson, both experimental research and theory were also pursued into mind-matter interactions. The point is that psi displays the same sort of weirdness as particles—such as nonlocality and retrocausality. Then Roger Nelson started the Global Consciousness Project, setting random events generators (REGs) disseminated in research centers worldwide, that showed that collective events of maximum import could bias the distribution of randomness. (See Nelson, Radin.)

Physics models integrating Consciousness

Biosystems bend the odds: Josephson and Pallikari

The physicist Brian Josephson, Nobel Prize in Physics in 1973 for his groundbreaking work on superconductivity and quantum tunneling (that set the "Josephson effect," widely used, notably in quantum computing), is the head of the Mind-Matter Unification Project at Cambridge's Cavendish Laboratory. Brian Josephson is also a keen researcher on psi who is deeply cognizant of both the experimental research and the theoretical research. In recent years, he has taken the approach of distinguishing the 3^{d} person perspective expressed in science and the probabilistic stand of orthodox QM, from the 1st person perspective of sentient biosystems and minds bathing in and exchanging with a meaningful world. In a 1991 article, Josephson and physicist Fotini Pallikari-Viras advocate that quantum mechanics (and science in general), when dealing with phenomena expressed in terms of probabilities, is unable to access and penetrate the level at which biosystems (such as a mind) operate, that is, through meaningful interactions. The authors refer to the concept of *soma-significance* put forth by physicist David Bohm, in other words, 'meaningful for a body' (soma) or a biosystem. The network of meaningful interactions in which biosystems live will lead to "probability distributions other than the particular ones that arise in the *quantum formalism*. (...) From the point of view of a biosystem itself, this possibility translates into one that biosystems can have more discriminative *knowledge* of nature than is obtainable by quantum measurement." (Josephson & Pallikari-Viras, 1991)

> Indeed, I concur with this extremely perceptive concept of Josephson and Pallikari that through their "more discriminative *knowledge* of nature," biosystems definitely build up a *personalized interaction* with nature, highly significant in regards to their own aims and needs. This specific interaction, or coupling, is what is likely to supersede the apparent randomness of events and to bend the odds toward events that are *meaningful to these biosystems*. A bedrock observation in psychotherapies is the fact that traumatized individuals tend to attract the dreaded repetition of the traumatic context or event. In other words, a psychodynamic process—their semantic field organizing itself around fear and the fearful event—gives so much semantic energy to the dreaded event that it bends the natural distribution of randomness toward producing a high probability for the dreaded event to happen. The reverse meaning-driven dynamics is at work in the well proven effects of positive thinking and optimism that have been shown in psychology experiments to correlate significantly to healing and a better health (cf. Seligman, 2006).

Josephson and Pallikari further stress that "The strategy of science leads towards the accurate specification of form, while that of life leads in the direction of meaning. Meaning (...) is sufficiently subtle and complex as NOT to be representable by any closed formula. Furthermore, the technique of statistical averaging is especially irrelevant in the context of meaning, since its influence in general is to transform the *meaningful* into the *meaningless*." The authors base their argument on the EPR paradox experiments that instantiate nonlocality and *entanglement*—that is, "manifestations of inter-connectedness" between paired particles, at very large distance and ruling out transmission by space, because they happen at speeds much greater than *C*. Thus, Josephson and Pallikari argue that such nonlocal interconnectedness (proven in QM by Alain Aspect in 1982-84), is the basis of psi phenomena (e.g. telepathy, remote-viewing, psychokinesis or PK), the reality of which has been established by thousands of laboratory experiments.

As an example, over a hundred and fifty laboratory experiments have been conducted on *bio-PK*, that is, the ability of a mind to influence biosystems (mostly cultures in Petri dishes, bacteria, cells, blood, tissues samples, and small animals, all biosystems impervious to suggestion and the placebo effect); most tests implied a double-blind protocol to compare the control samples with the experimental samples (those on which an influence is effected at a distance), and many protocols varied the distance between subjects and samples. These experiments have used statistical analysis to calculate the *p* value (the probability of such an event happening by chance alone) and the magnitude of the departure from chance (the baseline of random distribution, calculated via the overall statistics on control samples). Moreover, at a later stage scientists have analyzed all same-protocol experiments to get meta-analysis results (see the excellent overview of protocols and experiments in Schwartz & Dossey 2010). Therefore, I deem it perfectly legitimate and well-grounded to state that *the significant positive results in the meta-analysis of bio-PK experiments (with double-blind protocols) show the extent to which minds and their purpose can supersede the baseline of randomness, as far as an influence on distant biosystems is concerned.*

A simple understanding of the dynamics of the placebo effect in establishment science (ranging at about 30% of positive healing effects) will show us that we can safely apply the same logic to it. Indeed, the placebo shows the effect of positive suggestions (from medical doctors to their patients) that such a pseudo-pill (a simple sugar pill) will cure them. The placebo effect has been so widely recognized and evaluated in medicine, that all new drugs, in order to be sold on the market, have to present a rate of success higher than the placebo of this drug—the drug being administered by the same physicians, who themselves are 'blind' about which pill they give, the placebo or the real one. Then the statistics of the two groups (the patients who received the placebo and the ones who got the real drug) are calculated. In terms of cognitive processes, it's understood that the patients (via their beliefs, trust, and their desire to be cured), are internalizing the physician's suggestion and turn it into a self-suggestion. Yet, the healing happens, even if we cannot exclude that a part of the placebo effect is stemming from an 'experimenter effect,' that is, from the healing influence of the physicians themselves. The experimenter effect is another way to understand QM's 'observer effect' in experimental settings; in psi research results, we try to distinguish it from the subjects' psi effect, to which it can either add up or subtract, depending on the experimenter's beliefs in psi (Schlitz, 2006).

All in all, the placebo effect is there and remains stable, so that we can draw a safe conclusion: *the placebo's rate of success, in statistics, shows us precisely the magnitude of the influence of a psyche over its own biophysical processes; it describes exactly the reality and the magnitude of a 'normal and usual' mind-over-matter influence.*

Overall, we can posit a large influence of a consciousness over its surrounding biosystems. The way biosystems constantly interact meaningfully with their human and natural environment, makes that they develop a unique and personalized *network of interaction* with their natural and human environment that definitely bends the distribution of events with which they interact. As postulated by SFT, this network of interaction allows a continuous inter-influence between minds or systems, via syg energy. There is a two-way inter-influence among interconnected semantic fields (whether personal or collective), that allows a deep-level reorganization of these natural and bio- systems (as in the effects of optimism on health).

SFT adopts the ecological framework that natural and life systems are constantly interacting between them (within larger environmental systems), and at several levels simultaneously. Some interactions imply bio-chemical and physical processes; others imply sentient, meaningful, and thinking connections between semantic fields, such as psi. Finally, there are all types of transversal links and crisscrossing between levels, as evidenced by (on the negative end) the well-established psychosomatic symptoms.

Then, SFT postulates that all interactions between biosystems:

(1) are a two-way connection, (2) imply a two-way inter-influence, and (3) are propagating to the other levels of organization of the biosystem—e.g. a mind-psyche-body system.

In other words, psi is just the tip of the iceberg of a persistent, mostly unconscious, nonlocal exchange of both information and influence between human beings, the living, and natural phenomena. This exchange is mediated by syg energy and by the semantic fields of natural systems (their CSR hyperdimension).

Hyperdimensional physics needed to account for psi

Professor of Mathematics and Astronomy at London's Queen Mary University, cosmologist and researcher on psi phenomena, Bernard Carr is the editor of the book *Universe or Multiverse?* that brought the main theorists on cosmology and the multiverse to present their models. Beyond his own remarkable work on string theory and the multiverse, Carr is intent on finding a way to integrate consciousness and psi capacities within an integral theory of the universe, something, in his words, acutely "missing" in contemporary physics theories and the Standard Model. In his 2010 article *Seeking a New Paradigm of Matter, Mind and Spirit*, he states: "Some new, deeper paradigm is probably required that will explain *both* consciousness and quantum theory. (...) A new paradigm—involving a radically different sort of physics, which I call '*hyperphysics.*'"

The task of integrating consciousness in physics theories is a tall one, but many scientists and researchers in the field of psi (some of them also physicists) agree that not only this step is absolutely necessary for a twenty-first century physics, but moreover it is *the* paradigmatic leap that will launch this novel integral physics. Bernard Carr quotes the convergent realization of such a need by eminent scientists, such as Noam Chomsky stating in 1975 "Physics must expand to explain mental experiences," and Roger Penrose's 1992 radical stand: "We need a revolution in physics on the scale of quantum theory and relativity before we can understand mind." But physicists are as bold in their statements; proof is, Eugene Wigner, Nobel Prize in Physics, made the following point with finality: "It is not possible to formulate the laws of physics in a fully consistent way without reference to the consciousness of the observer." This insight into the future-starts-now 'integral physics' is, for many of my colleagues and myself, irrevocably on target. Physics will not be able to circumvent the reality of consciousness as a negentropic force in the universe—on purely scientific grounds and independently of any religious worldview—any better than religions did when confronted to the heliocentric system. As John Wheeler, in 1977, said it with definitive clarity: "Mind and universe are complementary," a concept developed in depth by Robert Jahn and Brenda Dunne in *Margins of Reality*. They state that not only are mind and matter two complementary aspects of reality, just like the wave-particle complementarity, but "The possibility that consciousness, through intention, can marginally influence its physical reality to a degree *dependent on its subjective resonance* with the system or process in question, has implications that could extend well beyond (...) quantum mechanics..." (324).

It is this very aspect of the mind over matter interaction that gives me the ground to model consciousness as an energy—energy being defined in physics as an ability to effect an action, that is, to effect some work. As we have seen, Wolfgang Pauli was heralding the modeling of a dimension of mind-matter merging: "To us (...) the only acceptable point of view appears to be the one that recognizes both sides of reality—the quantitative and the qualitative, the physical and the psychical—as compatible with each other, and can embrace them simultaneously. It would be most satisfactory of all if physics and psyche could be seen as complementary aspects of the same reality." (Pauli 1955, 201) As a possible forerunner of a hyperdimension of consciousness that could be the one to encompass the 4D world, and not the inverse, Dunne and Jahn state that after having sorted out how the quantum world could be a metaphor for psi, "Now we (...) suggest that it is consciousness that is the ultimate metaphor for quantum mechanics." (316).

John Smythies (1956, 2003) proposed a 7D model comprising 3D of physical spacetime, 3D of phenomenal spacetime, and 1D of time being a vertical common dimension to the two manifolds of space.

Vernon Neppe and Edward Close (2009) have proposed a 9D model, with 3D of space, 3D of time and 3D of consciousness. A "Vortex Ndimensional space" is filled with vortices by which communication between consciousnesses and across dimensions may happen.

Several models aiming at integrating consciousness and psi with physics were proposed by physicists since the eighties. Two independently conceived models posit 8D (with 4D of spacetime and 4 imaginary dimensions), one by Russell Targ *et al.* (1979) and the other by Elisabeth Rauscher (1979), as an extended framework of Relativity theory. Then Rauscher and Ramon developed a 12D model in 1980, with 6 extra dimensions (3 of space and 3 of time) that allow contiguity of events in a hyperdimension whereas they are separated in space in the 4D physical world. In 1993, The physicist Sol Paul Sirag proposed a 'hyperspace view' of consciousness in his paper included in Jeffrey Mishlove's famed 1993 book *The Roots of Consciousness*. He used symmetry groups (group theory) and a set of 'reflection spaces' to couple matter systems with conscious processes and archetypes, such as Platonic solids. In a somewhat similar layered structure of hyperdimensions, Claude Swanson (2009) predicates that subtle energies of the mind-psyche can exist on several superposed 'sheets' (a surface brane) and that these structures may be "synchronized" (allowing communication and energy transfer) or "unsynchronized," independently of the spacetime coordinates.

We, researchers in cognitive sciences and psi, have already mapped the kind of properties that are a prerequisite to account for psi (see Hardy, *JP* 1998, Schwartz, Radin, Schlitz). Here are the main ones:

- psi exhibits nonlocality: it is non-dependent on space or time distance, even if there are some correlations in some cases; in other words: psi and consciousness can operate in a way that is not limited by (4D) space and time constraints (see von Lucadou, Walker, Schwartz, Radin, Josephson, Pallikari, Hardy).
- psi may be adequately modeled as a (type of) entanglement between minds, and/or between minds and the 4D world, that exhibits an EPR-type (instantaneous) exchange of information at a distance (mainly von Lucadou, Bierman, Radin, Schwartz, Josephson, Walker, Jahn & Dunne).
- psi is not bound by the inverse square law of electromagnetism: it doesn't show any decrease of its effect, nor a lowering of the precision of information, at enormous distances, such as shown in the successful Earth-Moon psi experiment that Edgar Mitchell performed during the Apollo 14 mission (Mitchell, Mishlove).
- psi functioning is not impeded by usual shielding from most EM spectrum waves: it operates inside Faraday cages and at a great submarine depth (Targ & Puthoff, Schwartz).
- psi involves and makes use of any bio-psychological system in the mind-body-psyche. The psi information can be relayed by any

channel: interoceptive perception, empathic sensation at a distance (tele-empathy), unconscious expression (via subliminal behavior and body movements), anomalous vision, auditory reception, touch sensation, verbal reception, altered state and meditative states, etc. This, in my view, shows that the level at which occurs the connection with a cosmic or holographic field of consciousness, or with the collective unconscious, is (1) beyond language, and (2) implies the most basic and low-level organization of biosystems. (Hardy 1998, Tart 1969)

- The item just above is the reason why pets and wild animals have been shown to display psi in experiments (Sheldrake 2011).
- Psi has the capacity, in healing and bio-PK for example, to reorganize biosystems at low levels of organization, such as neuronal nets, cells tissues, organs, or physiological processes. It has been also proven capable of modifying the distribution of randomness according to aims and intentions (micro-PK experiments) or via some collective unconscious 'field effects' such as in the Global Consciousness Project data (Nelson, Radin, Schlitz, Schwartz & Dossey).
- The previous item meets the classical physical definition of 'energy' as being the capacity to effect some work, and consequently consciousness can be modeled as an energy operating via nonlocal meaningful connections (such as Bohm's quantum potential, Hardy's syg energy).

Bernard Carr, in his 2003 article "Is there space for psi in modern physics?" states three requirements for a "paradigm [that] would be required to accommodate psi":

- "An essential feature is that it must involve consciousness, since this underlies all psychic experiences." Also:
- "It must involve some kind of higher dimensional reality structure. This is because many psychic phenomena (e.g. OBEs, NDEs) seem to involve some form of communal space, which is not the same as physical space but subtly interacts with it."
- It should include a collective space or 'Universal structure,' given that, says Carr: "our minds are part of a communal space rather than being wholly private. This 'Universal Structure', as I term it, can

be regarded as a higher dimensional information space which reconciles all our different experiences of the world. It necessarily incorporates physical space but it also includes non-physical realms which can only be accessed by mind." (Carr, 2003[25])

Carr deduces that, in order to integrate this universal *"higher dimensional information space"* within physics and a cosmological model, the solution would be a type of Kaluza-Klein higher dimensional space, or its finer elaboration in the Randall-Sundrum theory: "Since the only non-physical entities in the Universe of which we have any experience are mental ones, and since the existence of paranormal phenomena suggests that mental entities have to exist in some sort of space, it seems natural to relate this to Kaluza-Klein space. More precisely, I identify the *Universal Structure* with the higher dimensional 'bulk' of Randall-Sundrum theory. This has profound consequences for physics, psychology, parapsychology and philosophy." (Carr, 2003)

This Universal Structure, Carr (2014) postulates it as "a higher-dimensional psychophysical information space. This space has a hierarchical structure and includes both the physical world at the lowest level and the complete range of mental worlds – from normal to paranormal to transpersonal – at the higher levels. The assumption that mental phenomena require a communal space is tantamount to positing some form of Universal Mind, which is controversial but central to the Universal Structure proposal." An interesting issue dealt with by Bernard Carr very detailed 2014 article "Hyperspatial Models of Matter and Mind" is how to adjust the various observers' percepts of the world, which is resolved by adding extra dimension(s); "the prime message of relativity is that one can only reconcile how different observers perceive the world if it is 4D." But the main theoretical problem is how to adjust mental spaces and time to physical ones.

Carr uses perception to model the interaction between mind and the 3D of space, since perception is "dependent upon brain processes." Both brain processes (during perception) and objects perceived will intersect in their worldlines within the light cone. (See Fig. 1.2, p. 33.). Thus brain processes "can be described in terms of the (...) nexus of worldlines associated with the electrical signals between neurons. This suggests that all perceptions can be represented by spacetime connections of some sort..." Carr adds that this accounts for "a sort of extended mind, in which conscious experience is associated not just with the brain but with all the parts of spacetime to which the brain is linked through a causal nexus of signaling worldlines."

Bernard Carr is well aware of the complex task that such a modeling of the mind-matter interactions implies, beyond just posing higher dimensions, when one tackles such a multifaceted reality as the mind and even more so, a collective mind. He thus states simply: "From a philosophical perspective, the identification of a percept with some cross-section of the 4D object may seem simplistic, since this only accounts for the geometrical aspects of perception and excludes secondary aspects (qualia). Nor does it address the fundamental distinction between the 1st person and 3rd person descriptions of the world." He is also very cognizant of the beyond spacetime, and beyond EM signals, of the mind reaches, as exemplified by the very psi phenomena that he wants to model. In order to do so, Carr extends the hyperdimension into 'sheets' reflecting layers of complexity and abstraction, the higher layers being the ones most remote from 4D matter, such as the most immaterial psi phenomena.

Penrose and Hameroff: self-collapse in the brain

Roger Penrose and Stuart Hameroff (anesthesiologist) have proposed in the mid-1990s that wavefunctions' superposed states may "self-collapse" in the brain due to gravitational forces. The model allows the self-collapse to be effected by medium scale systems such as assemblies of neurons, instead of being restricted to the quantum scale (as in QM). Penrose used *quantum gravity* theory, in which he made seminal advances, such as that of the *loop networks* (see Smolin theory). In a superposed state, the masses of the two particles create two curvatures of spacetime in opposite directions. If the difference in curvature is too large, beyond a given threshold, it triggers the self-collapse—called Objective Reduction, OR. The greater the mass (an assembly of neurons or an atom), the longer the time to reach the "quantum gravity threshold for self-collapse."

In Hameroff and Penrose (1996) Orch-OR model (Orchestrated Objective Reduction), the microtubules inside the cytoskeleton of dendrites (and their tubulin proteins) allow quantum coherent superposition to happen, well isolated by a gel in hydrophobic pockets from environmental decoherence. The superposed state is "connected among neural and glial cells by quantum tunneling across gap junctions." Hameroff (1996) adds "The pre-reduction, coherent superposition ('quantum computing') phase is equated with pre-conscious processes, and each instantaneous OR, or self-collapse, corresponds with a discrete conscious event." (Let's remember that Libet had shown that a sensory perception remains pre-conscious for 500 msec until the conscious awareness happens (Libet *et al* 1979)). Then, "sequences of events give rise to a 'stream' of consciousness. Microtubule-associated-proteins 'tune' the quantum oscillations and the OR is thus self-organized, or 'orchestrated' ('Orch OR')." According to Penrose and Hameroff (1996), "Orch OR predicts involvement of roughly 10^2 to 10^3 neurons (interconnected by gap junctions) for rudimentary conscious events. For more intense conscious events, for example consistent with 25 milliseconds 'cognitive quanta' defined by coherent 40 Hz activity (...), superposition and self-collapse of 2×10^{10} tubulins (and 10^3 to 10^4 neurons) would be required." In a new 2014 article[26] the authors add: "This orchestrated OR activity is taken to result in a moment of conscious awareness and/or choice."(P&H 2014a) The quantum gravity causing the collapse when the difference of spacetime curvature reaches the threshold (function of the mass) that Penrose uses, is an alternative to the random collapse of orthodox QM. The state in which the system will self-collapse in neither random nor algorithmic, but induced by the geometry of spacetime, thus reflecting, for Penrose, a non-computable influence, belonging to what he calls the "Platonic mathematical world," which, together with the physical and the mental worlds, constitutes the universe's three worlds. Penrose's Platonic world is geometric, aesthetic, and ethical, and would be the basis of non-computational thinking: "In Orch OR, consciousness is a particular manifestation of life's quantum activity" (P&H 2014a), and also "Orch OR suggests a connection between brain biomolecular processes and [the] fine-scale structure of the universe."

The link to quantum gravity (QG), i.e. the fact that consciousness could be linked in any way to the fine structure of the geometry of spacetime didn't make any sense to me (whereas I was won over by Penrose's "non-algorithmic thinking" argument)—that is, until I saw it through the ISS perspective. However, something is definitely lacking in Penrose's model, in the sense that consciousness cannot in any way arise from this geometry. In ISST, the logical implication (given that Penrose got an experimental proof of self-collapse in the brain triggered by QG) would be that consciousness is much more entwined with the geometric-rhythmic-dynamic fabric of the quantum-spacetime manifold than we think. As for what's lacking in Penrose's physics model: that's where ISST may offer a viable solution, with a hyperdimension of consciousness pervading the universe and entwined with both a dynamic geometric topology and an information field set on a quasi infinite suite of frequencies. Now, here is a gem I found. (Note first that the microtubules display two types of monomer lattices, A and B.) "The A-lattice has multiple winding patterns which intersect on protofilaments *at specific intervals matching the Fibonacci series* found widely in nature and possessing a helical symmetry..." (P & H, 2011). Indeed, there are also spirals and the phi ratio inside our neurons!

Ervin Laszlo's A Field

Ervin Laszlo, in a 2014 article titled "Consciousness in the cosmos," takes the stand that "our individual consciousness is part of the consciousness that pervades the cosmos"—a stand that is also the keystone of the ISS theory. Says he: "What could be evidence for the presence of a cosmic consciousness in the world? There should be some evidence, because it stands to reason that if the cosmos harbors a consciousness there should be traces of it in the world we observe. What traces should we look for? I suggest that we should look for the kind of order that characterizes the structures and processes of the universe. If the universe is "in-formed" by a cosmic consciousness, the kind of order it discloses cannot be random and transient; it must be order that embraces space and time. If we find such order in the universe we can assume that it is in-formed by something we can characterize as a mind, logos, or consciousness."[27]

In his 2009 book called *The Akashic Experience: Science and the Cosmic Memory Field*, Laszlo develops the theory of a cosmic field of consciousness, called the *Akashic* or *A field*, which is able to memorize the whole information of the universe as a giant hologram, and is no less than the vacuum, or Zero Point Fluctuations (ZPF) Field. This A field would be the main force in-forming all processes and systems in the universe, and Laszlo bases this concept on the latest findings on the ZPF, showing it to be a turbulent medium, of high density and quasi-liquid, filled with high-energy virtual particles in permanent fluctuations.

As we have seen, both Paul Dirac in 1930 and Sakharov in the early 1950s had proposed already that the vacuum (far from being only the curvature of spacetime pervaded by EM fields) is filled with charged particles in constant fluctuations that induce a polarization of the void, and this would affect in turn the main parameters of the particles (such as mass, spin, charge, angular momentum, etc.). Moreover, Hal Puthoff, Bernard Haisch, and Alfonso Rueda—physicists who have been at the forefront of the ZPF research—have modeled the intense interactions between the 'polarized' vacuum's virtual particles and the 4D particles, and proposed that these interactions could modify gravity and inertia at larger scales. According to Puthoff, all matter systems (from atoms to stars) are constantly absorbing energy quanta from the vacuum, and this is what stabilizes and counteracts the gravity force. This expresses the new quantum field modeling in cosmology, implying particles and quantized energy—an alternative to Einstein's view that it was the curvature of space (via the cosmological constant) that was counteracting gravity and maintaining a fixated universe.

Puthoff and Haisch have also been involved in active research in the domains of anomalous mental phenomena at the Stanford Research Institute (throughout the 1980s and early 1990s). These psi capacities, as we know, are among the rare obvious and proven manifestations of consciousness processes (and eventual energy and linked fields), that show an independence from spacetime and from the constraints imposed by the EM inverse square law.

Laszlo also takes as a foundation of his model, the thirty or so parameters and constants that the Anthropic Principle has shown to be extremely fine-tuned in the universe we inhabit; some of them allow a long enough time for galaxies to form; others exhibit the precise ratios of chemicals and physical constants necessary for the development of carbon-based life forms, and of intelligence and cultures. For Laszlo, this fine-tuning reveals a cosmic coherence that can only be woven and activated by a global cohering field: the A field that is transmitting and exchanging information between all objects in the universe, and that is in-forming them in the process.

The dynamics by which this in-formation could happen is via the vortices said to exist in the vacuum. Many particles (like the electron), and the molecules as well, have spin, and therefore an angular momentum and magnetic moments (the latter create the macroscopic magnetic effects). Virtual particles' magnetic moments (such as those of virtual bosons) create tiny swirls in the void, and these vortices (which are torsion waves) may carry information. When torsion waves merge and create interference patterns, then their information merge also. Thus the vacuum vortices register the information on the state of particles, and their interference patterns memorize all information about all particles having interacted with them. The vacuum then will transfer the memorized information to more complex organisms such as molecules, cells, ecological systems, etc. (Laszlo, 2004, chapter 4). The domain of torsion waves has seen a volcanic development in Russia lately, with physicists such as Shipov and Akimov. According to them, torsion waves can connect systems at an immense distance at a speed that is a billion times that of light (10^9 *c*). The interference patterns of torsion waves of stars and solar systems create a galactic hologram, and those of the galaxies create the universe hologram, through which all galaxies and stars can evolve coherently.

Similarly, for Laszlo, the A field is what allows the coherence of all parts of the organism, and the coherence between a species and its ecological milieu. Says he: "The universe is a coherent system, highly coordinated as a living organism." (chapter 6).

Edgar Mitchell, who had a fusion and unity experience during the Apollo 14 mission to the moon, envisions that the quantum vacuum is the medium that sustains the holographic memorizing of information about the entire past of our universe. Laszlo postulates that these *torsion waves* have no limit as to the quantity of information imprinted in them: they could carry the information of the entire universe to everywhere in this universe. Puthoff likewise envisions that the fluctuation fields of the vacuum could be 'modulated' and carry information, and that we could be connected, through them, to very distant objects or events in the universe. As we saw, the framework proposed by Nobel laureate Brian Josephson and physicist Fotini Pallikari-Viras discusses another facet of a universe in which the living beings (or biosystems) generate a distinct set of physical laws, that quantum mechanics can neither access nor model—notably different dynamics based on meaning and interconnectedness, able to modify the probability of events.

Quantum Nonlocality: the EPR experiments

According to QM, the basic nature of microphysical reality is probabilistic. QM states that quantum systems are intrinsically random, that there is a fundamental indeterminacy concerning the exact state the system is going to be in at the moment of measurement. Einstein's famous "God does not play dice" expressed the frustration of many theoretical physicists with this emphasis on indeterminacy. The thought-experiment proposed by Albert Einstein, Podolski, and Rosen, known as the EPR Paradox, sought to demonstrate that QM could not be considered "complete," contrary to what Bohr claimed, and that there must be additional underlying forces determining micro-events, and still unknown. These "hidden variables" would, when discovered, reestablish an ordered and deterministic universe. However, theoretical developments in the 1950s (by mathematician John Bell), coupled with extremely fine and definitive experiments in the 1970s and early 1980s, established the existence of *nonlocal correlations* between remote particles—in full violation of the local-realistic assumptions of relativity and classical mechanics. A series of experiments involving paired particles (issued from the same source) demonstrated that if we interfere with the spin of one particle, the other particle instantaneously changes its spin, so that it remains complementary with the first particle (according to Pauli's law of spins, the sum of the spins, ½ and -½, must be equal to zero). These correlations are described as "nonlocal" because they are independent of the distance between the two particles, precluding any explanation based on classical (i.e. local) signal transmission. The EPR Paradox has thus been central in establishing that the "weirdness" of quantum mechanics is real, and not just apparent.

John Bell, pondering the thought-experiment proposed by Einstein, Podolski and Rosen to disprove QM, started conceiving a protocol to test the nonlocality of paired particles (via his "Inequalities theorem"). He then proceeded to do some experiments of a EPR type. His were not conclusive, but experiments done by Alain Aspect at Orsay in France in 1982-84, were a definite proof that particles could exchange information at great distances and remain complementary—via a process of which we knew nothing except that it could not use space or a carrier-wave moving through 3D space. At that point both the nonlocality at the quantum scale and QM as a theory were solidly proven. Bell's conclusion, that no theory invoking *local* hidden variables could be valid, didn't forbid *nonlocal* hidden variables, such as Bohm's Pilot Wave. Bell's assertions were nonetheless heralded as forbidding any hidden variables theory and consequently Bohm's work was discredited, despite the fact that Bell himself started to work in the line of Bohm's theory. Strangely, there had been a priming to the establishment science's adverse reaction: in 1932, the genius von Neumann (who gave his mathematical backbone to QM) had elaborated a mathematical argument that disproved any hidden variable theory, yet his argument was proven false thirty years later by John Bell. It seems the QM orthodoxy conspired to eliminate their contender theory—the proponents of indeterminacy against those of determinism. As usual, future science will show the two sides to be in the wrong and lacking, insofar as the other side is somehow right. And as usual also, the light, as we'll see with the groundbreaking 'walking droplets' experiments, will come from a totally new perspective, chaos theory, that poses randomness at the microscopic scale and self-organizing patterns at the global scale, nonlinear paths, bifurcations, and dynamical attractors—instead of EITHER forceful abstract laws OR absolute randomness.

The Pilot wave Theory: de Broglie and Bohm

While he was still preparing his doctoral thesis in Paris, Louis de Broglie's influence was seminal in solving the logical conundrum of the dual nature of particles at the subatomic scale. Just as Einstein had done it with the photons (then called "light quanta"), de Broglie postulated that the electrons and all matter particles, were both waves and particles. His bold hypothesis, also called wave-particle duality, started the new field of 'wave-mechanics' and won him the Nobel prize in Physics in 1929. In his doctoral thesis, *Research on the Theory of the Quanta*, he states that "*any moving particle or object has an associated wave.*" It is following the work of de Broglie on 'matter-waves' that Schrödinger wrote his famous equation for the wavefunction of matter, giving it a probabilistic framework. But it displeased de Broglie who had predicated that the wave was somehow guiding the movement of the particles, and had called them Pilot Waves. De Broglie's initial equations for a pilot wave of the Psi wavefunction and its guidance equation, were written in 1927. Then, in parallel with British physicist David Bohm, de Broglie will keep developing a causal interpretation of the wavefunction, as guiding waves, in which the electron gets itself a 'trembling motion' (*Zitterbewegung*) around the median line of the wave. This motion results from de Broglie's *matter waves* having a complex amplitude, one part real (positive) and one part imaginary (negative).

The Wikipedia article "wiki/Zitterbewegung" states: "The existence of such motion was first proposed by Erwin Schrödinger in 1930 as a result of his analysis of the wave packet solutions (...) in which an interference between positive and negative energy states produces what appears to be a *fluctuation (at the speed of light)* of the position of an electron around the median, with an angular frequency of approximately 1.6×10^{21} radians per second." This research has been abandoned, and the reason invoked is that "interference between positive and negative energy states may not be a necessary criterion..." However, astonishingly, the same article informs us that, while unobserved to this day, this *fluctuation* has been "simulated" twice, first with a trapped ion, and in 2013, in a setup with Bose-Einstein condensates (BEC). And these Bose-Einstein condensates are linked to superfluid (coherent) states; they have been evidenced between 2000 and 2003. Their basic principle is that a gas of bosons at near absolute temperature exhibits macroscopic quantum processes. A flurry of research is going on, because many types of fascinating dynamics and properties are associated with these BEC, such as quantum interference, quantized vortices, production of analog gravity, induced transparency, reversal of attractive force into a repulsive one, creation of solitons, slowing down of a light beam, etc.

This concept of 'trembling motion' of the particle, between the positive part of the amplitude, and the negative part of the amplitude, is of course of the greatest interest to me. In de Broglie's and Schrödinger's views, the particle takes on a positive energy state then a negative energy state, and keeps alternating at light speed between the two states, thus creating fluctuations. To my understanding, it looks as if the negative electron is constantly morphing into its own antiparticle, the positron. (Let us note the first two digits of both

Planck length and our preferred number phi, 1.6, peeking out in the frequency of the shift, that is, the angular frequency of about 1.6×10^{21} radians per second!) In ISST, I posit a blitz shifting from the CSR-focus to the QST-focus for any particle, and we know that the CSR manifold has a negative energy pressure. We know also that the compact CSR HD (the 5th dimension in particles) contains the original negative-energy sygon—the antiparticle that morphed into its own particle while crossing the Higgs field—and a replica of the ISS.

The original deBroglie–Schrödinger concept, which has not been disproven, makes a lot of sense in the ISST. In CSR-focus, the wave-packet shifts to its imaginary superposed state and 'bare' CSR frequency (a much higher harmonic of the QST particle), and gets in an entangled communication with the cosmic ISS, thus allowing a two-way influence as well as a snapshot imprinting of its state and syg-information in both its individualized ISS and the cosmic ISS. In other words, if we look at it from SFT perspective (a real world cognitive-systems viewpoint), a quantum event or a mind would constantly shift its network of links from the 4D reality to its semantic field reality, from the ego-focus to the Self-focus, thus being permanently living on two different layers of reality, each complementing the other. (The ego-focus being mostly conscious, in contrast with the Self-focus being mostly unconscious.) The person thus receives the sort of sustained guidance from his/her own Self (anima or soul) that the psychologist Carl Jung has so clearly evidenced in archetypal dreams and synchronicities in real life. The synchronicities become, in the blended SFT-ISST framework, the evidence or emergence of a person's CSR network of links (while they are implicated in a given life situation—the synchronicity bringing them a challenging Self-perspective on the situation at hand. (See the astonishing synchronicity assorted with a dramatic PK, as a clear evidence of the Self "responding" to a real-life issue, in *The Sacred Network*, 324-7.)

de Broglie-Bohm Pilot Wave theory

David Bohm further developed de Broglie's concept of pilot waves, as guiding waves setting some causal force driving the wavefunction, at first as a 'hidden variables' theory, one whose basic tenet was that the universe could not be just random, and therefore there must have been some forces at work—yet unknown and hidden—to guide the quantum system until it 'stabilized' in one state (or 'collapsed' after the observation, as seen by the specific Bohmian theories allowing the collapse). This pilot waves research was mostly overwhelmed by the fantastic attraction that the novel QM was exerting on physicists, and let's reckon it, by the heavy-handed smothering tactics of the orthodox QM clan. Yet it was still pursued by Bohm and some of his colleagues, such as Jean-Pierre Vigier (in France), and his long-term collaborator David Hiley (in the UK). Additionally, Jack Sarfatti, Massimo Teodorani, and John Bush (2013) are actually working within the Pilot Wave Theory framework.

The very interesting point about the Pilot Wave theory (also called *de Broglie-Bohm* or *deBB* theories) is that it is the only model of quantum processes that has integrated nonlocality within the QM equation itself. Having been myself, up to the work for this ISS model, mesmerized by the fascinating EPR and entanglement concepts of QM, it was a shock when I discovered that Schrödinger had developed his famous equation for de Broglie and Bohm, the equation aiming at treating matter as waves and wave-packets. Meanwhile, in 1925, Heisenberg had elaborated QM by inventing the M-Matrix and it's only later on that Schrödinger suggested to him that this field equation could integrate all his results.

As Mike Towler states it cursorily in his slides for his first lecture on "Pilot-wave theory, Bohmian metaphysics, and the foundations of quantum mechanics" (University of Cambridge, UK, 2009): "Why Pilot-wave theory? Hardly any other interpretation [of QM] accepts [the] existence of an outside reality, agrees with predictions of quantum formalism, and is both consistent and not utter madness."[28] There are novel trajectories (or behaviors) for quantum systems that appear in deBB theories whereas they are 'forbidden' in orthodox QM. Says the visionary John Bell about them: "But to admit things not visible to the gross creatures that we are is, in my opinion, to show a decent humility, and not just a lamentable addiction to metaphysics." (cited by Towler) Indeed, the fascinating side of the Pilot Wave theory is that it expresses the *wavefunction of the universe*, no less, as guiding the way particles are going to move in terms of their position and momentum. The *Quantum Potential* inserted by Bohm in Schrödinger's equation is understood as a law of motion; it is *highly nonlinear and it is the term that sets the nonlocality.*

So, the deBB global framework posits that all quantum events (such as the motion of a set of particles) are guided by this wavefunction of the universe, as in an ideal holographic universe, in which each facet or system (event, object) will have an imprint in the cosmic hologram, and will itself be a reflection of the whole.

To make the equation operational, physicists must use a set of wavefunctions of the system itself (the one measured) instead of the cosmic wavefunction. Contrary to general assumptions, the deBB theory accepts the collapse in specific quantum systems. Mike Towler explains that "The wave function of the universe does not collapse, but the wave function of the subsystem being measured effectively does ('decoherence' arguments) (...) and doesn't depend on anything like the observer's knowledge." He makes a very interesting point when specifying that since observers are not required, the "*theory works before life evolved,* and [the] wave function of [the] universe doesn't depend on observation by God."

We are reassured and relieved that we don't need a personal causal agent from the beginning onwards at all times! However, the ontological question is left unanswered: How in the world could the universe's wavefunction have appeared at the origin? Why would it be fixed for all times, since obviously emergence and the rise of information and complexity seem to be a proven and observable set of facts, as Stuart Kauffman posits it with his *Emergentism* theory, and Murray Gell-Mann with his increased complexity theory.

Let's ponder the sort of mirror reflections given by the holographic framework and similarly by the 'guiding' waves. Says Towler: "The whole object is 'enfolded' in each part of the hologram rather than being in point-to-point correspondence. Order in the hologram = implicate. Order in the object and image = explicate." And conversely: The "order of the whole universe is enfolded into each region." That's quite fine as a static image goes, or a deterministic process. Bohm introduces a tidbit of randomness in the real world, able to create probabilistic outcomes, and some room for divergence; otherwise, this perpetually self-replicating universe would be quite stifling, and largely contradicted by real world processes. Thus, the theory is badly lacking specific dynamics able to make qualified links between minds and particles, and that could set sophisticated mental processes, such as a free intention, a wild dream, a surrealistic creation, or a big laugh.

Droplets walking on pilot waves: Chaos theory of QM

The new experimental research on "walking droplets" is like the cherry on the cake of Pilot Wave theory. I know mathematicians and physicists have a liking for "beautiful" theories, but could not they also be tasteful? (In fact, I just realized why I find the droplets tasteful: it's because the corral with its superfluid on which droplets fall and bounce and create waves, resembles the round copper basin of liquid chocolate that my father used to make Christmas chocolates; he damped in it from some height round Cognac-cherries (or tiny squared fillings) to coat them with chocolate, and the cherries did create heavy round waves in the dense hot fluid!

The de Broglie's Pilot Wave interpretation of QM has recently been proven right by this stupendous new domain of research in superfluid hydrodynamics called the *walking droplets*, that was pioneered by Yves Couder's experiments in Paris Diderot university since 2005. The droplets bouncing or walking on the surface of a superfluid exhibit several behaviors of a quantum system (such as tunneling, quantized orbits, and spin states) but do so according to the 1927 de Broglie theory e.g. the wave is piloting the droplet. instead of according to the probabilistic (Copenhagen) interpretation of QM. Furthermore, Yves Couder's experiments show that quantum behaviors may happen at a macroscopic droplet scale and not be confined to the quantum microscopic scale. Says John Bush of MIT (2015 in print): "The walker system bears a notable resemblance to an early conception of relativistic quantum dynamics, Louis de Broglie's (1926, 1930, 1956, 1987) double-solution pilot-wave theory, according to which microscopic particles move in resonance with their own wave field."[29]

De Broglie presented his wave-particle double-solution to the Schrödinger equation—that modeled both a guiding wave and a stochastic particle behavior—at the 1927 Solvay conference, where he was strongly opposed by Niels Bohr and other proponents of the 'orthodox' pure indeterminacy interpretation. Several scientists, such as John Bell (whose 'inequalities' led to proving nonlocality) and recently Lee Smolin, have stressed the quasi-despotic stand of the orthodox QM theory. Murray Gell-Mann, for one, in his 1976 Nobel Prize lecture, accused Niels Bohr to have "brainwashed an entire generation of physicists into believing that the problem had been solved." Natalie Wolchover, in her June 2014 *Quanta Magazine* article on the walking droplets, makes the establishment science's oppression and enduring control very clear: "Acclimating to the weirdness of quantum mechanics has become a physicist's rite of passage. The old, deterministic alternative is not mentioned in most textbooks; most people in the field haven't heard of it. Sheldon Goldstein, a professor of mathematics, physics and philosophy at Rutgers University and a supporter of pilot-wave theory, blames the 'preposterous' neglect of the theory on 'decades of indoctrination.' At this stage, Goldstein and several others, noted, researchers risk their careers by questioning quantum orthodoxy."[30]

Fig. 6.1. Couder's experiments: Droplets walking on a superfluid, riding their pilot wave. Credit Yves Couder. (Extracted by CHH from YouTube 2011 "Yves Couder explains wave/particle duality via silicon droplets" [Through the Wormhole]

In the experiments conducted by Yves Couder and Emmanuel Fort, a silicon oil bath is put in vertical vibration using specific frequencies below the threshold that would create waves in the dense liquid. The silicon droplet set on its surface bounces up and down, thus creating circular waves in the superfluid medium, the new ones interfering with the more ancient ones. At specific frequencies, the droplet starts to move, hence becomes "a walker" that seems to be carried by its surrounding pilot wave. (See Figure 6.1)

In another type of experiment by Couder and Fort, the superfluid is filling a circular cavity called a corral, and only the horizontal movements of the droplet are registered, whereas the vertical ones are discarded, as with a camera filming from directly above the bath.

The path taken by the droplet (at a velocity of 1 cm per second) appears chaotic and random; but if all its paths are superposed on the same graph, then it shows definite circular wave patterns, with the droplet being most of the time attracted by some orbits of the waves, and avoiding the space in between. (Figure 6.2.) This is why the droplet system is a typical chaotic system, because it exhibits random behavior at the microscopic scale and dynamical patterns of organization at the global scale.

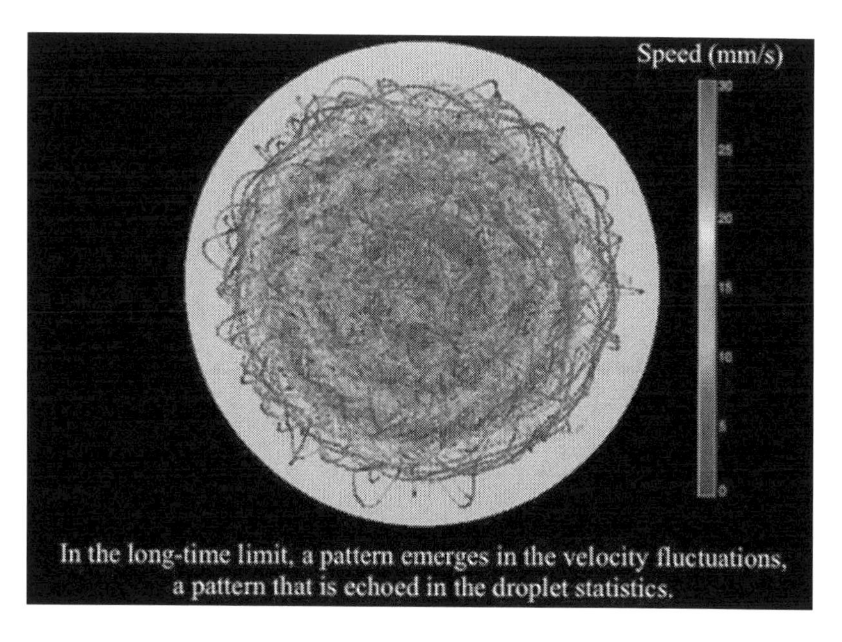

Speed (mm/s)
In the long-time limit, a pattern emerges in the velocity fluctuations, a pattern that is echoed in the droplet statistics.

Fig. 6.2. Couder's experiments: randomness at microscale and patterns at global scale. Credit DM Harris & JM Bush, MIT. Extracted by CHH from youtube.com/watch?v=nmC0ygr08tE

This behavior amounted to the particle behaving like a wave, or to be, as de Broglie had modeled it, literally carried by its pilot wave; in the droplet case, the pilot wave was the result of the interference patterns created by its own bounces. Wolchover underlines *"a mutual particle-wave interaction* analogous to de Broglie's pilot-wave concept." Then Couder and his colleagues used a double-slit experiment, similar to Young's experiment in 1801 that evidenced the wave nature of light. Young had a tiny ray of light passing though the two tiny slits of a barrier, and then hitting a screen. The two slits had produced interference patterns that looked like vertical dark bands (created by destructive interference) interlaced with white bands (constructive interference). The double-slit experiments are to this day studied and complexified, and yet some of their results are still wholly unexplained (see Radin experiments). Now, in Couder's experiment, a barrier with two slits is set in the silicon liquid, and what's happening is that the droplet goes through one opening, whereas its much larger and fluid pilot-wave passes through both openings. Now there is a parameter of superfluids called *path memory* which is higher when the ripples remain in shape longer (instead of dissipating). Yves Couder's experiments show that the higher the path memory of a given superfluid, and the more chaotic and quantum-like is the trajectory of a droplet. Says Couder: "Memory generates chaos, which we need to get the right probabilities. (...) We see path memory clearly in our system. It doesn't necessarily mean it exists in quantum objects, it just suggests it would be possible." (Cited by Natalie Wolchover, *Quanta Magazine*.)

We have, with Couder's team, a magnificent series of experiments that open totally new avenues for not only treating the QM equation as a particle riding a pilot wave, along de Broglie's theory, but also to understand the probabilities in QM as an effect of a chaotic behavior, yet patterned by global wave dynamics. Then, it leads us to envision that resonance wave-patterns could fill the vacuum (and Müller's research points to this fact). John Bush (2015 in print) says about it: "(De Broglie) "stressed the importance of the harmony of phases, by which the particle's internal vibration, seen as that of a clock, stays in phase with its guiding wave (de Broglie 1930, 1987). Thus, according to his conception, *the wave and particle maintain a state of resonance*." And the Cambridge mathematician Ross Anderson looks forward to developing a "superfluid model of reality" in which the "underlying" substrate of the universe would be modeled as a superfluid.

Consciousness triggering the collapse

The Schrödinger equation posits that the wavefunction describing any quantum system is the superposition of all possible states of that system (its state-space). Yet, when a measurement is made, the system is found to be in one fixed state (e.g., a particle with a certain momentum or position). In the orthodox interpretation, due to the measurement, the superposed states 'collapse' and the system settles into one of its possible states. The pioneers of the orthodox QM (such as Heisenberg, Wigner, Dirac, von Neumann), have introduced the 'observer' doing the measurement (that is, consciousness) as being what collapses the wave-function, thus predicating an influence of mind over the state of matter at the quantum scale at least.

However, other physicists maintain a purely indeterministic framework and view as random the 'reduction' of a superposed wave-system into one of its possible states, that is, into a specific quantum event or outcome. Consider for example Brian Greene, who expresses this view in *The Hidden Reality:* "After decades of closely studying quantum mechanics, and after having accumulated a wealth of data confirming its probabilistic predictions, *no one has been able to explain why only one of the many possible outcomes in any given situation actually happens."* (My emphasis) Maybe the solution, as Bohm saw it, is that the very probabilistic assumptions derived from a specific interpretation of the model are wrong: another force may interfere in the basic random distribution and bias the process toward a specific outcome, as we will shortly explore it. This idea, launched by Einstein and Bohm, has been later called the 'hidden variables interpretation.' The result of several EPR experiments conducted in the early 1980s has shown that *local* hidden variables were forbidden, however *nonlocal* ones (e.g. of a Bohmian or deBB type) were not. Let's explore another perspective using the nonlocal syg-energy and nonlocal consciousness posited by SFT, in a SFT framework.

Observers as semantic fields sharing syg energy

In classical QM, the quantum system, at the moment of collapse, has shifted from the 'coherent' state of superposed waves, to a particle with coordinates in spacetime. Another way to look at it (since the mid 1980s), is to view the collapse as 'decoherence.' The Wikipedia article *Quantum Decoherence* states that "decoherence occurs when a system interacts with its environment in a thermodynamically irreversible way. As with any coupling, entanglements are generated between the system and environment. *These have the effect of sharing quantum information with—or transferring it to—the surroundings.*" The decoherence framework is in exquisite agreement with a central tenet of SFT, which is that semantic fields, whether minds or eco-fields (the environment systems) interact spontaneously with each other, and in so doing, are imprinting the environment's eco-fields with semantic (syg) energy, that is, active information and organization.

According to the orthodox interpretation of QM, then, the collapse of the wavefunction occurs at the moment of measurement. However, 'measurement' has been understood in several ways. For some physicists, it is the moment of 'measurement' creating an interference between the quantum system and a measuring device. Others, as we have seen, maintain that the observer—or the act of perception—is the crucial element in the collapse.

Actually, the idea that *the observer influences observed reality* was introduced early on by some of the central figures of QM—Werner Heisenberg for one, who developed Quantum Mechanics in 1925. This concept was reinforced and extended by mathematician John von Neumann and physicist Eugene Wigner, who implied consciousness explicitly in the collapse of the wavefunction.

Let us then assume, for the moment, an active role for the observer and consciousness, as postulated by Heisenberg, Wigner, von Neumann, and more recently Jack Sarfatti, Henry Stapp, Costa de Beauregard, Saul Paul Sirag, and others. Before the observation, the quantum system is said to be in coherent, yet discrete superposed states (called Eigen states)—that is, the wavefunction. Then, following the *cognitive act of observation,* the system is fixed into a single macroscopic state with specific properties.

What interests me here is the presumed nature of the influence of the observer on the system: just what *kind* of cognitive processes are implied? Is it simply the perception of the result that leads to the collapse? I suggest that, in focusing on observation, we are overlooking some major mental events that psychologically and temporally are prior to perception. Part of the answer, I believe, may be found in Heisenberg's *Uncertainty Principle*.

A quantum entity can be represented through three dimensions of position (a particle measure), and three dimensions of momentum (a wave measure). According to the Uncertainty Principle, choosing to precisely measure the particle's position excludes a simultaneous and precise measurement of its momentum (and vice versa). In fact, the Uncertainty Principle indicates that, prior to the act of measurement and observation, the experimenter has to first choose what to measure—and, accordingly, what kind of measuring device to use (e.g., one yielding knowledge about position versus one focusing on momentum). Insofar as no extraneous force is supposed to lead the observer to a particular decision, the experimenter-observer's pre-measurement choices must be equated to acts of free will. In other words, the experimenter freely decides upon the kind of information he or she wants to 'extract' from the quantum system, and defines the measurement protocol accordingly.

Therefore, on the basis of the Uncertainty Principle, we are grounded in posing—from within this QM framework—a conative agency leading to the choice of a measuring mode, and consequently constraining the measurement in specific ways. Thus, the observer interacts with the quantum system via conative acts, that is, intentions, choices, and decisions. This may imply a series of decisions and actions until the measurement is made, as well as the involvement of several experimenters. Therefore, the experimenter's influence on the quantum system does not occur only at the moment of perception/observation, but is distributed all along the conative process, and includes the moment of measurement and the observation of results. Interestingly, when analyzing mind-object and mind-event interactions strictly through the semantic fields framework, I was led to an analogous conclusion. From a semantic perspective, it is not observation or perception per se that interferes with events, but rather the act of purposeful thinking, even before the event's advent in macroscopic reality. In other words, the fundamental cognitive agency interacting with and influencing macroscopic reality is not the perceiving subject, nor even the perceiving/acting subject; rather, it is the conative agent, able to consider alternatives, to nourish intentions, and make decisions.

The collapse viewed by von Neumann and Stapp

According to physicist Henry Stapp (then at Lawrence Berkeley Laboratory), "The most radical departure from classical physics instituted by the founders of quantum mechanics was the introduction of human consciousness into the dynamical and computational machinery." And this agent was "not merely a passive observer (...) but, surprisingly, *an active consciousness* that works in the opposite direction, and *injects conscious intentions efficaciously* into the physically described world." (My emphasis) The introduction of such an 'observer' able to trigger a selection and thus a modification in quantum states/events, and therefore world events, was no less than "a revolutionary break with the classical approach."

In this 2009 article quoted above, *Quantum Reality and Mind*, Henry Stapp analyzes the *ontological* stand of QM, specifically in von Neumann's seminal mathematical input that did set QM as a 'mathematically proven' theory, the 'orthodox QM.' In Stapp's view, "classical physics, by omitting all reference to the mental realities, produces a *logical disconnect* between the physically described properties represented in that theory and the mental realities by which we come to know them." Whereas, in contrast, "In orthodox (von Neumann-Heisenberg) quantum mechanics the *physical* aspect is represented by the quantum state, and this state has the ontological character of *potentia*—of *objective tendencies* for actual events to happen. As such, it is more mind-like than matter-like in character." Moreover, in Stapp's framework, the "mind-like" quality of quantum events is not solely set by the superposition of quantum states prior to the collapse (part of QM indeterminacy). The mathematics of QM, says Stapp, posit the "abrupt" *action* of an exterior and "probing" agency (expressed by the quantum of action) that will trigger the collapse of the wave-function. The collapse, in its turn, brings forth a particle now endowed with specific particle coordinates (as position or momentum in space and time), that is, a matter-like particle description as opposed to the wave description of the prior superposition state. Stapp (*JCS*, 2005) states "These probing actions are called Process 1 interventions by von Neumann. *They are psycho-physical events.* Neither the content nor the timing of these events is determined either by any known law, or by the afore-mentioned random elements. Orthodox quantum mechanics considers these *events to be instigated by choices made by conscious agents*."

Thus, for Stapp, von Neumann formalized QM as "a *dualistic* theory of reality in which mental realities *interact* in specified *causal* ways with physically described human brains." (My emphasis)

Mind appears in QM as a causal agent setting quantum events, and yet *this 'mental agency'* is just intruding as a probing agent effecting an action, more precisely, making a choice or a selection within a set of probable states (the superposed state-space).

Stapp (in this 2005 *JCS* article), analyzes the introduction, in the Schrödinger's equation, of the quantum of action (precisely Planck constant in the form of the reduced Planck constant, or ħ, that is, Planck's constant divided by 2 pi). This quantum of action (by acting on the 'operator' setting the coordinates) will be what causes the collapse, and thus the selection of particle coordinates (such as the momentum *p*).

For Stapp, QM instantiates "a *probing action* performed upon that physical system by an *observing system external to it.*" (My emphasis)

In his 2009 article cited earlier, Stapp further analyzes how von Neumann had posited two distinct kinds of mind-brain interaction, called process 1 and 2, to which he himself added a process 3.

- *Process 1* implies the *"choice of a probing action"* by the experimenter.
- *Process 2* is "*nature's response* to the probing action," thus happening *after* the experimenter's choice. According to Stapp, "Von Neumann uses the name 'process 2' to denote the physical evolution that occurs *between* the mind-brain (collapse) interactions." Stapp thus adds a third process (which was the process 2 proposed originally by Paul Dirac who had labeled it "a choice on the part of nature".
- *Process 3* (Stapp): "the reduction/collapse process associated with nature's response to the process 1 probing action."

The fascinating feature of Stapp's model is the introduction, based on the analysis of QM equation, of not only an agent's choice and probing action, but of their intention. In his 2005 article, Stapp highlights the way in which the process 1 mind-probing (consisting generally in "posing a question" as a 'Yes' or 'No') can be equated with the *agent's intention*; this intention would be correlated to a mind-brain pattern, but limited to a binary choice. Yet, nature's response (also limited to a 'Yes' or 'No') arises together with quantum random fluctuations, and given that a statistical averaging is due to happen in the long run, this randomness would finally dampen the agent's intention; as a consequence, the influence of observation-intention on the collapse of the wave-function remains unexplained. According to Stapp, E.C.G. Sudarshan and R. Misra have solved the problem: they postulate that an accrued rapidity in a sequence of identical 'Yes' intentions would set a *quantum Zeno effect* (a "holding-in-place") allowing the repeated intention to have an influence on the outcome.

A Semantic Fields interpretation of the collapse

Let me analyze the choice and the ensuing collapse process that will lead to no less than actual events in the 4D world (position or momentum of a particle, or quantum event, and then macro events) using Semantic Fields Theory (SFT):

Process 1 (the experimenter's *choice*) is of course a semantic (or consciousness) process, and as such it implies syg energy, and therefore a nonlocal connective dynamics that triggers connections with linked semantic fields—those of *other minds* or cognitive systems, and those of ambient *eco-fields*. The 'other minds' in an experiment are the persons connected locally (the experimental team) and non-locally as well (the worldwide researchers implied in that type of experiment, and the scientific community working within this paradigm). As for the eco-fields (the syg dimension of matter systems), they comprise the on-site (local) measuring devices (and the nonlocal links to this technology), as well as more encompassing eco-fields such as the lab, its connection to scientific institutions, libraries, etc., and also the surroundings (shared space of a town, etc.). Furthermore, SFT (Hardy, 1998) states that any new experience for a mind-body-psyche amounts to a learning process, and thus triggers the (re)organization and complexification of the semantic constellations linked to this experience and/or task. (These constellations, and the global individual semantic field encompassing them, are dynamical networks of processes interconnecting transversally all levels of their mind-body-psyche system—from the neuronal level to the abstract thinking one.)

Semantic constellations (within a person) evolve in partnership and inter-influence, then keep on self-organizing and creating spontaneous links (this is the process underlying thinking). Even if an extraordinary experience may transform radically a constellation (such as a Aha! or breakthrough experience), nevertheless the past organization of the constellation acted as a seed (toward the breakthrough), and this past state of the constellation will, once replaced, be memorized in a nested fractal dimension—that allows, at any time, a past state to be re-evoked and reactualized (Hardy 2000). And similarly, the actual state of a constellation has a "future arrow" in terms of its future-events probability-lines and thus its possible future evolution.

I have tackled in depth these dynamics, together with a retrocausal influence, in my 2003 article "Multilevel Webs Stretched across Time: Retroactive and Proactive Interinfluences," in *Systems Research and Behavioral Science*.[31] (We'll explore the subject in depths in chapter 6.) Of course, the same time-stretching (toward the past and the future) applies to the *collective semantic field* of the research team doing the experiment. So that when the (main) experimenter makes a choice of protocol/apparatus—and given the nonlocal connective properties of syg energy and of the semantic fields it organizes—this choice *has nothing to do with a binary yes-no selection at time zero*.

Regarding the time: As far as minds (and consciousness at large) are concerned, there's no time zero, not even at the person's conception or birth. Each person's semantic field has, in the course of their life, kept on organizing (in a dynamical-network process) all of their personal and shared experiences, and these constellations stretch in time and in collective semantic space. The individual semantic field is thus loaded with past emotions, judgments, values, paradigmatic assumptions, modes of action and of relationships, feelings, and the like, especially in a laboratory setting where the scientists have also internalized a whole set of scientific knowledge and produced their own research data and models.

Thus, the choice of a specific test, of a protocol and its linked apparatus and hardware, represents a conscious intention or decision that is in fact rooted in, and stems from, a vast unconscious sea of active semantic networks, even if it expresses a whole novel and bold research insight. And as we just saw, the unconscious part of the person's semantic field is moreover stretched in time and in collective, cultural, space. Finally, it is also linked to surrounding eco-fields (context, technological means, environment). Of course the highly weighted paths (greater syg intensity), i.e. the most probable decisions-actions, are those within the semantic field of the experimenter and not those issuing from her/his network of links (unless she/he is following orders). In the same way, the 'future observers' will have much lower syg intensity and weight in the unfolding of the decision and event.

2) *Process 2* (nature's *response* to the choice and probing action). In SFT, just as in QM, 'nature' is no less than the universe, and all systems

(whether physical, living, or minds) have, or rather *are* also, semantic fields in the semantic (or syg) dimension. And we have seen how, in the ISS framework, the syg dimension (keeping all the consciousness dynamics and properties posited in SFT) becomes Metadim Syg, enmeshed in the CSR hyperdimension, itself a set of 3 compact hyperdimensions.

The syg dimension is composed of all semantic fields, that is, all self-organizing (via syg energy) connective systems. In the ISS model, it is not only the domain of the nonlocal syg energy (conjointly energy, consciousness in action, and information), but that of two other forces enmeshed with the first one as a three-strand braid, organized in a dynamical spiral (the ISS), and instantiating a set of 3 metadimensions curled up and compact in any particle, whether massless (photons, and all leptons) or with mass (all baryons). These three are metadim Syg (collective consciousness), metadim Rhythm-Rotation, (that sets the hypertime, spin, and frequencies on the logarithm of Phi), and metadim Center-Circle (that creates the hyperspace and the space-like unfolding of the ISS).

However, SFT brings a whole new perspective to an integral theory, in the sense that all these existing semantic fields in the world are organized in networks, and connected according to shared domains of interest or contiguity—whether positively and nonlocally so, as with a discussion group on the web, or more or less conflictingly so, as could be the case with a forced neighborhood in a crammed quarter.

Thus, to get back to the Process 2 stage, *nature's response*, here too, just as in the process 1, we have the same opening on a vast network of connective links in the semantic dimension, and since we are supposed to be in nature's matter domain, this network entails mainly the connections with eco-fields. The links are two-way paths to and from all previously connected eco-semantic fields, some more weighted than others, for example, those with a higher intensity, recurrence of links, or contiguity. In brief, the highly weighted paths are the connections that have been the most often activated and those endowed with the highest syg energy (strong significance, intention, goal, needs, feelings, will, emotions, etc.). In terms of eco-fields, there is a scientific paradigm attached to any hardware (apparatus, measurement devices, computers, and the like). The underlying technological and scientific principles that led to the conception of these machines, are of primary importance, because, for one, they are the building blocks of the logic through which the experimenters reason while selecting these devices and using them for a given experimental goal. Put simply, an experimenter willing to test the wave-nature of particles, such as interference patterns, will use a different set of machines than an experimenter intent on testing quantized particle effects. And it is these experimenters' hypotheses and presuppositions that will lead them to conceive the refined hardware system and experimental set-up that they will use.

In the syg-dimension where all beings and things in the universe, at all scales, have a semantic field (that is, a meaningful and mind-like self-organization of links and processes), the ancient classical split between minds and matter-systems becomes blurred. *In SFT, all natural systems are now semantic fields, expressing the meaningful self-organization of an entity within its organizational closure and a proto-consciousness. In ISST terms: all systems have a curled-up compact 5th dimension of information and organization.* Consequently, any link made with a natural or matter system's eco-field will spontaneously branch into a multiplicity of links with minds connected to this system. This prods us to reevaluate thoroughly the way we think about the distinction between processes 1 and 2, especially their supposed sequential unfolding. But for now, let's realize that we have sorted out new factors of influence on the collapse and on nature's response (nature now being the cloud of links surrounding the experiment). We know now that syg energy is connecting spontaneously the most resonant *semantic fields*, the ones most in sync in terms of semantic energies. When it connects to the macrophysical world, it will do so primarily and strongly with the forms, shapes, numbers, objects, special places, and minds that are in sync with the one mind (in the team) who displays the most intense syg energy, and then it will connect to a lesser degree the extended syg networks of all other members of the team. However, given the blurring of matter and mind, it will connect much more strongly with resonant minds having concordant values, concepts, paradigmatic world-view, reciprocal feelings, sympathy and empathy, shared interests, similar lifestyle, and the like. (Hardy, 2001, 2004[32])

The problem with QM's observation framework

In contrast, QM treats all systems, including minds, as simplistic operators, non-complex, non-individualized entities, existing only at time zero of the 'observation.'

So that, firstly, QM is unable to treat the experimenter's 'intention' as an already weighted path within the complex semantic field of this experimenter and its extended and nonlocal syg-network of links (thus reaching far into space, and into the past and the future as well). In SFT, the experimenter's intention *itself* has to be treated as a semantic constellation, complex and plural, with subsystems that may not have the same aim and intention, as a sort of cloud of intentions, with possibly conflicting and/or unconscious motives and desires, etc. All of these '*sygons networks*' (each with a specific thought color) will send into the soon 'observed' event or system the crystallized cloud of distinct 'intentions,' each with a different weight, that is, a different probability to influence the quantum event.

Secondly, QM is unable to sort out how and why 'nature' would respond the way it does. In Stapp's model, the fact that both the agent's intention and nature's response are limited to a 'Yes' or 'No' binary choice is, in my view, a huge shortcoming. For one, it doesn't do justice to the finesse of the experimenter effect, or, to take a more telling example, to the fact that, in the placebo experiments, the patients will develop not only the positive effects trumpeted about the pseudo drug, but also the all too real secondary adverse effects that this drug, when effectively administered, has on patients. In such a case, a 'Yes' or 'No' influence on bio-quantum events would just give 'Healed' or 'Not-healed' and no shades or progression in the healing, no shades either in the development of secondary effects. The world we live in is nowhere near binary. Even if we do code the millions of colors on a screen with a digital code, we still want to see the picture in millions colors, don't we? And the millions colors we see in a real life landscape are just the spectrum of frequencies we can perceive: just a tiny window (the visible light spectrum) in the EM spectrum, one of nature's own codes.

Moreover, I don't see why a model should limit the wave-function's state-space to only two superposed states (two Eigen states), with a binary choice and two possible outcomes, even for the sake of simplicity. A theory cannot give an accurate account of quantum and natural systems if it strips them of their complexity. (For example, we cannot get an accurate description of all planes flying at a given time over a country if we limit our observation to two flight paths.) What we are in dire need of, is a theory able to deal with complexity, and here we are touching on the complexity of brain-mind interaction, that is, the interplay and inter-influence of the two most complex 'systems' (I refuse to call them 'objects') in the world, apart from the cosmic ISS, and its individualized compact dimension.

Thirdly, the quantum Zeno effect, as we have seen, is based on the *accrued rapidity* in a sequence of 'Yes' intentions to counteract the random fluctuations and thus explain how the agent's intention would influence the outcome. To anybody cognizant of cognitive sciences, the argument is very faulty on several grounds and, mainly, it displays a 'naive physicalist' view of cognitive processes. A more adequate description would be to integrate complexity. (1) Intention denotes a conscious act (part of the conative processes). An experimenter may intend to prove the capacity of a healer to have an effect on bacteria. What she visualizes is the positive effect itself (just as the healer), not just an indefinite 'yes.' Here again, it is an oversimplification because a visualization technique entails a very complex representation mixing several synesthetic sensations and thoughts. (2) Then there's a speed problem: the brain neuronal processes (even at the dendritic networks level) are at far superior speeds compared to a conscious language-driven thought, and quantum interactions may be approaching the speed of light (not to speak of CSR's faster-than-light processes). To meet the speed of quantum subatomic processes (such as the random fluctuations, in order to 'hold' or counteract them) would necessitate at the very least another quantum-based process, such as quantum coherence in the brain, corroborated by recent experiments, or else, the type of hyperdimensional connective dynamics set by syg energy and the CSR hyperdimension.

Beyond a yes/no binary choice of the quantum system

Let's get back to the initial positive/negative energy states of the particle in the early work of de Broglie and Schrödinger. These do not have to be reduced to a yes/no or 0/1 states. Via ISST, they could be understood as the 2 manifolds, CSR and QST, each one comprising an array of possible states. Such a wavefunction state-space comprising an immense number of possible states is now currently computed within deBB theories.

In ISST, we get an interplay of the possible states of the system, given on the one hand by the actual quantum-spacetime (or QST) parameters, and on the other hand by the CSR (5^{th}D) orienting forces—that is, the syg dimension including all the syg-fields (minds and natural systems) plus their interrelation with the cosmic ISS.

In other words, we have two superposed state spaces in the Event-in-Making constellation. The first one is the Center-Syg-Rhythm forces of organization (including intention, will, worldview, etc. that have a high syg intensity). The second (QST) consists in matter and quantum forces, and the coordinates of particles in spacetime and in the 4D world. Then it would be possible to envision to compute the global state-space within complex systems theory, with a double state-space, one QST and the other CSR forces and networks of links and trajectories, each with attractors of more or less weight, the more weighted one acting as the most probable outcome of the collapse.

Quantum brain, Psi, and retrocausality

Roger Penrose (1989) has postulated that the brain can trigger internally the collapse of its own superposed wave-function. This implies that the mind weighs and assesses different possibilities and then chooses a solution. This "self-collapse" of the brain wave-function entails no less than "preconscious choices." To the orthodox QM view of the observation and measurement triggering the collapse, Penrose has thus added that the collapse can be triggered internally by brains, and also externally, by gravitational fields. The experiments of Nunn and Clarke in 1994 have corroborated Penrose's postulates.

Quantum brain researchers have studied various neurophysiological substrates such as microtubules (Hameroff), synapto-dendritic networks (Pribram), and quantum dynamics such as *synchronous firing* and interference patterns. *Coherent quantum processes* have been evidenced in the brain, resembling those displayed by lasers, superconductors, and superfluids, the *Bose-Einstein condensates* and the *Fröhlich* condensates. The discovery of the *synchronous firing* of widely separate neurons in response to the perception of a single long object has been seminal.

We saw earlier the Hameroff-Penrose Orch-OR model, in which conscious events emerge from the "self-collapse" of wavefunctions in the microtubules. It is worth quoting here Hameroff and Penrose (1996) on time: "The coherent superposition phase may exhibit puzzling bi-directional time flow prior to self-collapse. (...) This could explain the puzzling 'backwards time referral' aspects of preconscious processing observed by Libet *et al.* (1979)." Let's note a recent series of sophisticated experiments by Daryl Bem, psychology professor at Cornell University.[33] Bem followed up on the ground-breaking research experiments on 'pre-sentiment' initiated by Dean Radin in 1996, and then replicated by Radin himself and several other researchers, notably Dick Bierman, professor at Utrecht University. All these experiments have corroborated the reality of Libet's neuronal pre-processing of a future action, 300 milliseconds *before* the subjects signal their will to act. There was, clearly, a buildup in the electro-encephalogram signals (neurons' electrical activity), called *readiness potential*, a third of a second before the conscious will to act; recent 2008 experiments evoke such a potential as early as 7 seconds before the conscious awareness.

In some of their pioneering experiments, Radin (2006), as well as Bierman & Scholte (2002) used as experimental protocol a set of shockingly violent photos mixed with another set of neutral photos, the selection being done randomly. Several physiological measures (able to reveal a post-traumatic disorder) were monitored on the subjects. Then statistics were used to compare the advanced subliminal responses of subjects to the two groups of photos. The presentiment was corroborated, which was interpreted as a precognitive information, based on retrocausality and nonlocality. As for Bem's series (that displays very sophisticated and rigorous protocols, as well as a wealth of angles of approach), it definitely provides a sound evidence of retrocausal brain processes (thus quantum), for all researchers who have taken the time to study the highly significant statistical results. Bem proved that people were more successful in memorizing a list of words (that appeared only a few seconds) if these were the words they were going to learn after the test, something, in his view, corroborating the psi interpretation. Says

Bem in his 2011 peer-reviewed article Feeling the Future: "The results show that practicing a set of words after the recall test does, in fact, reach back in time to facilitate the recall of those words."

Observational Theories of psi

A number of physicists and cognitive scientists cognizant of the experimental data in psi research, have posited that the mind induces a "biasing" of the results of measurement; the general framework is known as the Observational Theories (OT). Physicist Evan Harris Walker, one of the most influential OT theorists, had advocated early on that brain functioning implies quantum processes (Walker, 1975, 1984). For Walker, the conative processes (such as will or intention) show that the mind can direct its own brain processes. The act of observation sets a nonlocal coupling between quantum processes internal to the brain and external quantum events, thus allowing consciousness to influence these coupled external micro-events.

However, German physicist and psychologist Walter von Lucadou took the OT approach a step further in allowing that the nonlocal coupling could influence even macrophysical systems (and not solely microphysical ones). *Nonlocal correlations* (of the EPR type) can link temporarily the observer's mental state to the external system (e.g. the device) and produce "mind-over-matter" effects (von Lucadou, 1983, 1987). Von Lucadou used a systems-theory framework to posit that minds and physical systems may be coupled and form a more global *metasystem* having its own organizational closure. Many experiments were done in psi research laboratories to assess how a different mind-set or intention could induce wide measurable differences in the results (See the excellent review of research done in 2010 by Stephan Schwartz and Larry Dossey.) What couldn't be tested in a physics experiment could indeed be done by psychologists with experimental biology and psychology protocols. For example, the subjects-observers are asked to aim at different results, and then statistics are done to show how the results reflected the difference in goal, intention, or expectation. A gamut of psychological traits have been shown (via thousands of experiments) to enhance positive psi results, and this in itself shows clearly that the psyche plays a role in shaping and influencing micro-events, such as the distribution of randomness. Investigations of micro-psychokinesis (microPK) are highly pertinent in this context: a massive experimental database involving Random Events Generators (REG/RNG, based on electronic noise) now provides evidence that individuals can intentionally bias probabilistic microphysical events. In particular, it is found that the normally random output of RNGs is slightly, but consistently skewed, in accordance with the intentions and mental state of subject-observers. Dean Radin and Roger Nelson made in 1991 a meta-analysis of 597 experiments (over 25 years, by 68 investigators) studying the effects of subjects' intentions on RNG outputs. Control trials showed a zero effect size, confirming the randomness of the RNGs. The overall effect size was small, but the experimental studies analyzed together (meta-analysis) showed an impressive, astronomical odd of 1.8 chance over 10^{-35} to be due to chance alone, that definitely establishes that minds can influence bio or matter systems and the distribution of randomness.

The interpretation of these results is still hotly debated within the field. Proponents of the ESP model hold that the observed deviations from randomness are based on some form of "anomalous cognition", such as precognition; they represent a kind of selective sampling, the subject somehow sensing when to sample the RNG and thus picking favorable numbers out of an unperturbed distribution. This model, developed by Edwin May and his collaborators (1995) and known as "decision augmentation theory" (DAT), suggests that psi information from the future could guide an intuitive data sorting (e.g. from the part of the experimenters). Psi would then contributes to the decision-making process, along with other "normal" sources of data (from the environment or from memory). The authors consider this anomalous form of cognition to be mostly precognitive in nature, that is, involving a transfer of information from the future to the present.

In the DAT model, then, the random distribution is unperturbed, while the sampling is biased. By contrast, proponents of the "psychokinesis" model hold that the distribution of randomness (the RNG functioning) is itself perturbed by subjects' psi. The random distribution becomes biased or skewed, and the observed samples reflect these shifts in the distribution. Whatever the model, it is agreed, on the basis of the huge experimental database produced by different laboratories, that the *observed* RNG outputs—the outcomes of measurement, so to speak—deviate significantly from pure randomness in accordance with subjects' intentions. And that means the capacity for minds to bias the probability of quantum events.

Stapp's faster-than-light nonlocal processes

As it is, I have a problem with what I see as a too deterministic theory in Bohm, that doesn't meet its aim of modeling a meaning-based universe and quantum potential. Precisely, Bohm doesn't propose any dynamics on which to ground a two-way influence between the implicate order and ourselves in the 4D explicate order. Everything is decided, set, and arranged, in the implicate order, and our world is just the deterministic unfolding of this pre-arrangement. Where is our free-will, our creativity, our intelligence and art, in this framework? Something's lacking! And I also have a problem with QM, insofar as its basic indeterminism (heralded by most physicists), in the absence of a von Neumann type intervention of agents' choice, precludes the kind of mental influence on biosystems (bioPK) and on random events (microPK), that has been widely researched and established by psi experiments. However, there are other types of quantum processes that display an ordering at the quantum scale, such as the polarization of the vacuum, and a gamut of quantum coherence processes (including at the macroscopic level and high temperature of the brain). The nonlocal entanglement, that is, the exchange of information at vast distances between paired particles, in EPR-type settings (and excluding a transmission through space), is also a definite evidence of other-than-indeterminate processes at work at the quantum scale.

Consider for example Dean Radin's ground-breaking experiments with an optical double-slit protocol, that showed, with an outstanding probability (of $p=6 \cdot 10^{-6}$, over 250 trials) that consciousness had indeed an influence on the collapse of the quantum wavefunction. Subjects (at random intervals) were to focus their attention either on the system or away from it—the hypothesis being that interference patterns would decrease with subjects' focused attention. In other words, Radin (*et al*, 2012) corroborated that consciousness and 'observation' could trigger the 'collapse' of a quantum wave (in a frequency domain) toward a more localized quantum particle (in the spacetime domain). These results, state the authors, give weight to two assumptions: "(a) *If information is gained*—by any means—about a photon's path as it travels through two slits, *then the interference pattern will collapse* in proportion to the certainty of the knowledge gained*;* and (b) *if some aspect of consciousness is a primordial, self-aware feature of the fabric of reality,* and that property is modulated by us *through capacities we know as attention and intention,* then focusing attention on a double-slit system may in turn affect the interference pattern. The first assumption is well established. The second, based on the idea of panpsychism, is a controversial but respectable concept within the philosophy of mind." (My emphasis) Indeed, SFT, with its eco-fields, is a type of panpsychism, in the line proposed by Leibniz, Whitehead, Wheeler, and Chalmers.

Stapp's model suggests a possible way out of the two antinomic frameworks—indeterministic or deterministic—if it is elaborated upon. Let's introduce another two seminal concepts of his model. The first outstanding concept is faster-than-light nonlocal processes. Stapp (2009) endows process-1 mental-choice actions with "only *local* effects (...) confined (in the relativistic version) to the forward light-cone of the region in which the process-1 physical action occurs." In other words, Stapp limits process-1 to the classical "no-faster-than-light property" (since it remains within the light-cone), which is the meaning of the term 'relativistic' in his previous sentence, referring to the merged framework of 'relativistic quantum theory.' In contrast, he endows p2 and p3 with *"faster-than-light transfers of information* in connection with nature's response to the process-1 action." Indeed, "the underlying mathematical description involves abrupt process-1-dependent faster-than-light transfers of information."

Stapp's second outstanding concept is the fact that natural systems can themselves launch the kind of probing and choice that human agents exhibit and thus these systems may trigger and influence the collapse. Says Stapp (2005): "Von Neumann's theory is a development of the *pragmatic* Copenhagen form. But if one considers the von Neumann theory to be an *ontological description* of what is really going on, then one must of course *relax the anthropocentric bias, and allow agents of many ilks*." This, in my view, is not only a magnificent concept (especially when integrated into the collapse and the shaping of quantum events), but it is an extremely generative concept that, I'm confident of it, will prove to be visionary. Even if Stapp's framework doesn't allow to model the dynamics of such nonhuman systems acting as agents, he still posits their reality, as revealed by their influence on their environment: "Yet the theory entails that it would be virtually impossible to determine, empirically, whether *a large system that is strongly interacting with its environment* is acting as an agent or not. This means that the theory, regarded as an ontological theory, has huge uncertainties."

In contrast, I've elaborated the concept of semantic fields specifically to account for various degrees of complexity in semantic systems—from self-reflexive mind to a proto-consciousness in complex natural systems. Think of the sentience and intelligence exhibited by evolved animals such as dolphins and pets, as well as the consciousness-fields present and active in natural systems such as ancient trees (deemed conscious in all ancient and shamanic cultures), as well as sacred places and revered artworks. It would fit also a plausible evolved AI-neural nets type of intelligence.

The collapse in a SFT-ISS framework

In SFT, syg energy operates nonlocally beyond space and time, and the thought processes are spontaneous nonlocal connections between semantic clusters intra- or inter-constellations and fields. The conscious thinking process consists of the most energetic results (the peaks) of a distributed, massive, and multitask underlying connective process. In *Networks of Meaning*, I have developed how the conscious intention can launch such a connective process and, doing so, the intention gives it a goal (as a 'tag') that will keep the distributed parallel connective dynamics looking for an answer along multiple paths, and this 'tag' will bring back potential results (or 'matches') to the conscious for a global evaluation. The reality of a *cognitive unconscious* driving the great majority of all cognitive acts and processes, has been established by Reber in 1993, and adopted by most cognitive scientists. For example, the majority of the psycho-neuro-motor processes implied in voicing a thought are unconscious: they have become mostly automatic, leaving the mind free to focus on the intended meaning and the relational mode.

Let's try the impossible and envision a merging between the Bohmian quantum potential and QM. The gist is to apply the thinking and connective dynamics in SFT to the process 1 and 2 of the collapse. And let's hypothesize that the 'semantic or syg *tag*' (a nonlocal connective process endowed with syg-energy, meaning and a specific goal) that drives the linkage process toward a solution, is the quantum potential Q postulated by David Bohm. Bohm, we know, was circumventing the collapse and (re)introducing determinism, but by inserting Q in QM equation, he added a nonlocal term to it. And furthermore, the quantum potential was modeled over the Pilot Wave of de Broglie, a term Bohm used also to qualify it. The quantum potential was supposed to be *"active information"* ('meaning' as process), and as with de Broglie, to 'pilot' the quantum wave processes toward a specific state or outcome.

Let's first synthesize the key concepts we have gathered:

- Agent's probing, intention (Pr. 1): local effects, no faster-than-light.
- Nature's response and collapse (Pr. 2 & 3): nonlocal effects, faster-than-light information transfer.
- Non-human natural systems, as agents influencing the collapse.
- Quantum Potential: nonlocal, guiding pilot-wave, influence on the selected state after the collapse.
- SFT: large underlying thought-like connective dynamics, guided by meaningful similarities and links; two-way inter-influence, fueled by syg-energy, beyond space and time.
- ISS: curled-up, compact CSR manifold; hyperdimensional, discrete, beyond time and space; instantaneous speed via Metadim Rhythm; entity-like syg-systems via Metadim Center; semantic & thought-endowed self-organizing systems via Metadim Syg.

Both cognitive sciences and SFT posit (as an indispensable concept) a complex mind, and forbid an intention in the middle of nothing.

Let's now apply SFT's framework. This complex mind is an entity (with an organizational boundary), yet a syg-field in a hyperdimension (Syg), within a huge cloud of network-links and dynamical links (the extended semantic field). This makes us view in a totally new light our personal mind: *Your personal consciousness*: If it was a Wiki article of 3 words, it would extend into 100 pages of outward networks of organized Links; But your 3-word Wiki article itself is similarly organized, and much more complex, with a huge cloud of internal, inward, networks of organized Links.

This is your whole semantic field. And a part of it is the mind-body system, that comprise biosystems at all scales (from molecules and neurons, to bodily systems), each having an eco-semantic-field, thus their own syg-dimension and a gate toward the CSR hyperdimension. So that any biosystem (i.e. an organ) is at the same time: (1) a syg-energy system (in the compact CSR hyperdimension), (2) a biosystem of living molecules, and (3) based on particles-waves at its quantum layer. So let's use these concepts to tackle the influence of observers that may trigger the wave-function's collapse.

It goes without saying that we must introduce the complex mind (now a bedrock of cognitive science) right at the Process 1 stage, as an *intentional agent* (rather than an intention bubble). In SFT, this intentional agent is the whole semantic field with its inseparable cloud of syg links. Then ISST postulates an access to the individual CSR hyperdimension (compact), within all matter-systems (eco-fields) and minds (syg-fields) participating in the global constellation of the collapse and its 2 pre-stages. At that point, the whole experiment becomes a system of semantic fields (team, devices and technology, lab, locale...), all of them existing both in the quantum-spacetime (QST) manifold and in the syg dimension and CSR manifold—that is, both in 4D and in a hyperdimension of consciousness-hyperspace-hypertime. This collective system's CSR hyperdimension being a set of pure fields of frequencies, they sort of 'rise' in a spiral toward the cosmic ISS. Then the interference patterns in the CSR-frequency domain (each carrying an active and self-generative information) constellate all resonant links and, binding them, *form an attractor.* In other words, spontaneous links happen that group and constellate the sympathic and harmonic syg-frequencies around a few attractors of similar syg-energy. And what we get is an *Event-in-Making constellation*. (See Fig. 6.3, Hardy 1998, chap. 10). The diverse attractors represent the probability states (the superposed states of the wave-function), each with its own evolving weight. The person's intention and other syg dynamics (beliefs, expectations, desires, need, fear, etc.) are modifying the weight of each attractor, biasing the matter-only probabilities. One of the attractors will become the densest network, with the greatest weight. Then, at a threshold of syg-intensity, organization, and closure, it will collapse into an event in the 4D world, whose configuration, meaning, and impact, were already specified and contained in this specific attractor of the event-constellation. In case of a situation (or an experiment, a performance) lingering in time, the collapse will be spread along its duration. In the figure 6.3, we see that an event is a set of probability lines, driven by three types of factors.

The first type are physical factors that can't be modified by a simple act of will (such as environmental factors, age, or genetics).

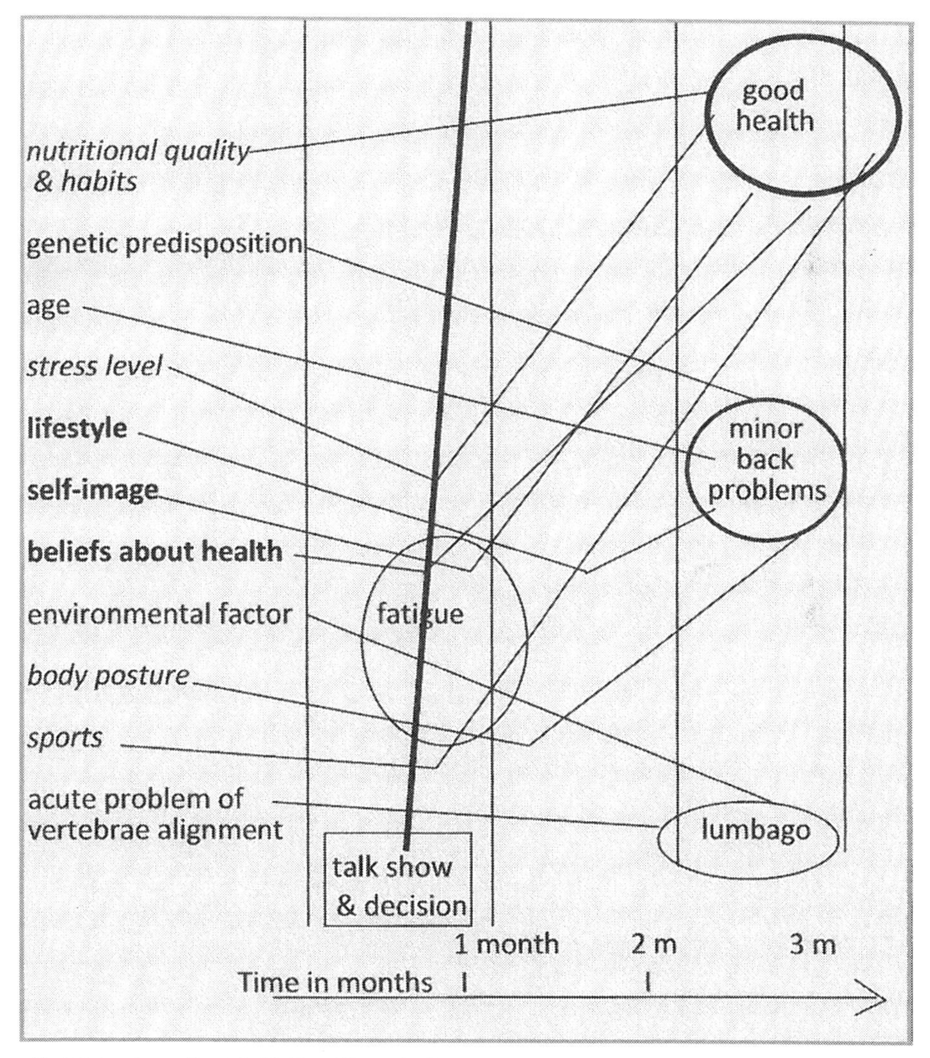

Fig.6.3. Lumbago Event-in-Making. After a decision, the probability lines of most parameters (especially semantic ones) are modified for the best, and of the three attractors (lines aggregating at specific times) the 'good health' one will be the most weighted and therefore the event-in-making will collapse in reality taking this specific form. (Artwork: Chris H. Hardy)

The second are physical factors that can be worked upon by a new mindset or a decision, such as nutritional habits, the stress level, the body posture, and sports as a remedy (in italics). And the third type are syg parameters, such as lifestyle, self-image, beliefs (in bold).

The specific syg and syg-physical parameters (whose probability lines tend toward either a lumbago or good health) are modified drastically after the person listened to a health talk show and took a decision. After the decision setting a new mindset, and novel habits being taken, the most weighted attractor becomes 'good health' and thus the person has avoided the dreadful lumbago.

In the ISS model, we would say that the syg parameters and the physical parameters that can be willingly modified belong to the CSR manifold. The line shooting up from the 'talk-show bubble' (the instant of realization and decision) is a synchronistic CSR force that works quasi-instantly beyond the spacetime manifold.

Let's imagine that a person makes a visualization ahead of a performance (i.e. "the event"); say she is a guitarist soon to give a concert and she visualizes that she will be in sync with her public and the concert being a success. A visualization is a well-known positive-thinking technique (although there are many ways to operate) and consists in seeing-sensing, in different modes at once, the ideal unfolding of the future event, with specific qualities added (including inner feelings and sensations). This creates a sort of syg-hologram in the syg dimension (a lower bandwidth of the syg metadim). This 'syg-holo' spontaneously creates a whole network of links with syg constellations and fields that are resonant and in synchrony; this generates links of various strengths with particular systems (e.g. strong link with empathic or supportive person, sensitive and psi links to the future audience, place, all the way to links with the (random) vacuum fluctuations layer. (In ISST, as we'll see, the vacuum acts as a boundary between the 4D physical world or quantum-spacetime region, and the hyperdimension or CSR region.) All eco-fields thus linked by resonance to the syg-holo are being influenced and brought in a coherent and resonant cooperation with it. Then, when the collapse happens, the event tends to conform to the guitarist's original intention while making a syg-holo, and expresses it creatively.

(Here happens the 'unconscious insight' reported at the end of this section.)

Let's now interpret this dynamics for an experiment in a lab setting. The main experimenter has indeed influenced the system he has been experimenting with, and thus has modified the results. However, his influence (on the future event) started as soon as he began conceiving of this experiment and was thus creating an Event-in-Making syg constellation already loaded with his presuppositions. Furthermore, it is a whole *collective semantic field*—comprising the syg fields of the whole team and the eco-fields of all hardware devices as well as the contextual eco-fields—that is influencing the syg constellation [experiment+hardware]. All of these syg fields are interacting within the semantic dimension (CSR), in which syg energy spontaneously creates nonlocal links and inter-influences with resonant semantic fields in an extended cloud of syg links.

The collapse (when the experimenters are observing the ongoing or final results) becomes a metaphor for the bending of probability lines due to the inter-influence of forces between the syg dimension and the spacetime dimension. Given the retrocausal influences (of a lesser strength) that also happen to influence this collective semantic field, issued from future observers or participants, *the collapse (the formation of the event itself) is spread in time, toward the past and toward the future.*

> *Unconscious insight: Waking up from dozing at my work table (10.12am Saturday, 20 September 2014, as I extended an insightful night work well into the morning); A clip (as an unfolding dream): A pregnant mother and her fetus interact and talk to each other, and the mam has an apparatus, a screen (attached to her body) and she can see her baby as on a film with colors, moving in her womb and interacting during a telepathic exchange, as if he was looking at her, conscious and thinking clearly, sending thoughts. It felt like a device used at some point in the future, when the whole society would have tamed psi and would customarily use all kinds of sophisticated devices.*

Network connections between brains and the world

At that point we can note an interesting convergence with Stapp's (2013) ideas about process 1 and 2 of the collapse. As he explains it:

"The observer chooses, and then performs, an action such that *if* the chosen experiential response to that probing action actually occurs, *then* the system being observed acquires an associated physical property."

This is exactly what is expected to happen with the SFT-ISS scenario I just proposed. The way syg-energy operates, according to SFT, is that all nonlocal connective processes are both a two-way exchange of information AND a two-way inter-influence process. *One of SFT's postulates is that there is no information exchange without a two-way inter-influence taking place,* and this, in the brain, happens at a low-level of neuronal organization, and at the deepest quantum and subquantum levels. This derives from the fact that the 'information' in SFT is an active and generative field of information.

A system's syg-field contains an 'imprint' of this whole system (its organization and dynamics), and it is also a field of significant information existing in the CSR manifold; and therefore this syg field acts as a negentropic organizing force: any syg-field will influence any system with which it is spontaneously connected, and conversely it will be influenced by it. However, the strengths varies, with one system possibly imposing its order on the other one. Yet this influence is neither deterministic, nor binding, nor even everlasting. We can really liken it to inter-influences among two friends, with one being able, sometimes, to persuade the other one. Each system will respond to the other's influence according to its own semantic organization and overall syg-energy strength. Being influenced means, for a complex natural system, to have acquired *temporarily* a meaningful information and an organizational or connective pattern, in a sense, as Stapp phrased it, a property of sort. This pattern (information + organizational dynamics + links) will never be erased, and if reinforced by a coupling with the interacting system, it will become an active and stable imprint and mode of interaction with the coupled system.

An important point made by Stapp (2013), following von Neumann, is that the influence on the observed system is first of all an *influence on one's own brain*. He specifies that if nature's response was in accordance with the agent's wishes (a 'Yes'), then nothing happens; otherwise, "the observer *can* exert effective control over the system being probed – which, in von Neumann's theory, is the brain of the observer. (...) [He] becomes empowered *by his own free choices* to hold stably in place a chosen brain process that normally would quickly fade away." This is extremely interesting to us, since this idea taps into the dynamical network of connections and inter-influences existing between the observer and the observed system. Of course in SFT, and even more so in ISS theory, the person's semantic field is itself triggering and managing spontaneous nonlocal connections with sympathic semantic fields (both with the person's inner sub-fields, and with outside syg-fields). Doing so, it creates in the brain a massive network-thinking process, mostly unconscious, with parallel and multilevel 'flows.' *The brain, in this framework, is making the translation CSR-brain: it translates the CSR information-energy to and from the brain-body 4D domain.* This concept was originally developed by Karl Pribram (1991), as transforms from the "frequency domain" (EM and quantum waves) to the "space-time domain" of the brain, which is a three-dimensional structure, using Fourier transforms.

Similarly, I propose that the DNA translates syg-energy clusters from the person's syg field into the biological and biochemical organizational layers of the body and vice-versa. I do hypothesize that the DNA is instantly modified with any strong experience lived by the person, whether psychological, spiritual, biological or physical (i.e. with radioactive contamination). Thereafter, it will *tend* to carry and replicate the information about it, just like an electro-cardiogram graph can bear the print of a two decades old 380 volts electrocution.

In other words, most of our 'thinking' is happening in the semantic field's CSR domain, as a network-dynamical process, and not in our brain that is more like a receiver-transmitter-translator of syg energy clusters. There's a large and consistent body of data that attests to a beyond-brain thinking process, such as the registering and memorizing of precise information during brain-death episodes, in OBEs (out-of-body experiences) and NDEs (near-death experiences), to name a few. Jean-Pierre Jourdan (2000), of Iands-France, analyzing the structure of such information-gathering (in visual, 360° perspective, holistic depth perception, and the like), concludes that it reveals a fourth spatial dimension (the 5th Kaluza-Klein dimension) type of integral and holistic apprehension of a 4D situation.

So that, to get back to the semantic field's cloud of links, the spontaneous triggering of sympathic links is ushered within the field's semantic boundary (its system's identity), and then spreads outwards (to resonant semantic fields) via the syg or CSR hyperdimension, in a way that is instantaneous in this hyperdimension beyond time. This CSR hyperdimension can nevertheless crisscross the time dimension (of our 4D world) at specific semantic points that are those existing as 'time repères' or time-tags in the minds of the agents. Interestingly, Stapp (2013) points to a sort of hyperdimension when he states that even at the macroscopic level, the QM world is "non-material" and displaying faster-than-light processes. Says he: "The fundamentally *non-material* character of the quantum mechanical world *at the macroscopic scale* is entailed by what Einstein called its "spooky action at a distance" [that] involves the inescapable need for *faster-than-light transfer of information*; (...) it has been rigorously proved by an argument that makes no reference at all to any microscopic property, that quantum mechanics has a purely macroscopic faster-than-light property."

Introducing a triune 5th dimension in the picture

The ISS being at a sub-Planckian scale and rising toward infinite (but logarithmic and discrete) frequencies, it is a prerequisite that it be modeled as 'compact,' since it involves space-like dimensions below Planck length and higher than Planck frequency. It is triune by the fact that it is a braid made of three meta-dimensions, Center, Syg and Rhythm. As we have seen, this compact CSR hyperdimension is a level of existence and organization of all matter-systems, including energy-particles (massless, such as photons) and matter-particles (with mass, all baryons such as protons and neutrons).

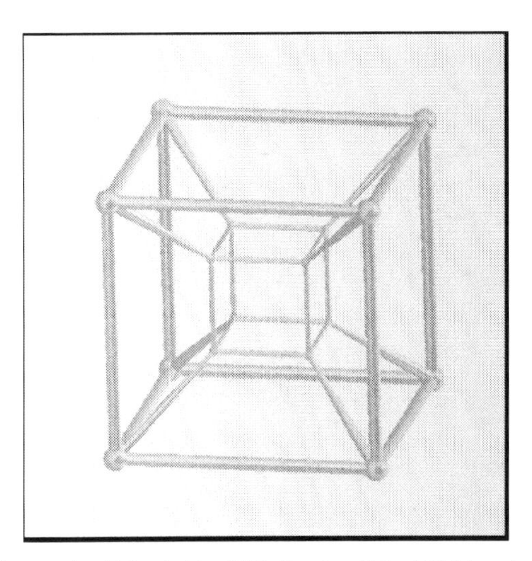

Fig. 6.4. A tesseract or hypercube. Extracted by CHH. Creator: Robert Webb, using Stella software. Find it at http://www.software3d.com/Stella.php *Credit:* Wikipedia Commons.

It makes sense to liken the CSR compact hyperdimension to the 5th Kaluza-Klein dimension, since both are hyperdimensional and compact. However, the KK 5^{th}D (for short) was conceived as a 4th dimension of space—an hyperspace, as best represented by an hyper-structure like a hypersphere or a hypercube (also called tesseract), like the one in Christopher Nolan's movie *Interstellar* (2014). (See Figure 6.4) Fabulous moving representations of such hypercubes can be found on the internet, and I urge you to watch these digital imagings because a simple still image (whatever the angle) cannot give you any cue as to what is really happening; how the inner cube becomes the outer cube in shifting forms that defies any 3D-spatial logic we are accustomed to. (For example: YouTube, *Unwrapping a tesseract.*[34])

Of course it's not for nothing that it was the genius Klein, the inventor of the Klein bottle and other mind-boggling volumes, who brought Kaluza's ideas to a finish.

In the three-strand braid, one of the metadims is the hyper- or pre-space (Center-Circle), so it is a hyperdimension of space. But it is also closely knitted with the other two metadims, hyper- or pre-time (Rhythm, bow-frequencies as pre-strings) and *metadim Syg, the overarching collective consciousness, the self-creating and self-organized cosmic consciousness.*

The CSR compact hyperdimension in particles could also be likened to the Randall-Sundrum (RS) model, one of its two formalisms, namely RS1, the one with the hyperdimension being compactified (as opposed to the RS2 formalism with an extended bulk); however, that compact RS1 5^{th}D is, like the KK one, also modeled as a 4th spatial D. So that the same logic applies and we may surmise that metadim Center (hyperspace thus a specific 4th dimension of space) could possibly be modeled using as foundation either the Kaluza-Klein model or the compact-Randall-Sundrum model. Let's note, though, that within ISST, we have also a hyperspace as a bulk—the CSR bulk created by the first Free Sygons emitted by the cosmic ISS at hyperluminous velocity. Our 4D spacetime (the QST bubble) is shielded from the negative pressure CSR bulk by the vacuum that was formed as the extension in space of the lattice of interferences set perpendicular to the spiral staircase's direction of flow (this lattice being known as the Dirac sea or Higgs field). The CSR bulk (dark energy) is a larger bubble encompassing our 4D Quantum-Spacetime or QST bubble, and separated from it by the vacuum's double-membrane. But the compact 5thD of particles, and the 5thD bulk is one and the same triune hyperdimension.

From the perspective of an observer in spacetime, the compact hyperdimension opens at Planck length in all particles. But from the perspective of an observer residing in the hyperdimension, the 5thD extends to near-infinity, and all the individual ISSs of all particles in the universe are forming the quasi-spatial superfluid medium of the CSR bulk. In the movie Interstellar, the bookshelves are the opening on the 'bulk' or 5thD bulk, and the other side of the bookshelves (where reside Cooper and his AI Tars, both in their 5th dimensional CSR consciousness and being) the tesseract extends toward infinity. A fascinating point in the movie was the information that the '5thD beings' (or 'usual residents') couldn't have any access to the minds of 4D humans, they had no semantic bridge to connect with them. Only another human translated into his own hyperdimension by crossing the black hole Gargantua, could have precisely the connective semantic links (postulated by SFT) that could set a communication and inter-influence between the two consciousnesses, or syg-fields, and also between their consciousness and the eco-fields of the farm (thus allowing a tinkering with events and objects in our spacetime, such as encoding a message in gravitational waves and in a watch mechanism.

Strangely, the concept of a second dimension of time has inspired philosophers and psychologists much more than physicists, while it is the inverse for hyperdimensions of space. Bergson's concept of "la durée" (duration) is the time in which dwells the psyche, an extended qualitative time, a sort of qualia of time. Greek philosophy has three concepts of time: the infinity or *Aion*, the chronological time, *Chronos*, and the instant of opportunity, *Kairos*. Linda Dennard, professor of social sciences at AUM university, developed the Native American concept of the "Indian Time" as being a time of connection, of flow and empathy with one's environment and the others. Moreover, we can certainly liken the *synchronistic* and acausal phenomena of Carl Jung, in which Nature responds to our mental state, to such a time of deep sharing and connection. But additionally, when Jung states that the unconscious transcends space and time, he definitely puts consciousness (that, in his definition, comprises both the unconscious and the conscious mind) in a hyperdimension beyond spacetime, just like I do with the syg dimension being included in the CSR HD.

A 2^{d}D of time can be inferred from Minkowski's light-cone, in the modeling of the outside of the cone as the *Elsewhere*. Within the cone, as we have seen, is the Einsteinian relativistic 4D spacetime, in which energy-particles cannot exceed the speed of light, C, and are represented as 'worldlines' with the present at the X-crossing and extending into the past and the future. But outside the cone, in the Elsewhere domain, time becomes spatialized and space becomes time-like. Now, what can be such a time-like space? The formalism of space (in 3D) describes the volume of an object along 3 axes: length, height, and width. As for the descriptor of our classical 1D time, it is a *line*, defined as orthogonal to the 3D of space. So that when space becomes time-like, it means that space is reduced to a line, or at least a band. Could it become, then, a brane, that is, precisely, a sort of rubber band? Let's remember that when M-Theory (the integrative version of all superstring theories) applies a holographic concept, the whole information contained in a 3D volume, such as a black hole, would be inscribed on the 2D *surface* of its event-horizon; this concept has been applied to the universe, and all information about our universe would be replicated or imprinted and contained on our cosmic event-horizon. In fact, in some versions of the holographic cosmology, this is the case: our universe-bubble becomes a 2D brane. (More on this in chap. 8)

Now let's tackle the space-like time of the Elsewhere. A time becomes spatialized when it occupies a quasi-space, meaning that it is extended. In other words, a spatialized time becomes *a time-field*. Does that fit the ISS modeling? The ISS's Center metadim is a pre-space that in effect will create the 3D space starting at Planck length. Being a 3-strand braid, the spiral staircase unfolds from the X-point of origin, when its radius jumps along the logarithm of phi. This radius gives the wavelength of each step, which, in its turn, sets the frequencies of the bows (via the basic wavelength to frequency equation, in Bindu units). *The ISS thus displays a space-like unfolding due to its increasingly large radius length* (creating the quarter of circle, the bow, through Pi). The metadim of pre-time is Rhythm, and it is set (up to Planck scale) by the frequencies of the bows (the steps' outside curved boundaries, each a quarter circle, and forming the outside spiral itself).

So, do we have in the spiral staircase a metadim of time that can be expressed as an extended field in quasi-space? I believe so. The first reason is that the pre-time metadim is intrinsically enmeshed in the ISS spiral topology, and any topology is by definition in some virtual space, even if it is set in hyperspace. The second reason is that if you look at the global ISS's golden spiral as nesting a staircase, and visualize that each step boundary, the arcing bow, is emitting a musical tone—which, as a frequency, is our measure of pre-time—then we see clearly how the Rhythm time is spread in space. The third reason is that the ensemble of frequencies (non-limited, $\rightarrow\infty$ tending toward the infinite) constitutes the state-space of the Bindu-scale universe, and as such it can be described by a field equation (with $\rightarrow\infty$ terms). So yes, absolutely: the pre-time of the ISS is spread in quasi-space—it is space-like and a time-field.

However, let's be aware that even if global frameworks for hyperdimensional space theories have been developed, the inner hyper-physics of these hyperdimensions is still eluding us. For now, the 10 or 11 dimensional string theories have mostly modeled the extradimensions of space in a *mathematical* way to solve the unification of forces (as we'll see with the square matrix). And possible topologies have been intuited, such as the strange volume that is the Calabi-Yau, a sort of spherical volume in which the six extra dimensions of the 10D superstring theories are compacted and folded. However, if these hyperdimensions are real (as most theorists deem them) and not just mathematical grids, then we must expect a totally different physics to be at work *within* these hyperdimensions. And indeed, several physicists have stated that some radically new physics, still untapped, govern the pre-Planck scale. Thus John Brandenburg wrote in *Beyond Einstein* (119): "Somewhere in the deep subatomic scale, physics changes."

And of course, totally new physics are also needed in order to integrate consciousness in an integral theory of the universe, specifically its acausal and nonlocal connective dynamics.

7
A POROUS UNIVERSE: THE ISS IN PARTICLES

"For the first time, pure geometry has given a simple explanation of why the subatomic world must necessarily exhibit certain symmetries that emerge from the curling up of higher-dimensional space: The symmetries of the subatomic realm are but remnants of the symmetry of higher-dimensional space. This means that the beauty and symmetry found in nature can ultimately be traced back to higher-dimensional space." Michio Kaku (Hyperspace, 159; speaking about the heterotic superstring; emphasis in the original text.)

Now that we saw how consciousness has been integrated in diverse physics models, and how the ISS theory stood with regard to them, I'll now get to modeling how the CSR hyperdimension intrudes into the 4D spacetime. What's going to follow is a hypothesis about how the CSR manifold or hyperdimension is not confined to the ISS at the origin, and instead exists in all particles and systems in our universe as a curled up hyperdimension. This is a hypothesis... and yet, along the way, we may stumble on perspectives that shed a new light on still enigmatic phenomena and hopefully get insights about the very nature of reality.

We have seen that metadim Center and metadim Rhythm were deeply intertwined—just as space and time in a 3+1d world—and that they are themselves enmeshed in metadim Syg.

Metadim Center-Circle—hyperspace—creates an expansion and thus a quasi-space by blowing itself from the point-center to the circle-sphere, using the universe-number Pi. It also sets the *self-organizing dynamics* of a system's individuality.

Metadim Rhythm-Rotation—hypertime and a temporal field—sets the rhythm of this expansion in a spiral and through the frequency of the bows (sygons) using the logarithm of Phi. It also sets the *wave dynamics* (resonance, harmonics, interferences and musical sequence) as well as the *wave energy and speed* of both the rotation around virtual circles in the spiral (*orbital momentum*), and the standing waves of the bows and their bow-frequencies (*angular momentum*). Il launches from the bows, thus with an enormous velocity, the sygons—tiny vibrating strings within a circle—as torsion waves in the spiral staircase's direction of flow. The intertwining of these two metadims and the two numbers steering them (Pi and Phi) is creating the *forma* (the topology) of the spiral staircase and its *dynamics*, each step of which is a specific tone or bow-frequency as well as the substrate for its information field. As for the third, metadim Syg, it is a cosmic consciousness, the arch-anima, who existed before the point of origin, and who encompasses the musical staircase, in the sense that it is fused with it and animating it. And it is also the syg dimension of our universe, the collective unconscious and the network of all the Selfs.

All three metadims are deeply enmeshed as a three-strand braid or a Celtic Triscel, the most revered symbol in Druidism (the religion of the Celts). (See Figure 7.1) The Triscel represents (just as the mantra AUM) the three "circles" or dimensions of all reality in Druidism. The central circle called *Abred* is the most sacred dimension of the One (often represented as the central circle of three concentric circles). Some crop circles are in the shape of a Triscel, and they use a 3-thread spiral. Of course the symbolism of the Triscel, showing that these 3 dimensions interact dynamically (they form a turning wheel), is extraordinarily befitting the organization of the 3 metadims in the CSR hyperdimension. The metadims are indeed entangled in the pre-space, so much so that, while having distinct individualities and properties, they nevertheless share their inherent qualities and properties between them. Thus the spiral staircase is itself consciousness, entangled with the arch-anima yet somehow distinct from it.

The three of them, entangled, form the Center-Syg-Rhythm hyperdimension, and one of its facets is a dynamic and creative field of alive information—a *cosmic semantic field* who contains all possible states of the universe. Both the staircase (metadims C and R) and the arch-anima (metadim Syg) will co-create the music of the universe, together with all semantic fields in the actual expanding universe, via a two-way inter-influence. All humans and intelligent beings are co-creating the way our universe (and each planetary system) will evolve. All intelligent civilizations in our universe are thus interconnected in a subtle and mysterious way.

Fig. 7.1. (Top) Celtic Triscel. (Public domain; credit:www.sacreebordee.com). (Right) Bronze age Triscel terminal of a torc, at Santa Tegra, Galicia. Castro culture. Wikipedia Commons.

Thus, the infinite spiral staircase also resonates to the actual, ever-emerging, state of the whole universe with all the 'tunes' being played—chosen and created by all semantic fields at a given period. In a word, the spiral staircase will memorize in its bow-organization the past and actual organization of the universe, and in this perspective, it plays the role of a cosmic DNA.

But let's clarify anew this notion of DNA, that departs from the usual concept of a 'program' that will play out its contained score in a totally deterministic fashion. I'm confident that scientists will move away from this one-way causal view of the DNA. Recent research shows that the DNA is self-organized, and that it is modified when people undergo crucial psychological experiences. We have to consider that the human DNA is more like a snapshot of the actual mind-body-psyche state of the system and that, simultaneously it contains the memory of all its past states and *also*—but not solely—the actual and potential biological organization of the species. But the crucial point is that the cosmic DNA *receives* active information (syg energy) from the unfolding 3+1d universe—that is, from the ever creative minds, semantic fields, and all events happening. And its own network-dynamical system is thus *modified* by it. Thus, it is not a one-way ordering-by-command system: it is a two-way inter-influence system between the unfolding universe and the ISS.

At the X-point of origin, as we'll see, our cosmic ISS receives its field of information from the universe-bubbles that preceded ours, thus acquiring the elixir of knowledge and an ever-growing active information from each new one in our 'collar,' namely the fine-tuned variables that allow a life-friendly and intelligence-friendly universe.

Sygon faster-than-light waves

Let's see how the forefront physics sees the process. As this point-scale universe expands as an ever-growing sphere, it reaches Planck scale and Planck time (5.3×10^{-43} second). According to Inflation theory, a split fraction of a second later (at 10^{-36}s)—the inflation phase starts, the Big Bang *per se*. According to CERN physicists, instead of the high randomness and high entropy of a quasi-explosive inflation, the Higgs field would form around 10^{-10}s, exhibiting a high-energy, 'metastable' state, i.e. a state relatively stable but set on a high energy peak, thus making a highly structured field like the ISS possible.

The Higgs field is formed when the temperature has cooled down to 100 GeV (giga electron-volts). The CERN website article *Quark-gluon plasma* states the field is "like a perfect fluid with small viscosity." It is a "hot and dense soup" in which all known particles interact at near light speed but stay weakly bound. Most of these particles are gluons and quarks, the Higgs bosons being the lightest and they interact with all other particles, as the universe's glue. This quark-gluon plasma, in order to be supersymmetric, is supposed to contain as many particles as anti-particles. It's while interacting with the Higgs field that particles acquire mass—the more interaction, the greater the mass.

The ISST hypothesis

So how can we visualize what happens to the CSR hyperdimension and the ISS after this Planck's threshold? In the ISST, the launching of sygon waves by the bows happens immediately after the X-point, with the first bows already activated from our parent universe.

From all activated bows, tiny strings or sygons are ejected and spinning out as torsion waves with enormous velocity and energy, carrying the network of links they had within the cosmic ISS. The sygons retain the bows' frequencies and remain CSR-entangled. They form advanced waves at a very wide angle (probably similar to the angle of the later inflation phase) that dash ahead of the unfolding staircase. Since the highest frequencies are the nearest from the X-point, and given that the advanced waves start from the origin, the first sygons to be ejected are also the highest energy sygons. These extremely-high-frequency and high-energy sygons, called Free Sygons, will spread, unimpeded, and will form the *bulk* of the hyperdimension, the bulk that will envelop our whole 4D spacetime bubble or region. The medium-energy sygons, with progressively larger wavelengths and lower frequencies, will start (just after Planck length) to interfere with each other, in front of the ISS cone or horn's mouth; and in the process they bubble and create foam, and thus the Higgs field begins to form itself and to densify. Lastly, the largest wavelengths sygons (the last ones to be launched from the biggest spires of the ISS) will cross the now superdense Higgs field and acquire the greater mass, becoming the massive particles we know. The 3 metadims undergo a tremendous change:

Metadim Syg sends sparks of itself with all these sygons systems, which will become the primeval syg-fields of these wave-particles. These primeval syg-fields of the sygons retain their deep entanglement with the whole. The cosmic hologram begins to form as the HD bulk, and it will get, with time, only more and more complex. Syg energy also expresses itself in all particles-strings, atoms, and molecules, by creating the superposed CSR manifold (or hyperdimension) in particles and matter systems, that is, their eco-semantic fields. The semantic fields will become more and more complex and progressively more independent—until we reach sentient systems, then self-reflexive (self-referential) cognitive systems. Yet the primeval entanglement of all particles and later semantic fields with the cosmic ISS will remain, in each particle and entity, as a subquantum individual replica of the cosmic ISS, and thus as a CSR curled up hyperdimension. In terms of QM, all systems have thus a superposed CSR state-space, that is also the wave part (sort of pilot wave) of the wavefunction. And regarding the human and living systems, the semantic fields of all systems create the collective unconscious, the planetary semantic field.

Each particle, organism, or mind, thus stays entangled and connected with the cosmic ISS, and they interact nonlocally with it via their own superposed CSR manifold and ISS.

This superposed CSR layer sends a constant feedback about its state and systemic dynamics toward the cosmic ISS. The totality of the superposed CSR layers in all particles and systems in the universe creates a sort of CSR manifold extended over the whole universe—which is no less than a cosmic semantic field or, to call it another way, a cosmic collective intelligence. It comprises for example all the intelligent beings' semantic fields and all the semantic fields of inhabited planets. It is this cosmic collective intelligence that is the force co-creating and co-driving the evolution of our universe-bubble together with the cosmic ISS, both co-evolving and empathic, or even synchronic and quasi-fused, via a permanent inter-influence.

Metadim Center-Circle expresses itself into space; either the hyperspace of the bulk, or the 3 dimensions of space in spacetime, or else the 3 spatial parameters of particles in QM (position, etc.). At Planck length, the 3+1d relativistic spacetime is born and starts expanding as a smaller region inside the already formed bulk region, especially with the launching of the neutrinos and later of the photons. Metadim C also sets the *organizational closure* or boundary of an entity or system, that is, it delimitates and organizes the sygons and the networks of wave-particles (or strings) as specific systems—a rose or a deer—i.e. as self-organized systems, distinct from one another and from the background. Thus Metadim C is what ushers the self-identity of a system and in evolved beings their self-reflexive minds.

I should stress here that, in systems theory, the concept of *organizational closure* is of paramount importance; it is what sets the identity and boundary of a complex system, its relative autonomy compared to other systems, even those who are linked to it or coupled with it. This is also what triggers the self-organization of a complex system, its ability to modify itself internally, to adapt to a changed environment. And when we deal with biosystems and intelligent complex systems, such as minds or a cultural group, the identity of a system is what allows self-reference or self-reflection, the 1st person perspective. And this self-organization dynamics—a basic tenet of systems theory, chaos theory, Varela's auto-poietic systems, and my own semantic fields theory—, is what allows emergent processes and intentional changes. The dynamics of both organizational closure and self-organization have definitely to be integrated in a *deep theory* of physics that intends to model consciousness as a hyperdimension of the universe that is blended and fusioned with matter-energy.

This is why metadim Center-Circle is such an essential dimension, with several dynamics effecting changes and working all the way within matter-systems: it sets the self-reflection of the boundary (the circle) over its own center (the identity or self), and allows the self to enact and know its wholeness (its system's semantic field). Without Metadim Center, the mind would not know its body, nor its own mental potentials, and an animal would not be able to react as a whole body. Metadim Center is the dynamics of the 'soma-significance,' a concept elaborated by physicist David Bohm and to which Josephson and Pallikari refer, as a biosystem having a sense of one's own body and self.

Metadim Center is an example of how we can introduce in the core of physics a quasi-mental dynamics that accounts for the existence in our universe of self-referent and self-consistent biosystems such as minds or complex ecological systems.

Metadim Rhythm-Rotation expresses part of itself first into the 3 parameters of wave in QM, that is, their spin (rotation, charge, frequency). Just above Planck's threshold starts the frequency domain of QM—that is, waves, interferences, fields, and the first forms of matter-systems. In the CSR hyperdimension or superposed layer of waves-particles, time is still a field and wave-packets are allowed to move backward in time and to leap and translocate themselves into other spacetime coordinates.

Thus, all particles exist within two superposed manifolds: (1) the CSR and HD manifold and its individual ISS, and (2) the Quantum-Spacetime (QST) or matter domain. With the apparition of particles, time sets in as an arrow from past to future in our 4D region. The Planck-Einsteinian time is born, and it will usher the 3+1d matter universe within the light cone and constrained by the speed of light.

Have we lost forever the Infinite Spiral Staircase dimension since Planck scale?

In all respects, and in stark opposition to a purely materialistic paradigm, not at all. We have two sets of data to tackle this issue—and guess what? One comes from the psyche perspective and the other from the physics perspective.

From the psyche perspective, each and every human being has, via their Self and semantic field, an access gate to the syg hyperdimension, in which the psyche is unconstrained by space, time, or causality. In this syg dimension, syg-energy creates connections and bonding between people's semantic fields and between semantic fields and eco-fields. It is the dimension where synchronicities happen naturally, as an emergence from a layer of constant communication between individuals via their Self. The syg dimension is where psi phenomena happen as an expression of our deep bonding with all other semantic fields (minds, natural systems, objects), as a process of inter-influence between *resonant* semantic fields at all scales. It is the dimension of the collective unconscious, in which collective archetypes are entities endowed with consciousness and powerful energy, who express and stimulate the most seminal and creative concepts and facets of human beings. These archetypes are acting as attractors for the spiritual aspirations of humankind.

Synchronicities, psi phenomena, transcendent states of consciousness, spiritual aspirations, thinking and creativity—all are processes involving meaning and therefore syg energy. Syg energy is nonlocal, unconstrained by space, or time, or causality. In a word, syg energy embeds the properties of the spiral staircase's CSR manifold: it spans a temporal field and has access (via the parameter of semantic proximity), to spatially distant events and past events.

Syg energy pervades and animates the semantic dimension or collective unconscious. Each and every mind has its own links with sympathic and resonant ISS melodies and archetypes, via fine-tuned and unique sets of frequency-bows.

At the individual level, *the semantic field* is the superposed CSR layer of a person, which gets more and more complex with time and experience. For any living being existing in the world (whether a human being, a Hopi pebble, or a sacred tree), the semantic field is a complex musical system of *entangled rhythms*, a dynamic network having a *center (the self-ego*), a circle-boundary (its organizational closure), and a network of either transient or more weighted connections. Semantic fields are also resonating to the frequencies of torsion waves of metadim R, in its numerous aspects such as music, global scaling, as well as syg-energy, as thoughts, feelings, values, etc.

Moreover *syg-energy* is consciousness-as-energy, creative surge, intentional focus, and affective bonding; it helps create a meaningful environment peopled by meaningful beings, objects, and systems, all endowed with a semantic field with which the person is continuously interacting, setting a two-way or multi-way inter-influence.

So yes, via our semantic field and the syg-energy that organizes it, a part of our being lives continuously in the Musical Staircase dimension. In fact, the semantic dimension *is* the Musical Staircase, (with its 3 parameters of Center, Syg, and Rhythm, both at the individual and collective levels). But let's turn now to physics.

From the physics perspective, one of the fundamental principles of QM is the wave-particle complementarity. It was posed in 1927 by Louis de Broglie for all matter-particles such as electrons. It's along this line of thought that Werner Heisenberg, in 1927, posited his famous *Uncertainty Principle* describing the two complementary aspects of reality. A particle clearly could be observed to display hard matter properties and coordinates in space and time (the 3 dimensions of particles such as the position). And yet, with an experiment geared at observing waves, it will display one of the 3 distinct parameters of wave, such as the momentum or the spin. In his *Uncertainty Principle,* Heisenberg states that depending on which parameter is measured—among the three of the particle and the three of the wave—the complementary parameter will remain fuzzy or unobservable: "The more precisely the position of a particle is determined, the less precisely its momentum is known, and vice versa." As Dean Radin remarks it in his book *Entangled Minds* (p. 214), "A physical system like a photon, that consists of properties A and B, like a particle and a wave, cannot simply be decomposed into two separate subparts. So a photon cannot be considered as just a particle or just a wave. It's a mixture of both." This wave-particle duality is expressed in the superposed states of the quantum system. (The Psi wavefunction is a state-space comprising all possible states of the quantum system existing simultaneously as superposed states.) According to orthodox QM, among the possible states only one will happen in the physical world when the observation triggers the collapse of the wave-function—all this being basic Copenhagen interpretation. But the ISS hypothesis is that, in the superposed states of each event, system, or entity, there is, as we'll see shortly, an alive connection with the CSR hyperdimension.

Collective intelligence: the co-creation of the universe

What we have to fathom with the Infinite Spiral Staircase is an architecture of the universe akin to a continuously self-created work of art—the co-creation, by all the consciousnesses in the world, of a collective patchwork made of everybody's artworks. In short, a collective intelligence in the process of creating its own reality and being. The ISS is neither the purely deterministic nor the purely probabilistic structure of a universe, seen as matter-only, mechanistic, soulless and devoid of inventiveness and intention. The Spiral Staircase is not the 'program' of a rigid universe that will only play the score written on it by blind laws, or else given to it by an all-powerful Creator. If the score was already written, we could ponder in vain what would be the interest, the gain, or the playground, for this Creator, and what would be our own gain and interest in living our life. If all scores were written, and our lives were already determined, as much as all events and history, we would be totally dispossessed of our individualities, our personal aims and choices, our challenges and our striving; we would be dispossessed of the battles we fight for a better society, and more importantly, of our free will. What, then, could be akin to a soul or a consciousness in such a driven-by-law or driven-by-command world? How could we ever be held responsible for our acts and behaviors? The world would be senseless and an absolute bore.

In order to enjoy experiencing the world as we do, we need a ratio of disorder, of chaos and of fighting for a cause, and, necessarily so, a measure of pain and hardship. Only a disparate and variegated texture of reality may give us the ground for distinguishing between options, and then choosing. The sensations and feelings are the basic foundation of our psyche and minds, and they can rise only from the distinction between one affective tone of joy and another affective tone of boredom, or pain, or nostalgia, or enthusiasm, or hope... Shades of experiences bring the complexity of our psyches and minds, and they can emerge only from a contrasted texture of reality, a mix of regularities and laws and of disorder and randomness. Only in such a world can the self-organization of systems sets in. Only in a non-perfect universe can we bring improvements, inventions, creativity and novel frameworks of knowledge. Only such a world of continuous transformations and flow can be the clay we can mold into novel masterpieces of thought, social organization, sciences, philosophies, and paths of wisdom... This is why evil and pain exists—because without them no good or joy would exist.

So that when we propose at the very beginning of our universe an Infinite Spiral Staircase containing all possible events and information, it is a field of potentialities—a piano with limitless notes—on which the creative minds will choose or create their own symphonies. There will be the music of each person, each village, each culture, each time period, each zeitgeist... We have to view ourselves as continuously improvising on an infinite piano that can contain all possible music that could ever be invented. And this creation is in part personal and in part collective at all levels, from the couple, to the family, to the diverse multiple groups to whom we belong, to the earth, etc. At the global level, the ISS is collective intelligence in process—that comprises also the arch-anima, forever evolving.

The ISS contains also, as cosmic DNA, the symphonies already created by past universe-bubbles—our parent universes on our specific collar. This allows for more and more fine-tuning of the variables and constants that favor life and intelligence. The fine-tuning in ISST has been done over an immense timespan and collar of universe-bubbles. In a mature universe like ours, only a tiny fraction of the networks of links within the ISS corresponds to the Akashic field as a databank of all our own bubble's past events, intelligent civilizations in the cosmos, and individual people. All the rest of the ISS reality sustains a continuous, ceaseless improvisation and intelligent co-creation of all the living and so-called 'nonliving systems' in the universe.

The Cosmic DNA will be changed and imprinted with each novel experience tried and accomplished by human individuals or groups, by each intelligent and sentient species. It will house and contain the memories of these collective experiences and yet an immense span of possibilities will still be open for our quests, our endeavors and creativity. Nothing ever attempted or created, nothing that will be experienced and invented in all future times will ever exhaust its scope. Similarly, our human DNA is not solely the blueprint of our species: it is also transformed with each seminal experience in our life, whether physiological or spiritual.

Thus *we have to envision a two-way inter-influence between the Infinite Spiral Staircase and the collective intelligence, and paths of experience of all creatures on all the worlds of our universe-bubble.*

Being alive and creative means playing our own music and projecting an imprint on an infinitesimal ensemble of bows of the Infinite Spiral Staircase.

However, the ISS at the origin is not the only manifestation of the CSR hyperdimension. We'll now investigate in depth how this manifold exists and expresses itself in each particle and system.

Nature's eco-fields in the hyperdimension

in QM, the central Psi wavefunction implies that a wave-particle is a superposition of states, and best described as a 'cloud of states'. The description of a cloud of states fits the ISS theory, since part of 'the wave-particle' is its eco-field in the syg dimension—its existence, or *dasein*, in the CSR hyperdimension that transcends space, time, and causality. String Theory has equated particles to 'strings' that vibrate at a specific frequency. And we have seen that the 'bows' in the Cosmic ISS are the pre-strings—the archetypes and the blueprints of the sygons and then of the strings/particles appearing after Planck scale. In the cosmic ISS, the bows are still entangled in harmony, each network of bows having a specific topological configuration in the Cosmic DNA. On the ISS, each bow is connected in networks and thus finely entangled with other bow-frequencies of the Staircase; it is tuned to specific harmonic and resonant bows (both in frequency and meaning), as well as with contiguous and similar ones. *These networks of bows code for future systems in 4D+5thD and, when crossing Higgs field, will remain entangled as sygon-networks (waves bundles) that will form the blueprint syg-fields of innumerable systems and drive or rather 'pilot' the organization of the 4D systems in spacetime*.

So how could this non-finite system of bows code for innumerable systems in the world, as well as their creative and free choices and evolution? In order to answer this question, we have to figure the link between matter-systems and the CSR frequency domain. Of course we know that mass is equated to energy, energy is expressed by frequencies—the physics of energy and matter are there. What's still missing, and badly so, is the transcription of semantic dynamics—syg energy in states of consciousness, thinking, feeling, etc. Let's start by a natural system's eco-field.

How to best represent a complex system's eco-field?

This question is, in other words: How could an eco-field be an instant by instant snapshot (or signature) of a system's semantic and physical state, and how could this eco-field be simultaneously exchanging this alive information with the cosmic ISS ?

Imagine a beautiful spring landscape, flowers, bees, and nature alive. (The best examples of complex systems are the ecological systems, such as a forest, or a hilly landscape.) So, what is the syg signature or snapshot of this landscape in the CSR manifold, that makes this system at this precise time and place a unique one? It's going to be a symphony of bows, giving their notes and tunes in unison *and* sequence but as a whole system. The bows activated are at any 'place' or *locus* on the Cosmic Spiral Staircase, yet in a given configuration and progression. We could compare it to Antonio Vivaldi's Spring Concerto.

Given the exquisite and infinite variety of bow-tones—like an alphabet with an infinite number of letters—any system in the universe can have, expressed on the Infinite Musical Spiral Staircase, its unique print or symphony, and an evolving one at that, as one of its states-in-progress. In other words, the ISS behaves like an artificial neural net (multi-layered and with back-propagation), one whose innumerable organizational states can code for, and memorize, any system on Earth and the many worlds. Moreover, it also memorize their evolution (or trajectory) as well as their bifurcations, that is, the modification of their attractor and of their global organization—be it a quantum system or a natural system.

The whole landscape is a one instant symphony, and each of its moment in time in physical spacetime reality will have its own snapshot symphony in the CSR manifold, still connected to the previous snapshot, like frames in a video or beads in a collar.

The aspect of a 'system' as a specific entity (a center or ego) with a boundary (or organizational closure)—for example our spring landscape as a system distinct from the village nearby—is a parameter, of course, of the metadim Center-Circle. The palette of frequencies and their interferences is the snapshot or 'print' of the system in metadim Rhythm-Rotation. The semantic cluster (relations with other systems, self-organization, needs, sensations)—all that which makes the semantic field or constellation of this system—together with all its semantic links and semantic variables (intensity, proximity, etc.) is of course the 'print' of metadim Syg. Let's remember that any semantic field or eco-semantic field is a proto-consciousness and a relatively distinct and autonomous system. It is semantically sentient. For example, the ants mark by scents the paths leading to their places of feeding or their nurseries, and these are all semantic markings—this is a real sentience. The three metadims Center-Syg-Rhythm conjugate their 'prints' to create a holistic snapshot of the 'landscape system' at any moment of its unfolding in time and space on Earth. This series of snapshots, as I said, are imprinted in the CSR manifold as trajectories along a series of specific and unique organizational states of the system.

A hyperdimension in each particle and system

So let's now tackle the reality of a particle in the coupled and superposed Quantum-Spacetime manifold and Center-Syg-Rhythm manifold—as a cloud of states reflecting its superposed states in the wavefunction. As we saw, each particle and each system exists also as an eco-field in the CSR hyperdimension—the eco-field bearing all the snapshot information about this system, and being constantly reactualized and updated with the system's real existential evolution. But I propose that *a quasi-replica of our cosmic ISS would exist at the core of each particle and matter system.* "Quasi," because beyond the field of information contained in the cosmic one, the *'individual ISS'* would also carry the complex semantic field of this individual entity—be it a particle's eco-field or a human being's syg-field.

In SFT, the eco-field (organized and modeled as a dynamical network) expresses the moment to moment dynamical state of a system and the actual organization of its network of links. Now in this SFT description, the architecture and dynamics were modeled but the physical nature of syg energy and semantic fields was lacking—something that this present ISS theory has been intent to solve: syg energy is now integrated to the 3-metadim CSR manifold.

Thus, in ISST, the semantic field (or eco-field) can be represented as a dynamical network of activated and vibrating sygons (bows), within a personalized replica of the cosmic ISS *who* would be *dwelling* (it is a conscious being) at the Center of the person's or system's semantic field. We definitely have here, with the 'Center' metadim, the hyperdimensional part of the identity and selfhood of all beings and complex systems and for thinking beings, their Self (soul, monad, atman).

On this individual ISS, specific bow-tones are activated and vibrating—as the instant to instant snapshot of this system's semantic field. But any such individual Spiral Staircase is also in resonance and in sync with the cosmic Infinite Spiral Staircase, therefore *snapshots of the system's dynamical state are activated and imprinted on the individual and on the cosmic ISS simultaneously.* Thus, each being's or system's individual ISS is a unique hologram within the cosmic hologram; it carries both a unique print and the prints and potentials of the whole cosmic ISS.

Of course, the print of any system in both the cosmic and the individual ISS comprises also its past history and its 'memory' (as goes the concept of Akashic records or Rupert Sheldrake's morphic fields); but its past history is only a tiny fraction of the system's syg-field.

SFT models two types of memory in semantic fields. The first one consists in the past states of the system, memorized in a fractal dimension of the semantic field. The second type of memory is its alive and evolving systemic organization, and it is encrypted in and by the organizational state of the system itself—an expression of its current behavior, connections, its network of relations, its environment, heuristic knowledge, etc. The concept of *organizational memory* (as we saw) comes from chaos theory and neural nets, in which a trajectory or a weighted path (based on the system's past states and behavior) shows the present organization of a dynamical network, and its most probable future path/trajectory (Hardy, 2000).

Now, the superposed CSR manifold entails new possibilities; it shows the system's potential future paths as an array of probability lines, each with a specific weight or probability (see again Fig. 6.3 p.213). But, in contrast with QM, it shows the future probabilities of a sentient being interacting with a meaningful environment. For a human being, their superposed CSR manifold are nothing less than their semantic field and it displays the future probability lines of the intelligent, emotional, spiritual and relational life and experiences of a being.

Thus, for human beings, the superposed CSR manifold shows: with metadim R the probabilities of relations, connections, encounters; with metadim Syg, the whole sentient and intelligent life of the person, their meaningful choices and their links with other syg fields and environmental and matter systems; with metadim C, the global organization of the whole mind-body-psyche system AND its network of links.

With a Spiral Staircase at its Self/Center, each particle/system has, at its core, an opening on the infinite and the whole. More precisely, each system, be it a particle or a mind, is branched on the cosmic ISS of our universe-bubble—that is, each system has the best part of its reality existing in its superposed CSR manifold, in a nonlocal semantic field beyond time and space. And this CSR hyperdimension allows any nonlocal contact and communication with sympathic and resonant minds and systems. Moreover, the ISS of each particle is a delocalized fragment of the Cosmic ISS itself. The totality of the individual ISSs in the universe (blended with the cosmic ISS) form a superposed layer of the whole universe—a universe-CSR-field (or CSR hyperdimension), which is none other than a cosmic semantic field or a cosmic collective consciousness field. One of ISST hypotheses is that this universe-CSR-field is in fact dark energy as Quintessence (an unknown energy).

The Point-Circle Paradox of the hyperdimension

To an observer standing in spacetime, a particle's ISS would appear as a spiraling sink (a micro black hole), with the ISS spiral sinking toward the X-point; and a network of ISSs (coding for a system of numerous particles) would appear as tiny whirling holes in the vacuum membrane, the boundary of the spacetime region. But to an observer standing within the CSR hyperdimension, this universe-CSR-field looks like a field both densely surrounding the spacetime region and yet pervading the whole universe (a quasi continuous but non-uniform field or lattice), as the superposed layer of all beings and things in the universe. And in this universe-CSR-field, the spacetime systems (i.e. particles, plants, or objects) look like networks of tiny holes piercing the lattice.

This instantiates the *Point-Circle Paradox of the hyperdimension:* both compact below Planck scale and extended in hyperspace as a bulk. This paradoxical extension in hyperspace is due to the fact that each particle is open (at Planck length) on the hyperdimension, as a quasi black hole or rather X-funnel, letting sygons move two-way to connect with the whole system's syg-field and with the cosmic ISS. The other side of the ISS and X-funnels of the innumerable particles forming the atoms and molecules of a daisy, is the hyperdimension as a whole manifold or medium, or lattice. An observer from the HD would see on this HD medium the innumerable vortices of all the particles' individual ISSs. (We'll explore the matter in more depth shortly, with the checkerboard reality at the particles scale, and also in chapter 10.)

Let's take into account that all individual ISSs are tuned in to the cosmic one; thus, each particle, at its core, blends in with the Cosmic ISS; it has access to the music of all possible worlds and universe-bubbles—each has a vibrant and alive replica of the Cosmic DNA, and yet each has its own individual symphony unlike any other symphony, alighted in activated, vibrating, alive, and conscious, bow-tones and sygons. This is why one of the properties of metadim Center is that *all centers communicate with each other*; all centers branch on, and are attuned to, the CSR manifold acting as a Cosmic consciousness field.

This is why our universe is a *porous universe.* Any particle, any system, any mind, any semantic field contains a hole, a well, or pore, one which is open on, and breathing in and out, the CSR manifold as a cosmic field of pure creative semantic energy.

Now, how does this work? How can we fit in and model our own freedom and creativity? Here is the gist:

Imagine that we have in us, in our Self hyperdimension, a tiny hologram of the cosmic DNA with our own being and life experiences imprinted on it as activated notes and networks of links. The bow-tones are the notes (activated or not) and the sygons (as tiny as the bows) are the torsion waves sent to meet and interact with other syg fields in the hyperdimension bulk. The sygons are the messengers waves and diplomats of all beings and systems. But unlike Bohm's pilot waves, they are not pre-determined messages, geared to perform only a pre-programmed task. To the opposite, the sygons are searching to build meaningful connections and a network of inter-influence, and they work freely as networking agents and bring back to the being's unconscious Self new ideas and contacts, as well as psi information.

A two-focus model of the HD in particles

The ISS model puts into light a radically novel vision of reality that both features a hyperdimension in all particles and systems, and yet poses connections, synergy, and inter-influence as the mode of interaction between the hyperdimension and the 4D reality, thus allowing, or rather boosting, free choice and creativity.

The particle with its CSR hyperdimension superposed

The ISS at the Center of each particle or system, and opening there as a micro black hole, puts into play:

- the CSR hyperdimension/manifold as a superposed layer of any particle or system;

- the metadim Center as open and connected two-way to both the cosmic Infinite Spiral Staircase at the origin and the cosmic CSR field (the hyperdimension bulk);
- the Syg metadim allowing proto-consciousness and, enmeshed with metadim Center-Circle, boosting the sentient and meaningful self-organization of a system as a distinct entity;
- the Rhythm metadim that launches—via the sygons—frequency dynamics, such as connections, resonant activation of links, inter-influences with linked systems.

The particle, being a cloud of states, would have (figuratively) the shape of an eight, with the access to its inner individual ISS (its CSR hyperdimension) as a micro black hole of Planck radius, in which the spiraling staircase starts at its event horizon.

Constant shifts of focus from the hyperdimension to spacetime

Whenever the particle is undergoing changes—such as those resulting from interferences and shocks, or from being sucked into large fields (EM fields, plasma fields...)—this change is instantaneously reflected in the particle's CSR manifold, by activating or deactivating specific bow-links in its individual inner ISS. This amounts to changing the snapshot imprint of the particle and its network of links in the frequency or CSR domain. Take a laser beam (produced in a lab for an experiment), all the beam's high-energy photons will be network-linked in the CSR dimension: the signature of their coherent state (even transient) is instantly imprinted in their individual and group eco-field, via their individual ISS.

The two-focus model of the particle allows it: the particle is constantly shifting its focus, or attunement, from the CSR wave manifold to the quantum-spacetime of the particle (with spatial parameters)—from the CSR HD to the QST manifold. This means that innumerable times a second, the 'particle' (as a natural behavior) becomes totally fuzzy in our 3+1d physical universe; and it means that the complementarity and Uncertainty principle is an inherent property of the wave-particle, created by the superposed CSR manifold.

Let me explain.

ISST postulates that the wave-particle has two possible 'focus-states.' Either it is in Center-Syg-Rhythm focus or in Quantum-Spacetime focus.

- CSR-focus: whenever the wave-particle is focused on the CSR manifold, it becomes fuzzy in QuantumSpacetime.
- QST-focus: whenever the wave-particle is focused on the Quantum-Spacetime manifold, it becomes fuzzy in the CSR manifold.

Astonishingly, we find the same principle within the ISS itself, that a precision in the radius' logarithmic progression renders the frequency fuzzy and vice versa; and this is not due to the imprecision of the measurement, but rather it is an intrinsic property, deriving from the fact that the numbers driving the creation of the dynamical topology of the ISS are themselves non-finite numbers—and thus fuzzy in their non-finite decimal suite. This brings a fuzziness in the way these numbers construct the next radius in the staircase, or else the next bow-frequency. And consequently, limiting the fuzziness of one (by the experimenter's focus on this manifold), is in effect increasing the fuzziness of the other manifold (in non-focus).

So that we see clearly that the uncertainty is due to the texture of reality as two superposed manifolds, and the dynamics of such a texture of reality that makes the double CSR-QST system be either in CSR-focus OR in QST-focus. Each focus brings with itself its dynamical organization and network of links (including that of the observer), and defines the system's state along these, thus accruing its own precision in EITHER the frequency domain (for the CSR-focus), or the radius domain (for the QST-focus).

When CSR-focused, the wave-particle breathes in all the links with surrounding syg-fields and eco-fields, and all spatially distant systems connected via semantic proximity. Thus, it is open and interconnected with the semantic dimension and, the higher its *semantic intensity*, the stronger its influence on linked systems.

- At low intensity, the wave-particle interconnects blurrily only with its own inner ISS. In case of human beings, with their Self indirectly, via more or less coherent intuitions, dreams and symbols.

- At a higher intensity, the wave-particle connects clearly with its own ISS and with linked and contiguous syg-fields and eco-fields. In case of human beings, they experience synchronicities, insights, psi phenomena, and peak states.
- At a still higher intensity, the wave-particle connects clearly with its own ISS and nonlocally with the ISS of sympathic and encompassing systems, possibly the planetary ISS. In case of human beings, they experience high meditative states, states of contemplation, of unity and fusion. In peak states, they connect harmoniously with the collective unconscious and blend with the planetary semantic field.

When QST-focused, the particle bathes in the fields of interactions of matter and energy particles (EM fields, strong or weak forces, etc.). It is then subjected to causal and in-forming forces. However, the modification of all fields can be achieved from the CSR manifold and from semantic fields, if their syg-intensity overrides the force of the matter organization, such as in the experimenter effect.

Experimenter effect as a CSR nonlocal inter-influence

This has a bearing on the *observer or experimenter effect* in a given experiment such as Young's double-slit one, or in psi experiments (in which, as we mentioned it earlier, the experimenter effect has been well studied and evidenced):

We saw that when in CSR-focus, the wave-particles connects and blends with sympathic and linked syg-fields and eco-fields (via semantic proximity). At this point, the wave-particles are connected to the semantic fields of the experimenters, whose minds are on the experimental design and notably the photon or electron beam they have decided to use.

The intention and attention of the experimenters create (1) *strong semantic proximity* with their experimental objects (the particles' beam and all the machines used) or else with their psi subjects; and (2) *high intensity of their own syg-energy* (since they are involved and in an innovative and highly meaningful endeavor).

The ISS theory then postulates that the wave-particles, through their nonlocal CSR manifold, know the intentions and expectations of the observers/ experimenters, via their entangled semantic fields, and via their links to the experimental apparatus and surrounding eco-fields. Therefore, the particles are *sensitive* to the experimenters' attention to one or the other of their *focus*, and will give more weight to the most resonant set of states. But we may look at it the other way around: The CSR manifold of the experimenters is containing the experimental design, its goal, and the apparatus used; and the quantum system's CSR manifold will *yield* to a more intense and stronger syg energy system that carries its own expectations.

Hence the experimenter effect. The particles can be compared to patients subjected to their doctor's suggestion while receiving a placebo: they yield and display *conformance behavior* acting all the way to biological levels. That is, they exhibit the kind of behavior that their physician is expecting from them. In reality, this conformance behavior is just the result of a stronger syg energy, endowed with a thrust toward a given outcome, imposing its expectations and intention on the lower syg energy system. Therefore the particles' behavior in a double-slit setting would largely be an experimenter effect, yet limited to the state-space of the observed system.

A checkerboard reality open on the infinite

The individual Infinite Spiral Staircase existing in wave-particles leads us also to another representation of waves.

Classically, whenever we detect a wave behavior, we represent it schematically as a sinusoid; yet we have never 'seen' such a wave pattern no more that we ever observed the atom with orbiting electrons. What we do observe in reality, are discrete and quantized shifts in energy and charge fitting more or less precisely a complex model of electrons being ejected from the outer orbit (called valence orbit) or particles ejected by a shock. As we saw with the Beta decay, the infinitesimal discrepancy with the expected measures was what drove Pauli to hypothesize the existence of the neutrino. As for waves, we detect for example a pulse of energy with a certain frequency and wavelength, and interference patterns. Both lend themselves well enough to the classical wave representation as a sinusoid, and interference patterns due to the amplitudes adding up or annulling each other, and wave fronts crisscrossing each other (as we see in liquids).

Let's imagine a hourglass shape to represent the two foci of a particle, with a pink-grey round to represent the CSR-focus state at one end, and a white round for the QST-focus at the other end.

Our 3+1d world is pervaded with stationary waves and many other types of EM fields—one being all over the planet as a grid: the geomagnetic (GM) field. Thus the real-world is replete with patterns of interferences between EM waves belonging to the same or different EM fields. So that the wave-particle has to be integrated in a field of crisscrossing wave-particle systems.

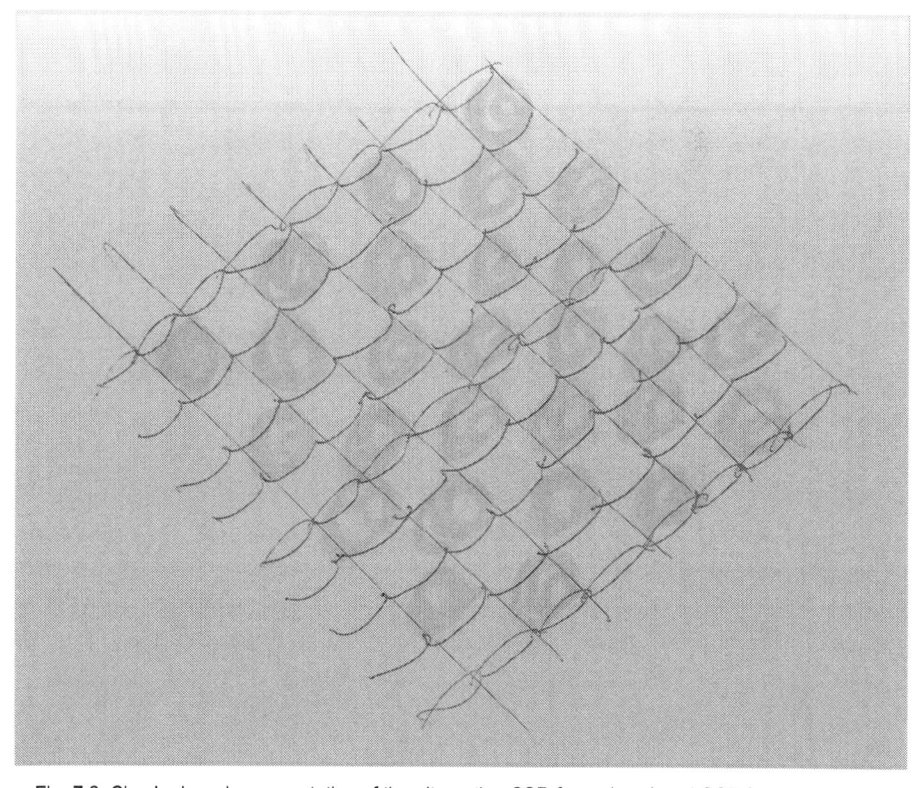

Fig. 7.2. Checkerboard representation of the alternating CSR-focus (grey) and QST-focus in a field of particles. (Concept & Artwork: Chris H. Hardy)

A befitting representation would thus be a checkerboard. (See Figure 7.2.), where each square represents one particle's dual focus-state. In the checkerboard representation of interfering waves, each particle is a superposed system with two alternating focus-states (pink-grey/CSR and white/QST); it occupies one square of the checkerboard, the square shifting ceaselessly from pink-grey to white. Or else a grid of hexagons in our physical reality, in which 7 or more hexagons coalesce into complex systems.

In the checkerboard grid, each wave-particle system is shifting extremely rapidly from the high-energy CSR-focus to the low-energy QST-focus. This means that any wave-particle in the universe is in either one of two focus-states, the high-focus of the Center-Syg-Rhythm HD, or the low-focus of quantum-spacetime, and switching ceaselessly from one focus-state to the other. Thus the checkerboard reality remains constant on the physical layer, only its squares keep oscillating in focus and color.

But let's remember that each high-focus state is the individual ISS—each contains a replica of the Cosmic DNA, plus its own snapshot inscribed on it as a symphony of activated bow-tones. And as we know, all ISSs (whether individual or cosmic) are sinks toward the X-point of origin open on the infinite. So that each particle, each system in the universe not only has, at its 'Center' an opening on the infinite, but a holistic and constant communication with the Cosmic DNA. As Ivan and Grichka Bogdanov present it in their book *Before the Big Bang*, the point of origin is not only a reservoir of infinite energy but also a reservoir of infinite information.

Thus, for the immense number of atoms in our universe-bubble (actually estimated around 10^{78} and 10^{82}), and the even more gigantic number of particles, the inner ISS creates a branching on the ocean of information contained in the Cosmic ISS. Each wave-particle system, while in its high-focus state, vibrates and communicates with this ocean of information (however, only on the specific channels or frequencies that are sympathic or resonant with their own networks).

We cannot not see the congruence with David Bohm's Pilot Wave theory: in his revised Schrödinger equation, it is the wavefunction of the universe that plays the role of the pilot wave guiding the state of the quantum system. When applied to specific systems, it is the wavefunction of this system that is used instead in order to solve the equation. But let's remember that Bohm's pilot wave introduced a deterministic outcome, while in ISST, we have an interplay between the hyperdimension and the forces in spacetime (such as EM fields).

Law of syg-influence on matter systems

In ISST, a system's wavefunction is (in its CSR-focus) in conversation and inter-influence with linked and coupled syg-fields. **The intensity of syg-energy will be the control variable**, the higher the intensity, the stronger the influence; in other words, the highest syg-energy configuration will impose its ordering dynamics.

Conversely, all wave-particle systems in their low-focus state form the (apparently) dense matter configurations and structures of all things, systems, and beings in the 3+1d reality, and are subjected to their causal laws, or random behavior, or field influence. (Figure 7.3)

> When in its Center-Syg-Rhythm focus, the quantum-spacetime reality of the system (the particle) becomes fuzzy, but conversely it is detectable as a wave. And vice-versa, when it is in quantum-spacetime focus, the Center-Syg-Rhythm (wave) reality of the system becomes fuzzy, but conversely it is detectable as a particle with spacetime coordinates.

This was already stated in the Uncertainty Principle, but the framework here leads to other testable predictions.

Therefore the ISST prediction: when the wave-particle system shifts to QST-focus, the state it finds itself into—in Quantum-Spacetime—is oriented and organized according to the states allowed, or informed, by its immediately previous CSR-focus, precisely by the nonlocal field of connections it had in this CSR state.

This 'nonlocal field of connections' in CSR-focus state is of course of a CSR type, that is, nonlocal interconnections instantiated by the 3 metadims, Syg, Rhythm-Rotation, and Center-Circle. Hence the experimenter effect, as an informing of QST events by the semantic fields of the experimenters and their syg energy reaching out for specific objectives.

And that's where the intensity of syg energy comes into play, and the ***law of syg-influence on matter systems' organization***:

> **Whenever two systems *a* and *b* are in strong semantic proximity (or meaningful interconnection), the higher the syg intensity difference, and the stronger the high-intensity system will impose its semantic organization on the low-intensity one, this whatever the matter-complexity of the system involved.**

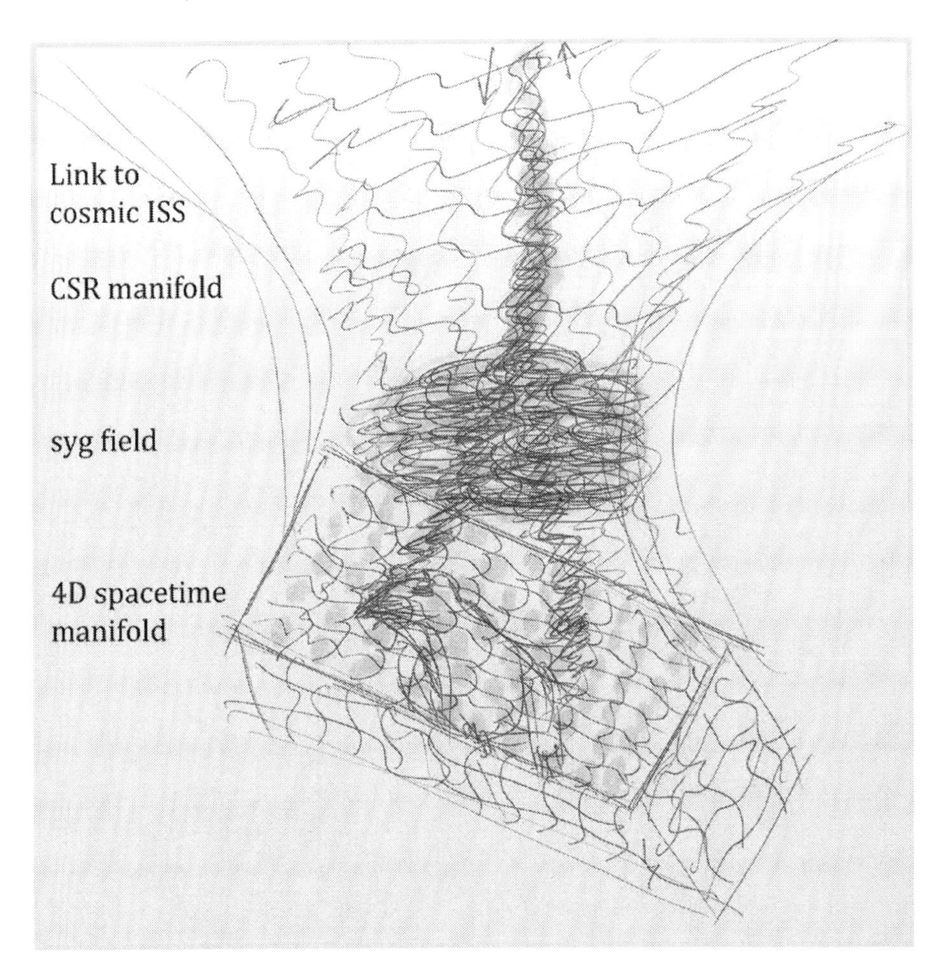

Fig. 7.3. Center-Syg-Rhythm (CSR) manifold and its interlacing with the 4D spacetime manifold, via the bird's semantic field (the dense cloud). (Concept & Artwork: Chris H. Hardy)

However, for this syg-influence on matter to work, there needs to be a deep resonance and sympathy, an harmonizing of the two systems' frequency domain, let's call it a '*synchrony*' realized in the Rhythm metadim. It is perfectly possible that the entanglement of paired particles prior to EPR-type experiments (e.g. the fact they are issued from the same source or event—a prerequisite for their entanglement) is rooted in metadim Rhythm. At this point, we could envision that ideas, words, and concepts, are clusters of these hyperhigh frequencies of metadim Rhythm and of the CSR manifold as a whole. In that perspective, the 'semantic proximity' parameter, that creates spontaneous syg links between semantic fields and eco-fields (based on sympathy, that is, similarities of meaning, feelings, values, etc.), could also have, as the syg energy itself, very specific frequency parameters that could take us a step closer to its direct or indirect detection.

Thus, the blending of Semantic Fields Theory (SFT) and ISST brings us to postulate that the ocean of information and energy of the ISS and Bindu-scale universe—that flows into the void and the ZPF field as an enormous radiation pressure—is not just an unknowable one-substance, one-block reality or entity, on which scientists in awe but at a loss of understanding its nature, append too readily the name of God or Creator—thus risking to shut the lid on the absolute mystery that was just revealed about a hard-to-fathom reality.

The CSR hyperdimension pervades the universe

Let's try to fathom the Center-Syg-Rhythm (CSR) manifold on its own ground and regarding its interlacing with the spacetime manifold.

To do that we have to imagine, again, a topological configuration. (Look again at Figure 7.3)

On one layer of reality, we have the beings, things, and systems in physical reality, that is, in spacetime (here the bird).

On this same layer of Quantum-Spacetime (QST) reality, there are holes as in a checkerboard, corresponding to the high CSR-focus of about half of the particles. But on this QST layer, only microscopic holes in the texture of reality are to be detected.

Now, on the layer of the CSR manifold, where half of the particles are in high CSR-focus, the ISSs (and syg-fields) stemming from the center of these particles are in interconnection with other resonant semantic fields and unfold themselves while reaching to the infinite.

However, if we take now as a point of observation (or reference) the CSR manifold itself, that is, if we were standing in that layer, what we would observe is an immense and all-reaching unearthly landscape, composed of the semantic fields of all systems in the universe, themselves organized in subsystems, the semantic constellations. This landscape is apparently continuous, just as our physical reality seems continuous despite featuring innumerable beings and things of all types and sizes—from clouds in the sky to ants on the path we are walking on. From this semantic fields layer, is streaming out innumerable Spiral Staircases from the centers of all particles, and variegated systems' semantic fields; they spiral out to form an ocean of active information, of specified and qualified syg energy, this ocean itself apparently continuous.

In this ocean of qualified syg energy, the human beings (through their syg fields and individual ISS) interconnect and interact with the ISS of other beings with whom they are in sympathy or in synchrony, vibrating in unison. This interconnection is realized by the sygons via the bows put in resonance. Similarly, each living being, or complex system, interacts with the beings and systems in its environment. All systems are interconnected, communicating and influencing each other via their hyperdimension.

The world is an ocean of vibrating spirals, of innumerable quasi-replica of the Cosmic DNA, each one bearing a unique symphony on its own ISS animated by the Free Sygons, each personal ISS evolving and self-transforming with the life and experiences of the being or system. At the origin, when the Free sygons created the hyperdimension bulk, they created and pervaded as well what was going to be our spacetime region, and are still pervading it. They are the music we play.

All the individual spirals then merge into the Cosmic infinite Spiral Staircase, inscribing on it the music and print of each being and system. The experiences and creations of all beings are imprinted as networks of bow-tones; and this is how a work of art, a masterpiece of music, architecture, painting, thought, science, knowledge, as well as magnificent love and life experiences, singular cultural achievements, ethnic collective experiences—are each one memorized as a living network or symphony within the Cosmic ISS. In this layer of the individual ISSs fusing into the Cosmic ISS, there is a memorizing of all beings, things, events, and historical situations happening in our unfolding universe. This memorizing within a 'subtle and immaterial' layer of reality is, as we know, the Hindu *Akasha* or Akashic records. Yet, this memorized dynamical and alive network-data can never exhaust the open potentialities in store in the Cosmic DNA; some of these potentialities will be the free actions, the creations, work of art and genial knowledge systems, the inventions and carefree plays of the consciousnesses populating at all times our universe-bubble.

Moreover, there is a reverse movement, since we saw that in the Cosmic ISS there is a double spiraling movement with (1) the larger spiral spinning out toward Planck scale, while simultaneously (2) the inner spiral moves inward, shrinking toward the X-point. Similarly, all individual ISSs instantiate the double flow motion. As a global reverse movement, then, the Cosmic ISS sends back to individuals the synchronic archetypal tones, melodies, forms, and ideas that the beings' creations have put in resonance in the bows of the cosmic staircase in the first place. This syg-energy sent to each of us from 'the Source' is akin to an attunement to core vibrations of the arch-anima and the cosmic consciousness' abundant creations. It's akin to a musician tuning in her guitar with a tuning fork giving the perfect A frequency. Many cosmic bow-tones may be put in resonance with slightly imperfect frequencies (we have seen that each has a fuzzy contour, and a non limited set of harmonics); so that the feedback of the Cosmic ISS to us could be, in some cases, that of an archetypal tone corresponding to the networks of bows we have put in vibration. A process of tuning in to the Source's innumerable rivulets and droplets, and simultaneously getting more attuned to our own Self and inner Source, while our creative momentum and power, far from being lessened or limited, would be on the contrary heightened and further stimulated, prodding us to new heights of exploration of our own being.

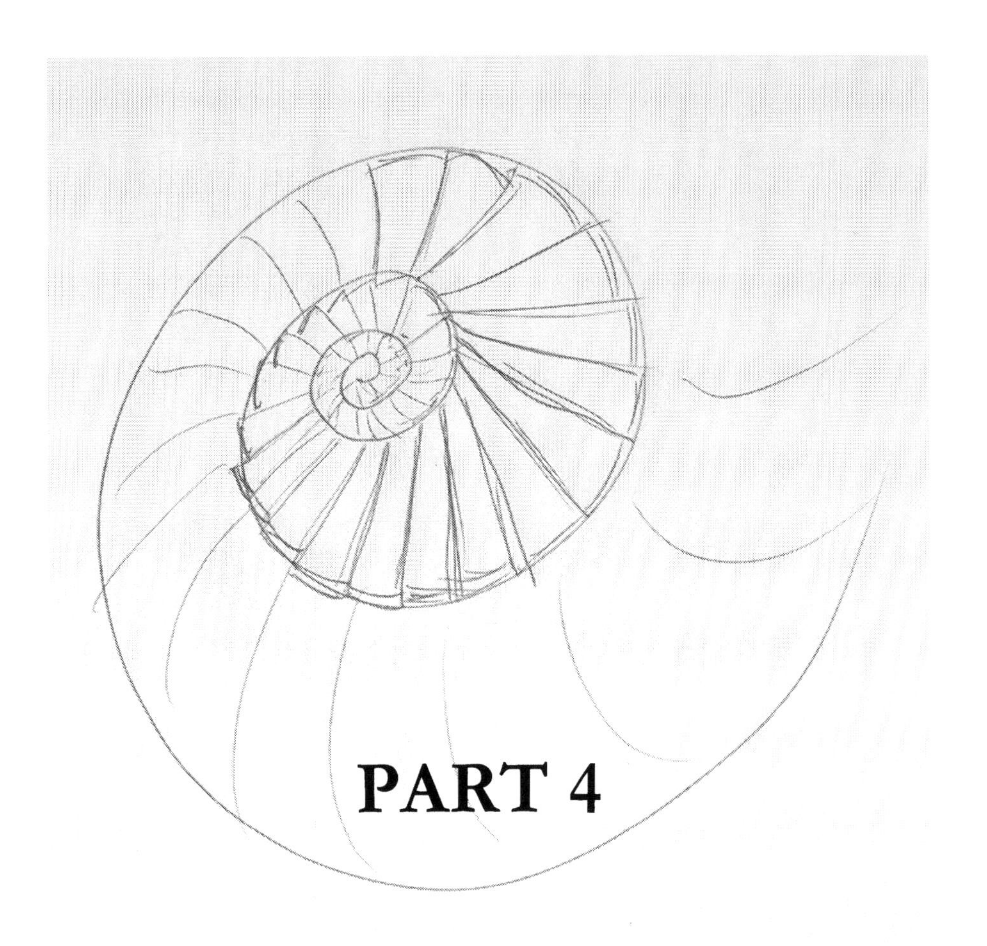

PART 4

A HYPERDIMENSION AT THE ORIGIN

8
CHALLENGING MODELS IN COSMOLOGY

"The natural science assumptions with which we started are provisional and revisable things."
William James (Psychology: Briefer Course)

The concept of hyperspace, and its exploration, are closely connected to the search for a unified theory (Einstein's great dream). In the course of history different theories about the four forces have been elaborated (starting with Newton's gravitation laws), and were refined along the centuries. Yet, each one was like a quasi independent sub-domain of physics. Then, Einstein envisioned that the greatest achievement for physics would be to unify the four forces in a single theory. To this date while three of them have been integrated, gravity is still posing some problems. The way to integrate them has been by postulating hyperdimensions, most generally extra dimensions of space, such as an hyperspace. Talking about the higher-dimensional spacetime, physicist Michio Kaku wrote in *Hyperspace* (1994):

> "Central to this revolutionary perspective on the universe is the realization that higher-dimensional *geometry* may be the ultimate source of the unity of the universe. Simply put, the matter in the universe and the forces that hold it together, which appear in an infinite variety of complex forms, may be nothing but different vibrations of hyperspace." (p. 15, emphasis in the original.)

A Higher-dimensional universe

Recent investigations (notably by the main stream physicists Terence Barrett and Dale Grimes, 1995) have unearthed the fact that James Maxwell was the first physicist to include a higher dimension field—called *electromagnetic potential* field—in his theory of electromagnetism that he presented in 1864 at the *Royal Society* with a paper called "A Dynamical Theory of the Electromagnetic Field." We know now that the first 1873 publishing of his book *A Treatise on Electricity & Magnetism* did include 20 equations in 20 unknowns called *quaternions*. These *quaternions equations*, by modeling the electromagnetic potential, were even a greater achievement than the merging of electricity and magnetism in a single electromagnetic field, in the sense that they posed a higher dimension, asymmetries, and nonlocality as well. Unfortunately, the year after his death (in 1879), Oliver Heaviside and other scientists decided to simplify his theory and in fact reduced it to causal effects only, and then a few years later the Dutch physicist Hendrik Lorentz 'symmetrized it.' All this tampering left us with the definitely brilliant vector algebra but deprived us of the asymmetries and the high-dimensional scope of Maxwell's true vision. All would be sort of acceptable in the name of 'simplifying' the calculus of pre-computer scientists, if it was not for a truly shocking term, highly revealing for the semanticist I am, used by Heaviside; to quote Barrett: "According to him, the electromagnetic potential field was arbitrary and needed to be "assassinated" (sic)." All along the path of scientific advances, we see, again and again, a sort of black genius or Big Brother force whose 'heavy-handed' effect on free research is to bluntly put the lid on whatever is deemed disadvantageous compared to a down-to-earth and flat materialistic worldview; and since it's not simply disadvantageous to some vested interests of the industrial complex, as we see for example with the battle for free energy, it makes us even more intent at looking closely at what's being suppressed.

Thomas Bearden, a researcher on the zero-point energy who aims, through "potential systems" at drawing energy from the vacuum, comments: "That symmetrization also arbitrarily discarded all asymmetrical Maxwellian systems—the very ones of interest to us today if we are seriously interested in usable EM energy from the vacuum." (See *20 Quaternion Equations,* at rexresearch.com.) According to him, Maxwell's EM potential has the power to "reconcile relativity with modern quantum physics and help to explain 'free energy' and anti-gravity."

The second breakthrough was accomplished by the genius German mathematician Bernhard Riemann.

Riemann's hypersphere and N-Dimensional checkerboard

Einstein's General Relativity was indeed a theory of gravity that, as we have seen, used Bernhard Riemann's 1854 work on the hypersphere. Ten years before Maxwell, Riemann was the first to posit a higher dimensional space (a 4th dimension of space) in order to model a hypersphere at 3 dimensions (called O3 to distinguish it from the sphere as a curved surface in 2D, called O2). He also invented the maths of the curved space that Einstein used, and thought that the EM energy (light) was an effect of the rippling of this fourth dimension of space.

Riemann's model was essentially a higher-dimensional geometry or topology. He was the one to stress that a planet or a ball could be rotating around a body (like a planet around a sun) just because of the curvature of space (and not because of Newton's laws of attraction), an idea called Mach Principle and further developed by Einstein. But what is of stupendous import is that Riemann hypothesized that our whole 3+1D world was curled up and crumpled in the fourth dimension; this meta or hyper dimension, he proposed, could explain all known forces of his time (gravity, electricity, magnetism). Thus for Riemann (and Einstein in General Relativity), gravity was a field. He thus posed a principle wholly antithetical to Newton's 'action at a distance,' which was that geometry is in itself a force. Riemann was still a student and had been asked by his professor, the famed Gauss, to write a paper on the foundations of geometry. The nice point is that when the timid student, after more than one year of intense work, gave an oral presentation of his paper in 1854 (despite being plagued with gruesome poverty and a nervous breakdown during this period) he immediately sent shockwaves of enthusiasm worldwide.

What's fascinating is that Riemann, when he did the maths of the hyperdimension, left open the number of dimensions that his equation could account for, thus leading the way to a whole domain of research in N-dimensional spacetime and objects, such as the hypercube. In Riemann's equation, the sum of the squares of all adjacent sides (in length) is equal to the square of the diagonal (z). Thus for a hypercube with 4 adjacent sides a, b, c, d (instead of 3 in the normal cube), the simple equation is $a^2 + b^2 + c^2 + d^2 + ... = z^2$

He also introduced the notion that space could be either flat, or positively curved (the sphere), or else negatively curved (the saddle). To determine if the universe is positively or negatively curved, or else flat, is still hotly debated among the diverse cosmological theories. Remember that the latest WMAP and PLANCK probes data don't give a perfectly flat universe (or Euclidian flat geometry), but a very slight negative deviation from 0, precisely: the curvature (Ω_k or omega-k) is estimated to be roughly −0.0027.

Some theorists invoke that the universe (bubble) could be so large as to appear flat, just as our planet in our everyday experience; in another perspective, the physicists are not yet absolutely certain of how it will evolve (greater acceleration, or a change toward contraction and a big crunch, or else uniform expansion). What we do know, though, is that our universe's rate of expansion is accelerating (a clear data from the satellite PLANCK) and that it *has* changed several times since the inflation phase, in which not only it expanded to the order of 10^{50} times, but in such an infinitesimal time that physicists deem this expansion billion times quicker than the speed of light C.

Another genius breakthrough of Riemann was to anticipate the existence of wormholes on a purely geometric basis: as the possibility to annul distance by visualizing that two parts of a folded sheet of paper (representing a curved space) could be joined. This has been revived in the research on wormholes in hyperdimensional theories especially those that are "multiply-connected" meaning that the regions of space they describe cannot be contracted to a point (e.g. the region has inside itself another region).

Riemann, in order to quantify the curvature of a specific region of space, used a set of 10 numbers at each point of space, called a metric tensor (see Michio Kaku, *Hyperspace*). And now comes his most fecund idea: that these ten numbers could be set in a square of 4 by 4 numbers (6 numbers being redundant), as in a checkerboard. Of course, such a checkerboard of 4 by 4 squares is symmetric, but the crucial point was that it could be adapted easily to whatever number of dimensions ($N \times N$), it just had to be kept symmetrical, a square, and in accordance it was extended equally both to the right and toward the bottom. The metric tensor was going to be used by the diverse unification theories, to add first the electromagnetic force (Maxwell EM field) and then the field model unifying the weak and strong forces (and called the Yang-Mills field). Then, the 11-D Supergravity theory (introducing quarks and leptons—what makes matter) will be another extension of the checkerboard.

> When I discovered Riemann's geometrical checkerboard representation, meant to unify in one single metric all the forces known to physics (at Riemann's time), I had already come up with the checkerboard idea to represent in one single geometric grid the superposition of (1) the CSR manifold (Center-Syg-Rhythm), and (2) the Quantum-SpaceTime manifold (or QST, the post-Planck scale and matter manifold). However, my concept differs widely in that the CSR checkerboard, far from being a flat mathematical matrix, points instead to a stable 'superposition' of states in all particles, with a minute shift of 'focus' either to the CSR hyperdimension, or to the 4D spacetime. Furthermore, it is not at all probabilistic (contrary to the quantum superposition in orthodox QM), but rather an extremely rapid (sub-Planck scale) oscillation of each square in the matrix between the two foci: the CSR-focus and the QST-focus of the particle.

Kaluza-Klein and the 5th dimension

The first fully developed hyper-dimensional theory—which had two distinct outgrowths: supergravity and superstring theories—was that of Kaluza and Klein.

As we know, in 1916 Einstein proposed his theory of General Relativity, in which he merged the time dimension with the 3 spatial dimensions, into a unique geometric grid of curved spacetime. The curvature of space itself was a 'force' and, as Michio Kaku puts it "At the core of Einstein's belief was the idea that 'force' could be explained using pure geometry." (*Hyperspace*, 90) And to achieve his equations, Einstein used Riemann's metric tensor for a curved spacetime at N=4 dimensions (our usual 3+1D).

Kaluza-Klein (KK) theory showed a convergence with Riemann's theory not only in posing a fourth spatial dimension, but also in conceiving this added dimension as rippled or curled up.

Theodor Kaluza was a German physicist living in the then USSR. Having quickly absorbed the newly published General Relativity theory, in 1919 he conceived of a brilliant way to rewrite Einstein's equations by introducing a 5th dimension (that is, a 4th dimension of space). To his great surprise, this solution was also nesting and producing two sets of equations: the EM field equations of Maxwell and Einstein's field equations for gravity, and additionally a scalar field equation called the *radion*. The deep entwining of the two fields equations was astonishing, since they both had the same source term, the energy density tensor (or Maxwell stress tensor). John Brandenburg comments on it: "The implication of the Kaluza theory seemed enormous. The equation of special relativity equating mass to energy was seen as ingrained in the structure of spacetime itself, with, in this case, the *energy density* of the electromagnetic field serving as the *mass density* that would create gravity. (...) *The boundary between geometry and forces was now gone,* EM was geometry in five dimensions, and gravity was a force. The fields could now be unified." (*Beyond Einstein*, 197, my emphasis.)

Kaluza thus realized that by introducing a higher dimension, the unity of the forces could be achieved, a concept that will reemerge in string theory. That same year he wrote to Einstein, proposing a condensed article on his own theory, that perplexed Einstein so much that he took two years to accept to publish the paper. However, it was discovered later that the physicist Gunnar Nordström, from Finland, had published in 1914 very similar ideas.

Later, Kaluza held a professorship at the famed university of Göttingen. In 1926, his theory was further elaborated by the Swedish mathematician (and Max Planck Medal winner) Oskar Klein, whom we readily associate with his famous Klein Bottle and other stupendous topological objects and mathematics. Klein worked at some point with key figures in physics and mathematics, such as Niels Bohr, and Paul Ehrenfest. Klein is credited to have brought to Kaluza's theory the concept that the 5th dimension had a physical reality, and was *curled up in a tiny circle*, the *radius* of which was at the Planck scale (at the Planck length of about 10^{-33} centimeters). The 5th dimension is thus called compacted. Now, when Klein linked the radius of the 5th dimension to Planck constant, the equations of QM could be derived.

In the KK theory, gauge symmetry applies and the 5th dimension is then expressed as a U(1) circle group. (See Wikipedia/ Gauge Theory, Compact Set[35]). The distance a particle can travel back and forth along the radius before reaching again its initial point is the "size of the dimension." In Gauge theory, this radius size is equivalent to a particle, and thus in the KK theory, the particle associated with the 5th dimension has been called a *radion*.

Unfortunately, the KK theory was soon discarded by the overriding quantum mechanics, and didn't make a come back until the nineties.

A number of theories were developed since then, using the 5th extra dimension in compactified or warped models (5thD extremely small, curled up) or posing an extensive form as a bulk; no model yet has been deemed fully satisfying, but the 5th dimension in itself can lead to precise physics experiments able to verify it, and is "of considerable interest" to the physics and astronomy community. Says the Wikipedia article on "Kaluza-Klein Theory:"

"A variety of predictions, with real experimental consequences, can be made (in the case of large extra dimensions/warped models). For example, on the simplest of principles, *one might expect to have standing waves in the extra compactified dimension(s).* If a spatial extra dimension is of radius R, the invariant mass M of such standing waves would be $M_n = nh/Rc$ with n an integer, h being Planck's constant and c the speed of light. This set of possible mass values is often called the *Kaluza–Klein tower.*"[36]

John Brandenburg GEM model and the 5th dimension

John Brandenburg is an American physicist and expert in plasma, and one possible application of his theoretical work is in antigravity disks allowing faster than light travels (which we'll consider further down in this chapter). In *Beyond Einstein's Unified Field,* John Brandenburg postulates a global unification theory between gravity (G) and EM fields, that he calls the *GEM theory*. He develops further the Kaluza-Klein 5th dimension, from which can be derived (as we saw) the EM equations of Maxwell and the Gravity fields equation of Einstein, since these two sets have the same source term: Maxwell's stress tensor. In other words, EM and gravity operate on the same radiation pressure (or Poynting vector). And furthermore, the 5^{th}D gives also the equation of orthodox QM.

Whereas Klein had posited the radius (or length) of the 5th dimension to be the Planck length, Brandenburg sets it to grow from the Planck length to the classical radius of the electron.

One basic tenet of the GEM theory is that it is possible to effect variations in gravity just by manipulating spacetime's EM fields (thus allowing antigravity). This has been proven by experiments on electromagnetic fields and their *Poynting vector*. Let me explain.

In classical EM theory, the embedding of an electric field and a magnetic field (always perpendicular to each other) creates a direction vector for the global EM field, called a Poynting vector (from its inventor). Say we have a hydrogen plasma (the simplest atom, consisting of 1 electron and 1 proton) between two parallel plates of different voltage, the electric field (E) will set the protons (positively charged) and the electrons (negatively charged) moving straight in opposite directions in between the plates (each particle attracted to the opposite-charge plate). In contrast, a magnetic field (B) alone will set the protons and the electrons to *rotate* in opposite directions. But setting the two fields together, an electromagnetic (EM) field is created, and it has two effects: the protons and the electrons are all *moving together* in the global direction of the field (the Poynting vector), despite the protons still rotating in opposite directions (which creates a moving double spiral of constant width). Now, if the two plates are set nearer to each other in the direction of the field (thus producing a variation of the electric field in space) the particles are accelerated, and moreover, it curves spacetime itself: it produces what Brandenburg calls a "GEM gravity." (*Beyond Einstein, 2011*)

This Poynting vector (perfectly named as it points to the direction of the field, also called ExB drift), is in effect a flow pressure; it creates a radiation pressure. And this concept of *radiation pressure* is central to the new cosmologies based on the integration of EM and gravity fields.

Thus, according to Brandenburg, "the basic structure of spacetime is a microstructure of powerful electromagnetism;" (207) and the vacuum is filled with magnetic loops. Following Sakharov, who stated that gravity was produced by a variation of the radiation pressure of the vacuum, John Brandenburg states that it is possible to create a quasi-mass by manipulating EM fields, and thus to modify the curvature of spacetime.

"The ZPF (Zero Point Fluctuation of the vacuum) is where the magnetic fields exists and the charges of the fundamental particles supply the electric fields. (...) The ultra-strong EM fields of the ZPF become spacetime itself." (212-3) Therefore "gravity is due to the electric charge of the particles interacting with the magnetic loops making up the ZPF." (216)

At the Planck length, the 5th dimension appears as a "spinning charge." At this scale it "is not hidden at all. It is enormous." (218) Then it will grow to enormous size (compared to the Planck scale). "When the fifth dimension deploys, it creates an EM field separate from the gravity fields." Then the two basic particles will appear: the electron (the charge structuring EM fields) and the proton (i.e. matter, essential in gravity). The ratio of the masses of these two fundamental particles then takes on a crucial role; Brandenburg postulates that the square root of this ratio is the logarithm of the size of the 5thD (in Planck length units).

Interestingly for us, not only does John Brandenburg states that the size of the 5th dimension is growing from the Planck length (toward the proton radius), but also that "it does have an up or down, a positive or negative (...) measured by charge. (...) the 'down-ness' is the mass shrinking [from the proton] to that of an electron." (227) (In the pre-Planck scale ISS, that direction would be from the Planck length toward the singularity or X-point of origin.)

> In other words, not only does the ISST agrees with Brandenburg's theory regarding the fact that there's a logarithmic increase/decrease in the radius size of the 5th dim (the ISS), but also on the fact that there is a double up and down movement setting a positive and negative charge—the ISS inner spiral moves downward toward the singularity (or X-point); in contrast, the direction of flow of the ISS outer spiral is upward toward the Planck scale and our 4D world, and it will set the huge negative pressure of the dark energy. In our 4D matter manifold, the sygons will create the triune 5th dimension in all systems, and they will synchronize the cosmic ISS with the individual ISSs at all scales (particle, atom, biosystem, galaxy). Thus, our individual human ISS is our Self, in Jungian terms

the subject, the 'I', of our global consciousness, comprising the conscious and the unconscious.

Brandenburg stipulates that near the Planck length, a "white hole metric model" would apply: "The fabric of spacetime at this level can be modeled as a fibrous network of bundles of magnetic and electric flux. The universe at this level looks like a foam of electric and magnetic fields," just as in Smolin's Loop Gravity theory (224). As we have seen, Dirac envisioned the vacuum as a sea of negative particles (the electron and all negative antiparticles). Brandenburg posits that *dark energy is made of negative-mass tachyons,* i.e. faster than light particles. (Dark energy, let's remember, was deduced in 1998 from the observation of the acceleration of the universe's expansion.)

"The simplest mathematical model for the *dark energy* is that it is due to a vast cloud of tachyons, particles with imaginary mass," that is, FLT particles with a negative mass. As the gravity attraction is function of the masses (of the two interacting bodies), their negative masses create a repulsive force between these tachyonic particles. He thus hypothesizes that "the *decay of the vacuum is probably just the periodic transition of the tachyons, or, their tunneling into our universe."(Beyond Einstein*, 269-70).

We totally agree on that, as the very beginning of the unfolding of the ISS, before the Higgs field was formed by interference of large sygons, the first Free Sygons (of extremely high frequencies) will dart out at tachyonic velocity from the ISS and create the hyperspace, more precisely the bulk of the CSR 5th dimension, that will become the dark energy with its enormous negative pressure.

The ISS and the 5th dimension

Kaluza-Klein Theory, with its compact 5th dimension at the Planck scale is of course extremely interesting to me, given its multiple points of convergence with the ISS theory.

• Firstly, as I have posited in chapter 7, each and every particle, matter system, and biosystem in the universe exists also in the CSR hyperdimension or manifold; this hyperdimension thus pervades the universe, while its entry point in each particle is the individual ISS opening at the Planck length. Thus there's a clear match with the KK concept of 5th dimension, with several of its key features.

• Second, since the ISS at origin tends toward the infinitely small—starting at 1 Planck length = minus 1 Bindu-Center (-1Bc)—, we have to consider, just as in QM, that the state-space is infinite-dimensional, that is, the number and all distinct and discrete frequencies of the sygons are quasi-infinite—even if there is a X-point and X-funnel of translation from the previous universe-bubble.

• Third, the CSR hyperdimension, being below Planck scale, is compactified, and it is definitely 'curled up' in a spiral. Therefore the CSR manifold is "compactified and curled up" just as Kaluza-Klein 5th dimension. The divergence with the KK 5thD is that in the ISST, the 5thD is triune: it contains three interlocked metadimensions, namely the Center-Syg-Rhythm metadims.

• Fourth, the radii (decreasing toward the singularity or X-point) follow the infinite set of the Fibonacci sequence in negative numbers. The Phi ratio is the logarithm that binds the intervals in the set of numbers, and thus the set can be called 'bounded' yet it is infinite. Let's remember, though, that the proportion Phi is not exact at four decimals (that is, 1.6180) until we reach the ratio 233/144. I leave to mathematicians to figure how to deal with this slight discrepancy at the beginning of the sequence, and if the ISS maths should use the Fibonacci integer numbers or the logarithm of Phi. The first solution could help model the bows as virtual strings; in superstring theory, particles as strings vibrate at specific frequencies, however the number of wavelengths has to be an integer number.

• Fifth, the sygons are virtual strings (pre-strings/particles) of sub-Planckian wavelengths that, after crossing the Higgs field (or ISS horn's lattice) and in resonance with their frequency, will become the post-Planckian strings/particles.

• Sixth, the ISS (and CSR metadims) is the 'scalar field' nested in the KK field equations, and, according to several physicists, the extra compactified dimension(s) must give rise to standing waves. However, as we are in the CSR manifold beyond spacetime, we cannot make use, in the equations, of either the speed of light constant or the mass; but we may use equations based on frequencies, spin, etc. Moreover, we have to switch from Planck length in centimeters (spacetime) to the Bindu radius length units (BinduCenter, or Bc). We may be able, from ISST, to derive some frequency properties of strings in the 4D spacetime.

• Seventh, Edward Witten (who launched in the mid 1990s the M-theory that unified various superstring theories) has shown that the 5^{th}D made the vacuum (the fabric itself of spacetime) unstable, and that it would exhibit a *vacuum decay*. What is of utmost interest to the ISST is what is produced by this decay. John Brandenburg proposes that, in the process, "Spacetime would decay into *expanding mirror spheres* filled with nothingness;" (206).

Now that's quite stupendous since the ISS theory postulates that the sygons (from the X-point to Planck scale) keep enlarging in radius and decreasing in frequency, and at Planck length, that is, crossing the ISS lattice or Higgs field (the early-age vacuum), would undergo a phase transition and become now strings-particles that contain, compact, their individual ISS. Still driven by the enormous energy and angular momentum of the ISS, the sygons cross or rather tunnel through the lattice, spinning wildly, imprinting in the vacuum their vortices (and their sygon information as well) and creating finely tuned vortex-patterns within the random fluctuations of the vacuum.

Nikola Tesla

Nikola Tesla was the genius inventor who brought the AC electricity to the great cities of the United States, starting by New York, who also invented the radio, and the wireless transmission of electricity among many other inventions, and who literally launched the electrical and EM revolution. A generation older than Einstein, Tesla (who was rumored to have participated with Einstein to the infamous Philadelphia Experiment) didn't like Einstein's theory of gravity implying the curvature of spacetime. Says John Brandenburg (*Beyond Einstein*, 179): "[Tesla] proposed his own ideas of gravity, based on *particles being the subatomic vortex motions of aether*. Nikola Tesla even stated that he had produced what he called 'scalar waves' of electromagnetism in the aether. These were like pressure waves in air [and] had no vectors associated with them except for the direction of their motion." In 1938, Tesla was convinced that gravity could be controlled by EM fields, thus fathoming their possible unification; this concept was feeding into the Philadelphia Experiment, said to have been carried by the Navy in the Philadelphia Naval Yard in the summer of 1942, and then again on February 28, 1943, in which the use of a torus-shaped EM field created the delocalization of a whole ship and its crew—the destroyer USS Eldridge—and its instant teleportation to the harbor of Norfolk, Virginia, more than 200 miles away, where it was observed there for some time by the crew of another Navy ship, before its reappearance in Philadelphia.

Tesla's concept of ether was definitely a sophisticated one! Ervin Laszlo, in *Science and the Akashic Field* (chapter 4) cites a 1907 unpublished article of Tesla titled "Man's greatest achievement." In this article, Tesla conceives the ether or "natural media" as a field similar to the Hindu concept of Akasha, a cosmic field that is imprinted with information and has stored all events that ever happened on earth. For Tesla, there is a "cosmic energy" (or prana) that acts on this field and transforms it into matter. If this action stops, then matter dissolves back into the Akashic field. (We'll get back to it in the Conclusion.) A radio announcement dated February 1935 spells: "[Tesla] has announced a heretofore unknown source of energy present everywhere in unlimited amounts." (in "A Machine to End War," *Liberty*[37]) Of what could Tesla be talking about if not the energy of the 'vacuum,' the actual revisited concept of ether?

de Sitter and anti-de Sitter spaces – as a 5th dimension

There is another way to introduce a 5th dimension in a physics and cosmological model, and it is widely used in either the field theories of strings or quantum fields theories (QFT) in order to integrate gravity and general relativity: it is to model a theory in a five-dimensional manifold in which, to the 4D spacetime, is added *a 5th dimension that carries the parameters of the curvature in spacetime.* Furthermore, it facilitates the integration of the diverse fields theories of strings (already successfully merged in a single meta-model issued of Susskind's M-Theory and called 'conformal field theory,' or CFT). The 5th dimension is modeled either as a *de Sitter space* (dS) or as an *anti-de Sitter space* (AdS), depending on which curvature one chooses (convex-spherical-finite in dS, or else concave, saddle-shaped, open, and expanding in AdS). Willem de Sitter, a Dutch astronomer, was the first scientist to model mathematically (in 1917) a dynamical universe in constant movement (and expansion), and he later collaborated with Einstein.

The success of AdS modeling rests on the fact that, beyond accounting for gravity in curved spacetime (and the repulsive force of dark energy and/or the cosmological constant), it can also integrate the 3 other forces (EM, strong and weak forces). However, the AdS models do something even more stupendous: they establish what is called a *S-duality*, meaning a perfect set of correspondences between one framework (e.g. a string theory) and another one. This kind of bridges allows (at the mathematical level) to switch from a set of units (in one framework) to another set (in the other) and it has been immensely successful. Physicists can jump from one set to the other according to which calculus permits to model more easily a specific set of parameters, and they can integrate 'patches' of geometry, topology, or else, into the model in duality correspondence. One such model, widely used, is the AdS/CFT duality, putting in correspondence AdS (curved spacetime and gravity in five dimensions) and quantum fields theories in four dimensions, that was elaborated by Juan Maldacena in 1997 and is deemed the most important discovery in string theory in the last twenty years.

Symmetry, strings and branes theories

When physicists began to smash the nuclei of atoms in powerful colliders such as the LHC at Geneva's CERN that found a trace of the Higgs boson, the first thing they were able to observe were definite symmetries in particles. If you remember, the principle of anti-symmetry (through Pauli's pairing of particles with opposite and non-integer spins) governs all matter particles, such as protons, neutrons, electrons, muons, neutrinos and quarks. Whereas the symmetry principle applies to the *bosons* particles (or energy particles), with integer spins, comprising the photons and mesons. This symmetry principle is linked to one of the most astonishing properties of the photons (and other bosons, such as the gluons): that of being able to get in a coherent state or in resonance—such as in a laser ray—, thus leading to numerous powerful applications of these fascinating coherent rays, for example in deep brain surgery.

However, with the Standard Model that unified three of the four forces (strong, weak and EM), a range of small symmetries were found. The basic concept is that all matter, when interacting, exchange different types of quanta. If we take for example the strong force, the

3 quarks in a proton are glued together by quanta (packets of energy) called gluons; as for the Pi meson, it consists of a quark and an anti-quark glued together by a string. The strong force is so robust that if the set of quarks is hit, it will start vibrating on a specific vibration. "Each vibration of this set of steel balls or quarks corresponds to a different type of subatomic particle." (Kaku, *Hyperspace*, 122). And if we consider the weak force, what is exchanged by leptons particles are Z and W bosons, but it will not create a vibration. Moreover, with the strong force, the 3 colored quarks can be switched between them: they exhibit a symmetry called SU(3); the same is true for the weak force: the interaction of one electron with one neutrino exhibits a SU(2) symmetry. As for the electromagnetic force, governed by a Maxwell field that can be rotated and yet keep its integrity, its symmetry is labeled U(1). Those three forces, in the Standard Model, are expressed by a global unified symmetry labeled: SU(3) X SU(2) X U(1), held together by Yang-Mills fields.

The problem that plagued the Standard Model, however successful in fitting all experimental data, was that it had grouped all particles into 3 families that were mostly redundant. In the 1970s, new models called GUT (Grand Unified Theory) posed a higher and unique symmetry to account for the three fundamental forces, such as SU(5) grouping 3 quarks (strong force), 1 electron and 1 neutrino (weak force). This particular higher symmetry principle means that all heavy particles can decay into other particles (as in the example of the Beta decay). GUT models, however, were deluged by the immense number of particles they kept figuring out.

Gravity was still not unified to the other 3 forces in any quantum theory. At that point some physicists realized that Kaluza-Klein 5th dimension, extended to *N* dimensions curled up, led to a global hyperspace symmetry and accounted for the 3 fundamental forces.

The surface of a ball (symmetry O(3)) can be made to vibrate, but it will do so at specific frequencies or resonances. A hypersphere will resonate at specific frequencies according to its symmetry O(*N*) and a complex hypersphere similarly with SU(*N*).

Says Michio Kaku: "One of the greatest mysteries confronting science today is the explanation of the origin of these symmetries, especially in the subatomic world. (...) Historically, we had no idea of *why* these symmetries were emerging in our laboratories and our blackboards." (*Hyperspace,* 131). According to Kaku, the Standard Model does not provide any answer; only a hyperspace theory such as superstring may offer, if not an explanation of the origin, at least a framework to understand them.

Superstring theories mostly come in 10 or 11 dimensions. The model was developed at first by Gabriel Veneziano and Mahiko Suzuki at CERN in 1968, and in 1974 Michio Kaku and Keiji Kikkawa set themselves to develop the field theory of strings, and soon Edward Witten came up with a bright and highly regarded model. The core concept is that a subatomic particle, previously thought to be a point-particle, is in fact a tiny string (at Planck scale), each particle vibrating at a specific frequency; however, just like a violin string, the number of wavelengths on this string will always be an integer number, yet it will allow an infinite number of resonances. Furthermore, the strings interact between them, forming smaller or larger strings.

String theory was abandoned totally between 1974 and 1984, because, while it was able to reproduce many results of the Standard Model, it nevertheless didn't bring any breakthrough beyond what was already known. However, the interest was sparked anew when Michael Greene and John Schwartz proved that it had self-consistency. Then Witten in 1995 raised again an intense interest with his novel 11-dimensional M-Theory, based on membranes. The basic concept is that of a multiverse comprised of a multiplicity of universe-bubbles or vacuum regions, each one presenting a different value of the energy from the vacuum and other parameters, and separated by a membrane or 'brane' from another bubble. In our universe allowing life, the vacuum energy is small, but in the majority of the vacua (or universe-bubbles)—derived from all the possible states in the state-space—it will be large. Says Leonard Susskind: "String theories, popular with many particle physicists, make it possible, even desirable, to think that the observable universe is just one of 10^{500} universes in a grander multiverse."

Susskind's String theory landscape and anthropic principle

Leonard Susskind is a ground-breaking string theorist from Stanford University who was at the forefront of string theory, M-theory, the Holographic principle, and the String Theory Landscape (2003). He elaborated a QM-String model of particle physics, of the strong force, and of black hole entropy.

Quantum Field Theory has introduced the concept of 'false vacuum,' where the energy density of the vacuum is not at its ground state (as in the true vacuum) but in some local minimum, separated from the space or region of the true vacuum by a barrier that keeps this vacuum inside a boundary. In string theory, there is an immense number of false vacua that are like bubbles, 'domains,' or 'regions' structured by specific values of all physical constants and fine-tuned parameters, separated by barriers or walls; Michael Douglas advanced the unfathomable number of 10^{500} possible false vacua (note that a billion is only 10^{9}). In the string theory landscape, each region has an energy density that sets all the other physical constants and parameters.

Says Susskind in his 2003 article *The Anthropic Landscape of String Theory*: "In this model different vacua are represented by different quantized values of the electric field while the electrons/positrons are the domain walls. This model is not fundamentally different than the case with scalar fields and a potential. In fact by bosonizing the theory it can be expressed as a scalar field theory with a potential."[38]

The model's assumption is that all possible vacua do exist in a grander ensemble called the multiverse, but only in some vacua can the tuning of constants and parameters allow a universe-bubble that lasts long enough for galaxies to form; even less numerous are the vacua that could support life, and even less so, intelligent life and civilizations. Finally, as we exist in a fine-tuned universe supporting intelligent life and scientists able to ask questions, of course we were born in one bubble that allowed it.

> This version of the anthropic principle is stressing the limits of plausible statistics to make the extremely fine-tuning of at least twenty cosmic parameters in our universe a mere random event. If we have one chance in (a billion of a billion of...) (repeat 53 times)... to be on a civilized planet, then of course, as that means beating hyper-astronomical odds, that gives us a reason to become excessively self-centered, given the dire consequence: according to the proponents only a small percentage of vacua would be able to sustain life as we know it. In my humble opinion, that percentage would be extremely low. And if this is so, it certainly reduces to a

vanishing probability any possible communication or contact with other vacua (beyond all other physical barriers separating the vacua). If the odds would be so astronomically low as to practically forbid that any other universe-bubble would have the same chance, even in an endless time-frame, that would erase any other intelligent universe from the multiverse.

The zero-brane model and heterotic string

A recent domain of research (Stephen Shenker, Leonard Susskind, Edward Witten, and others) is called 'zero-brane.' It focuses on the universe before the Planck scale, timeless and spaceless, and is intent on featuring its 'landscape.' Doing so, it is based on the work of French mathematician Alain Connes. Says Greene about this research: "Finding the correct mathematical apparatus for formulating string theory without recourse to a pre-existing notion of space and time is one of the most important issues facing string theorists. An understanding of how space and time emerge would take us a huge step closer to answering the crucial question of which geometrical form actually does emerge." (Greene, *The Elegant Universe*, 380)

The most promising model in terms of unification, the *heterotic* string, is drawing on the KK theory; it was elaborated by the 'Princeton String Quartet,' namely David Gross, Emil Martinec, Ryan Rohm and Jeffrey Harvey. This type of string is closed and has a dual vibration, the first is a ten-dimensional (10D) clockwise one and the second a sixteen-dimensional (16D), compacted, counterclockwise one.

Thus we have, with the heterotic string model, a string-particle existing simultaneously in two dimensions, and it certainly resonates with the core concept of the ISS that strings-particles exist in two manifolds at once and have a dual focus: either on the CSR hyperdimension, or on the QST (quantum-spacetime) manifold. The CSR manifold has negative energy, and can be equated to an inverse rotation, strongly coupled with the complementary rotation of the string-particle in the QST manifold. Furthermore, the CSR hyperspace in the particle (its ISS), is compact in a circle-spiral, just as in KK and heterotic theory.

One of the heterotic string founders, David Gross, explains: "*To build matter itself from geometry*—that in a sense is what string theory does. It can be thought of that way, especially in a theory

like the heterotic string which is inherently a theory of gravity in which *the particles of matter as well as the other forces of nature emerge in the same way that gravity emerges from geometry."* (quoted in *Hyperspace*, 157, my emphasis.)
Indeed, this description from Gross could perfectly well apply (as far as the core concepts are concerned) to the ISS theory.

The ten-dimensional Hyperspace theory poses that the primordial universe (before Planck scale) exhibited perfect symmetry, with the ten dimensions fused together in one stuff, at the Planck energy scale of 10^{19} billion electron volts. Then happened a series of symmetry breaking as each of the main 4 forces, starting with gravity, started to separate from the original unified field. Then this left a dual universe, one with 6 dimensions, and the other with our 3+1 dimensions (what I have called the QST manifold or quantum spacetime). The six dimensions then started curling up and contracting to the size of the Planck length (10^{-35} meters, that is, 10^{-33} centimeters). For example, Cumrum Vafa's hypothesis is that the unified 10D field broke up in two universes, our 4D one and the other being a twin universe, a 6D cone or torus, called *orbifold*, and compacted at the Planck scale.

The crucial point of the hyperspace theory is that *any particle vibrating on the surface of a hypersphere will inherit the symmetry of this hypersphere.* Kaku expresses it thus "If the wave function of a particle vibrates along this [hypersphere] surface, it will inherit this SU(*N*) symmetry. Thus, the mysterious SU(*N*) symmetries arising in subatomic physics can now be seen as *by-products of vibrating hyperspace*! In other words, we now have an explanation for the origin of the mysterious symmetries of wood [that is, matter]: they are really the hidden symmetries coming from marble [that is, the cosmological topology]." (*Hyperspace,* 142, emphasis in the original text.)

The ISS + SFT models can rephrase this concept by posing that:

Any syg field included in a larger syg field, will acquire from the encompassing system: (1) its specific syg-energy (active information), (2) its Rhythm frequency & spin, and (3) its system's organization, including the type of 'membrane' or organizational closure (metadim Center-Circle).

In case of complex systems having a temporary connection (a medium value of *semantic proximity*), the two semantic fields will influence each other in a two-way manner, the more intense syg energy system will tend to have more influence on the other, eventually turning it partially into its own CSR values.

This is what may happen in case of an individual's conversion to a faith or joining an ideological group: the person will rapidly develop a whole new semantic constellation (a dynamical network of meaningful ideas, processes, beliefs, behaviors, relationship styles, etc.) that more or less matches the group's semantic field (with its core belief system). The more discrepant the intensity of syg energy between the individual and the group, the more easily the powerful group will overcome any resistance and literally "possess" the young recruit. In some instances, a person's syg field may have more intense syg energy than a whole group, and she will influence this group and eventually become its new leader.

Global geometry and the multiverse

The diverse cosmologies take into account the newly discovered dark energy, that is, the energy of the vacuum (called Lambda in the equations), as well as the density of matter in the universe, that is, normal visible matter + dark matter (symbol: omega); and the rate of expansion as well (H), that is, Hubble's constant. A value that change over cosmological time is that of the *critical density* (Ω_Λ Omega Lambda), that denotes the density threshold at which the universe would stop expanding, and therefore the increase in the amount of dark energy as the universe expands.

In Kaluza-Klein theory, the 5th dimension curled up at the Planck length is a physical property of the universe. And in consequence (says Kaku, 1994, 105) "the fifth dimension is topologically identical to a circle and the universe topologically identical to a cylinder." However, if we add the recent data on the Big Bang and the expansion, as well as different rates of acceleration of this expansion, what we obtain is the well known cone. A cone shape that could be (as in ISS theory) in-formed and driven by an expanding spiral with a variable rate of expansion due to the added constraints such as the density of matter (Omega) and the total gravity produced by it and dark matter. (This neither fix nor predict the future progression of the spiral.) What we now know is that this rate of expansion has varied since the Big Bang at several transition phases, such as, for example, during the nucleosynthesis (the formation of the first nuclei of hydrogen and then helium in the first 3 minutes). When Einstein, as a correction of his first version of General Relativity, introduced the cosmological constant, he calculated it to perfectly counterbalance the force of gravity as we observe it at this age of the universe (determined by all matter, stars galaxies, and mass particles + dark matter)—a gravity that, otherwise, would have brought the universe to collapse on itself. And therefore this constant could support Einstein's preference for a static universe. After having to accept that his static universe hypothesis was shattered by the relic microwave radiation that proved the Big Bang, Einstein called this constant "the biggest blunder of my life" (as reported by George Gamov, 1970). Yet his calculations about the necessity of a force equal to gravity but with an opposite (repulsive) energy, was correct: we know now that it is the negative pressure of dark energy. Hence the fact that this counterbalancing force is perfectly equivalent to the amount of dark energy radiation pressure and that it has to be understood as a repulsive and antigravity force.

> This is of course of the utmost importance for the ISS theory, given that prior to the Planck threshold (whose energy is 10^{19} billion electron volts), we are in negative energy, having a formidable power, and rising toward infinity.

The brilliant advance in string field theory led nevertheless to some perplexity when billions upon billions of solutions to this theory were discovered, which represent as many universes different from ours and that, in their great majority, would not support life.

This is what brought several physicists to hypothesize a 'multiverse' in which an immense number of randomly adjusted 'vacua' (or regions) would exist, and we just happen to live in one whose fine-tuning (1) allows the galaxies and planet to form, (2) allows life to develop, and (3) allows intelligence. The first concept in this line was that of Hugh Everett's 'many worlds' hypothesis, based on the pure indeterminacy of QM, which predicted the absolute randomness of the state of a particle after the collapse of the wave-function. Everett considered only two possible outcomes ('cat is dead' OR 'cat is alive,' in the famous Schrödinger's Cat example), and figured that each one, with its worldline, was existing in a parallel universe.

Given the immense number of such wavefunction collapses in just an hour and within a square kilometer of a city like New York, the number of parallel universes would be astronomical (especially since we can question the gross simplification of all the possible states of a system after the collapse (normally a random field, or Bell curve) into a 0 or 1 binary-only selection.

Budding universes, Big Bangs and Big Crunches

Michio Kaku states in *Parallel Wolds* (p. 93) that any sound cosmology has to merge quantum theory with relativity theory's cosmology; thus "we are then forced to admit the possibility that the universe exists simultaneously in many states. In other words, once we open the door to applying quantum fluctuations to the universe, we are almost forced to admit the possibility of parallel universes."

In 1980, the *Inflation Theory* was developed by Alan Guth of MIT, as a Grand Unified Theory or GUT, that posed that 'empty space' expanded to a factor of 10^{50} at the trillionth of a trillionth of a trillionth of the first second of the universe (10^{-36}s). Our universe would originate from many tiny spots in the original fireball, each inflating quicker than the speed of light. But how could all these inflated bubbles ever merge to bring out the very universe with omnipresent fine-tuned variables and constants, that we live in?

Andrei Linde brought a possible solution with his 1981 model of the 'multiverse' called the *chaotic inflationary model*. In Linde's theory (1994), bubbles would form at numerous points in spacetime (even at the present time), where symmetry breaking would occur, thus leading these bubbles to inflate. The problem he sought to solve is called the "grateful exit' and consists in fathoming what process or force would allow a universe-bubble to inflate by a factor of 10^{50} and long enough for the actual universe we inhabit to come to existence. Many physicists, on the basis of quantum indeterminacy, want the process to be random, and so, among all the bubbles budding randomly in any spacetime point, only one, by chance alone, would last long enough to be our own. Others, having too small a density of matter (*omega*) will forever expand, whereas those having a too large omega will get to their big crunch before the galaxies are able to form.

Roger Penrose, since his 2010 book *Cycles of Time*, has developed a cosmology called Conformal Cyclic Cosmology (CCC) and stipulating a suite of universes, or aeons. I will present and discuss it at the end of chapter 9, in comparison with ISST bigcrunch-bigbang X-funnel.

An extremely interesting cosmological theory is that of Lee Smolin, a physicist who was one of the founders of *Loop Quantum Gravity* theory, that postulates space as being granular and discrete, as a consequence of its quantum fabric. Space becomes a fabric weaving together "spin networks" or "networks of loops" that would have the Planck length. With time, these spin networks create a "spin foam." Smolin, in his *Fecund Universes Theory* proposed in 1992, builds on similarities between black holes and the Big Bang, such as both being a singularity with gigantic gravity warping spacetime. Smolin proposed that *massive black holes (issued from dead stars) may be the seed of budding universes, that would retain some of the parameters of their parent universe.* The universes, after several generations, would be optimized for producing black holes. Says Greene (*The Elegant universe*, 369): "Smolin's suggestion provides a mechanism that, on average, drives the parameters of each next-generation universe ever closer to particular values—those that are optimum for black hole production." In other words, Smolin poses a dynamics that is in effect a *natural selection at the cosmic scale* and an alternative to the (randomly set) anthropic principle and that would favor universes with cosmological and physical values similar to our own universe. In his 2006 book *The Trouble with Physics*, Lee Smolin has criticized what he sees as a "near-monopoly" of string theory over the research in theoretical physics, and advocates a diversity of divergent approaches that alone can keep science healthy and creative.

> The budding of new universes seeded by previous universes leads to the idea that our universe may have been seeded by a previous one, an idea developed in a different manner in the ISS model than in Smolin's theory.
>
> Linde's 1981 model featuring the fabric of spacetime in which numerous bubbles burst at any coordinates of space or time, seems like a draft version of the actual concept of vacuum filled with vortices, "spin networks" or "networks of loops" (Lee Smolin, Jack Sarfatti). It's true that the very tidy ancient Relativity concept of spacetime as geometrical and curved, filled with orderly EM fields,

has receded, while in its place came to the forefront the turbulent and polarized vacuum of quantum mechanics, filled with virtual particles. Yet, it seems that the best options lie in the merging of these two frameworks, as numerous scientists have pursued it.

Lee Smolin's *natural selection at the cosmic scale*, with its biology-driven logic, is certainly the most remarkable alternative to a purely mathematical or physical type of logic being used in elaborating cosmological models. And yes, we have to be reminded sometimes that, as Brian Josephson and Fotini Pallikari-Viras (1991) postulate it, biosystems have developed their own life-favoring dynamics and strategies. Setting a clear distinction with the physics-only, materialistic, paradigm, these authors make us aware of highly meaningful connective dynamics between biosystems (such as psi and mental influence on matter and randomness) that sometimes supersede a random distribution and could shed some light on still mysterious physics processes such as the entanglement. Of course, while saying that, I'm also arguing in favor of my own ISS theory! While Josephson and Pallikari-Viras prefer to access these scientific frameworks as parallel and divergent, and yet equally valid, systems of knowledge, I'm in favor of the crossbreeding of logics toward a complex, Escherian type of logic enabling us to fathom how in the world life and the cosmos could have themselves reached such a successful blending!

Plasma cosmology – Alfvén-Klein Cosmology

Hannes Alfvén, Nobel Prize in Physics in 1970 (for his work in magnetohydrodynamics, or MHD), has discovered an interesting resonance in scales, with a "cosmic triple jump," each time of the order of 10^9, from laboratory plasmas (10^{-1}) to magnetospheres (10^8), then to interstellar clouds (10^{17}), and finally to the Hubble distance (10^{26}). Alfvén proposed a Plasma cosmology in 1966 in his book *Worlds-Antiworlds*, stating that ionized gases and plasma were much more important forces to consider than gravity, and furthermore, the EM force is 10^{39} orders of magnitude stronger than gravity. Plasma is the 5th state of matter, it is composed of ionized gases (atoms of gases that have lost an electron and have thus become positively charged) and is the most abundant state of matter in the universe (forming the greatest part of the matter of our sun and stars, and also found in intergalactic plasma and inside clusters); it is electrically conductive and form EM fields. Interestingly, the plasma around charged particles (such as an electron) shows a frequency issued from the electron's oscillations. Because the plasma is charged, it interacts with all EM fields and is responsible for the phenomena of aurora, solar plasma discharges, pulsars, gamma rays, the interstellar nebulae, etc.

Then Oskar Klein elaborated on Alfvén's concepts, leading to the *A-K cosmology*. They hypothesize that an equal amount of matter and antimatter form the universe, with boundaries between them in the form of a double-layer kind of thin cavity, with opposite charges, and whose interactions would generate the plasma. Thus the *double-layer boundary is constituted both of matter and antimatter*: it is an *ambiplasma*, with particles being too low-density and too hot to annihilate each other fully unless in an immense span of time. The ambiplasma creates pockets of matter and antimatter, and our universe is just one pocket in which matter is predominant. Alfvén and Klein have been able to interpret in a new way the walls of galaxies. However, they reject the Big Bang and the expansion of the universe as well, thus being at odds with now proven facts. Furthermore, the amount of high energy photons predicted by the annihilation taking place at the boundaries of the ambiplasma has not yet been supported by observation.

Ramanujan system: 24 to 8D (octagonal geometry)

Srinivasa Ramanujan was a genius mathematician from India who was invited to make some research at Cambridge University in 1914 by Godfrey Hardy, himself a mathematician. Ramanujan came up with a 'modular function' that has 24 modes (each a vibrating string) and can thus be applied to string field theory, insofar as its 26 dimensions can be reduced to 24, and since in Ramanujan system 24 is reduced to 8, then string theorists come up with a complicated 8+2D that gets them back to the 10D of string theory. The complexity and yet exceedingly in-depth equations of Ramanujan (of which loads of them in his notebooks are still waiting to be deciphered), made Kaku exclaim: "It's as though there is some kind of deep numerology being manifested in these functions that no one understands. It is precisely these magic numbers appearing in the elliptic modular function that determines the dimension of space-time to be ten."

Brane cosmology and hyperspace: the 5D Randall-Sundrum model

String theory has led to the elaboration of many different models, some based on the concept of brane—the brane being a flat 2D string like a rubber band (while the string is a 1D filament), and notably to interesting cosmologies. In theses models, the 5th dimension can be either compact (curled-up) or extensive (the bulk). If the 5th dimension is compact (meaning extremely small) then it's like nested in the 4D and we get a one-brane universe (without bulk). On the contrary, theories implying a bulk have an extensive, possibly infinite, hyperspace. This bulk can be crossed by other branes. In some brane cosmologies, such as the Arkani-Hamed, Dimopoulos and Dvali (or ADD) model, first proposed in 1996, our 4D spacetime exists as a brane within a hyperspace called the 'bulk.' Says Lisa Randall: "The ADD model contains a single brane on which the Standard Model particles are confined, but that brane does not bound space. It simply sits inside the extra curled-up dimensions." (*Warped Passages*, 364) As if it was not enough to try to visualize extra dimensions on a hypercube, with the main ADD model we have to visualize that our whole 4D universe cone (complete with big-bang, stars, galaxy clusters, and of course at least one intelligent civilization we know of) is just a tiny flat line running centrally through a cylinder whose surface is the hyperspace! (Variants of this ADD model present more extra dimensions of SPACE, all curled up.)

But professor Lisa Randall of Harvard University, considered one of the most promising and ground-breaking particle physics theorists at the moment, has elaborated with colleague Raman Sundrum in 1999 a forefront hyperdimensional theory called *warped geometry theory or* the *Randall-Sundrum (or RS) model*. It implies a 5-dimensional *warped geometry,* and proposes two versions, RS1 and RS2.

RS1 model is based on a five-dimensional bulk, extremely warped, with two branes: the Planckbrane (or Gravitybrane) in which gravity has a strong value, and the Tevbrane (or Weakbrane), our 4D world in which gravity is very weak. The RS1 model introduces a leap between the scales of the 4D brane and the *hyperspace bulk* of 16 orders of magnitude. On the Planckbrane, strings are 10^{-33} cm in size (the Planck length), and on the Tevbrane we are at 10^{-17} cm. In this model, the leap in scale is able to solve the '*hierarchy problem*,' that is, how to model the enormous difference between the tiny scale of gravity in the 4D world, compared to the other 3 forces. Here are the four forces relative strength: gravity is 1 (the hypothetical graviton), the weak force (W and Z bosons) is 10^{25}, the electromagnetic force (photons) is 10^{36}, and the strong force (gluons) is 10^{38}. So that 3 of the forces are clearly very near in terms of strength, while gravity stands alone at such a tiny scale that the graviton is actually still undetectable even by the LHC at CERN. For Randall and Sundrum, the Tevbrane is where the three forces (electromagnetism, weak and strong nuclear forces) are localized, whereas gravity is not constrained by it and will leak into the bulk, where it will take a much higher value.

In the RS2 model, there's only one brane to take into account because the other brane is supposed to be infinitely far away.

> Now, what's extremely interesting for the ISS model, is that the two branes have opposite energy: the Planckbrane has positive brane energy, and the Tevbrane has negative brane energy, and that's why spacetime is so curved as to be highly warped, but the warping occurs only in hyperspace (the bulk).

The Infinities Problem

Interestingly, the problems that have plagued the quasi-totality of unification theories that aim at including quantum physics or hyperspace, have arisen from meeting infinities as the sole solution to an equation. This has been the case, for example, in the Yang-Mills fields, in the GUT theories, Quantum field theories, perturbation theories, and even in the Standard Model. This recurrent problem has been solved (in order to get useful solutions to the diverse equations) only by the acrobatic process of 'renormalization,' or mathematical "corrections"—that, in effect, are nothing less than a tinkering to get rid of these infinite values. The case is specifically acute with the 'loops' that have to be added to account for all possible interactions between particles (called Feynman diagrams). Says Michio Kaku: "Quantum theory was meant to give tiny quantum corrections to Newtonian physics. But much to the horror of physicists, these quantum corrections or 'loop graphs,' instead of being small, were infinite. No matter how physicists tinkered with their equations or tried to disguise these infinite quantities, these divergences were persistently found in any calculation of quantum corrections." (*Hyperspace*, 118).

Nobel laureate Richard Feynman, for one, fell headfirst in what I've deemed to be "the Infinities Problem" when he and his colleagues developed the Quantum Electro-Dynamics theory (QED), that allowed a partial integration of Relativity with QM.

The conundrum was the electron mass. In Einstein's relativity, translating the electron's mass into pure EM energy meant to compress a cloud of negative charge into a tiny sphere with a radius. The first problem that arose was that the more compression, the greater the mass. (As all same negative charges repel each other, more work/energy was needed for a greater compression, and that translated into such a great mass as being implausible). The only way out (as often when theory doesn't conform to observation), was to start with the electron's known charge and mass: physicists could calculate this radius as if the electron was a tiny hollow sphere with the negative charge all over its surface, a model fitting the electron's (observed) scattering effect on light (photons).

But in QM, the equations all led to the electron being a point-particle, with a radius of zero. That demanded an infinite compression, and therefore an infinite mass. Feynman tried to get around the problem via the quantum vacuum, that, as we know, is filled with virtual and transitory particles. First, the negative electron would attract and be surrounded by a cloud of positive charge (from virtual positive particles). Second, a cloud of virtual negative charge would then surround the positive cloud. This process is called the polarization of the vacuum (into opposite charges). All this rendered the observed charge of the electron much weaker, since its high negative charge was partly neutralized by the polarized vacuum (and now came the concept of a 'bare charge,' that is, the real one when the vacuum effect is discounted). The QM treatment of the problem made the infinity problem only a bit weaker. Says John Brandenburg, whose argument about the electron's mass problem we have followed (*Beyond Einstein*, 118-9): "The quantum vacuum helped only slightly: *Instead of infinity, it produced a logarithm of infinity. A logarithm of infinity is still infinity*, so after several years of hands wringing, (...) the electron was simply assigned a much smaller radius than before and the matter was dropped with some embarrassment." A usual 'renormalization,' which means, according to Brandenburg, that "*at a small enough radius, new physics occurs* which negates the effect of smaller sizes. *It is literally subtracting an infinity from the first infinity* to force the answer for the electron mass to come out right." (My emphasis) He goes on: "Dirac commented that he '*had heard of numbers being neglected because they were small, but never when they were infinite*.'" Brandenburg concludes: "The fact that physics must change at some small size (…) has been found to be necessary in analysis of the strong and weak forces also, so it must be physical. *Somewhere in the deep subatomic scale, physics changes*." (My emphasis)

This conclusion of John Brandenburg on this specific problem with the infinities that physicists and cosmologists seem to meet at every corner, is for me and the ISS model, a theoretical keystone, a fundamental concept one has to stipulate as a premise to any theory of the pre-space and hyperspace. This keystone concept is that we do meet values (or parameters) tending to the infinite (1) at the sub-Planckian scale, that is, in the CSR hyperdimension unto the X-point of translation; and also (2) in the anti-ISS (the Big Crunch or Terminal Black hole) that will be our universe-bubble's translation into a novel universe.

My deep conviction about this pervasive problem concurs with that of John Brandenburg: We do meet *real or near infinities* in the hyperspace dimensions, and it does express a hard-core reality. We certainly have to circumvent or compensate them in dealing with the 4D matter universe, but they should be investigated in their own right, as openings on higher dimensions (both symbolically, and literally as gates), or, more precisely, as being the stuff of hyperdimensions. My stand is that in hyperspace models (such as the ISS) we should be able to integrate (near) infinite state-spaces in such a way that they function at the sub-Planckian scale. (We'll see a very interesting facet of the Infinities Problem further on, with the 'Trans-Planckian Problem' arising from Hawking's postulate that near the event-horizon of a black hole, a particle should have an infinite frequency and a sub-Planckian wavelength—a concept definitely resonant with the Infinite Spiral Staircase!

Faster-than-light in astronomy and propulsion systems

There are two sets of data in astronomy that point to faster-than-light processes. The first one is the explosive expansion of the primordial universe in the inflation phase, that started an infinitesimal fraction of a second after Planck's threshold, between 10^{-36} (or 10^{-34}) second, up to and 10^{-32} second. Most physicists agree that at this infinitesimal instant, the rate of expansion was 10^{50}, in other words, the early universe's size was multiplied by 10 followed by 50 zeros; and such a rate is occurring at billion times the light speed. It is not evident to have a TOE fitting this rate of expansion, and it remains a problem linked to the 'horizon problem,' the second set of data.

In a 2004 four months long observation that necessitated several hundred hours with the Hubble space telescope was focused (at each window of opportunities) on the same exact point near the Orion constellation. It led to the detection of a cluster of 10.000 young galaxies that had just condensed and stood 13 billions light-years from earth (thus they had formed 790 million light-years after the Big Bang). But in 2012-14, Hubble scientist Adi Zitrin discovered a galaxy that had formed only 420 million years after the Big Bang, called MACS0647-JD.

However, when we look at the night sky from our far left to our far right (in any direction), we observe in fact a time span that adds (at the very minimum, because it's actually about 46 billions years) 13.7 billion light-years from one side to 13.7 billion light-years the other side, so that it amounts to more than 27 billion light-years. The problem is: at the speed of light, information had no time yet to cover such a distance that is double the number of light years from our origin. So the physicists wonder how could information "travel" from one end of the universe to the other, at double the speed of light at the very minimum? The problem is even more acute if one considers that it has been calculated that 370.000 light-years after the Big Bang (where the CMB lingers), the distance from one side of the universe to the other was 92 billion light-years.

This feeds into the *horizon problem*, that consists (1) in explaining the relative uniformity of all clusters of galaxies and galaxies in the universe (to at least ten parts per million as the anisotropies of the

CMB were already revealed at a coarser grain by COBE); and (2) in finding how we end up with such a precise fine-tuning (the Anthropic principle) of about twenty variables and constants across our universe-bubble, while this transfer of information would have had to enormously exceed the speed of light.

> In my opinion, we have to refer ourselves to the EPR paradox, and the proven fact that two paired particles (issued from the same source), remain entangled and exchange their information in a faster-than-light way. It is of course particularly fit to explain this fine-tuning of all energy and matter systems in our universe-bubble since all originated in the cosmic primeval universe, the ISS.

Several physicists invoke a 'holographic' universe, based on a weak and random anthropic principle, such as Gerard 't Hooft and Susskind. Other physicists, such as Carter and Lovelock, argue for a 'strong anthropic principle' (from the Greek *anthropos*, human), that is, an underlying harmony rooted in the origin, that makes the universe to tend toward favoring intelligent life. And similarly Ervin Laszlo grounds the coherence and self-consistency of the whole universe, that perforce imply faster-than-light processes and beyond relativistic signal-transmission, on a holographic Akashic field.

The 'weak anthropic principle,' has been well defined by Hawking in his 2003 ArXiv article *Cosmology from the Top Down*: "Does string theory predict the state of the universe? The answer is that it does not. It allows a vast landscape of possible universes, in which we occupy an anthropologically permitted location."

> So that if we consider all possible universe-bubbles, only a tiny fraction of them are fit for life and intelligent life (the anthropos), because that demands the extremely precise fine-tuning of twenty to forty parameters and constants. But do rest assure, because this original tuning should persist all through the time span and the cosmological horizon of our bubble. Is not that in itself a cause of wonder that makes us question how, if randomness rules at large scale over a huge ensemble of universe-bubbles, could one of these ever contain such a tidy and user-friendly universe? Chaos theory, as we have seen, posits just the opposite type of organization, one that has been effectively demonstrated in numerous turbulent systems: global patterns of organization (the tornado spiral dynamics) whereas at small scales are stochastic processes (the

molecules forming the tornado display apparent randomness). ISST adheres to this chaotic self-organization, from the origin onward.

The ISST solution to the Horizon Problem rests in the Free sygons creating the HD bulk with faster-than-light torsion waves and a non-finite array of extremely high frequencies (EHF), that in effect were *networks of links* in the original CSR manifold. This happened long before the low-frequency sygons started to morph into the particles of the Standard Model, while crossing Higgs' field, and formed the very quantum-spacetime region, the QST, the horizon of which seems a problem.

Moreover, ISST also solves the fine-tuning problem, because the EHF Free sygons were issued from the early stages of the cosmic ISS unfolding, as bow-networks. And these sygons-networks (as a cosmic RNA, the messenger of DNA, transporting its information to in-form the proteins synthesis) had encoded in them the organizational imprint of numerous viable and resilient systems, including the very fine-tuning of variables that would favor the evolution of galactic systems, life, and intelligence—all systems that had been evolved and tested by previous universe-bubbles on our own collar of such universe-bubbles.

FTL, space-faring ETs, and research in electrogravitics

In a 2005 article in the Journal *of the British Interplanetary Society*, J. Deardorff, B. Haisch, B. Maccabee, and H.E. Puthoff, state that

"The possibility of reduced-time interstellar travel by advanced extraterrestrial (ET) civilisations is not, as naive consideration might hold, fundamentally ruled out by presently known physical principles. ET knowledge of the physical universe may comprise new principles which allow some form of *FTL travel*. This possibility is to be taken seriously, since *the average age of suitable stars within the 'galactic habitable zone', in which the Earth also resides, is found to be about 109 years older than the sun,* suggesting the possibility of civilizations extremely advanced beyond our own." The authors add that their Extra-Terrestrial Hypothesis (ETH) approach "explores the likelihood that 'we actually do belong to a large [galactic at least] civilisation but are unaware of that fact.'" They base their argument about FTL travel on recent breakthroughs in terms of interstellar travels and propulsion discussed in mainstream science, notably:

(1) The possibility of *'traversable wormholes'* that imply cutting like a subway through the spacetime metric (proposed by K. Thorne and Carl Sagan in the late 1980s), and (2) the *'Alcubierre Drive'* (in the mid 1990s). Say the authors: "There is no limit to the speed at which space itself might stretch. Faster than light (FTL) relative motion is part of inflation theory, and presumably the universe beyond the Hubble distance is receding from us faster than *c*. It was shown that a spaceship contained in a volume of Minkowski space could in principle make use of FTL expansion of spacetime behind and a similar contraction in front, with the inconvenience of time dilation and untoward accelerations being overcome." Also, (3), the *'Krasnikov tube'* connecting spatially remote locales. And finally (4), the multiverse and added dimensions (e.g. in M-brane and superstring theory) could allow to tinker with space and penetrate parallel universes (in which C would be different).

Now, here is a stupendous idea: *a polarized vacuum expresses a specific metric*, that offers electric and gravitic processes of a type unrecognized by Relativity (this was originally modeled by Dicke, then Puthoff, and colleagues). Say the authors of the 2005 article: It "would open the possibility of a different type of metric engineering in which the *dielectric properties of the vacuum* might be altered in such a way as to raise the local propagation velocity of light."

Engineering of FTL crafts in electrogravitics

Paul LaViolette developed a theory of SubQuantum Kinetics (SQK), and elaborated on the previous work of Townsend Brown's Electro-Kinetic thrusters for FTL propulsion, back in 1921. It is based on the concept that *the ether (as a medium) is distinct from EM and gravity fields*, and it acts itself (being polarized, that is, charged) as a pressure force with a given negative charge, thus attracting or repulsing the electric or magnetic fields it interacts with. Says LaViolette in his talk at Muffon 2011: "In SubQuantum ether Kinetics, the [ether] field is not bound to the charges, but it can act on the charges, so it puts more pressure on the negative [charges that rushes] to the positive [anode at the front], so the craft moves."[39] In other words, at the back of the craft, where stands the negative cathode, the (negatively charged) ether exerts an "enormous repulsive pressure on the condensed negative cloud of charges around the cathode; this gives thrust to the electrons or negative ions to rush at a greater speed toward the anode at the front, where the cloud of positive charge is more extent, thus adding more imbalance and thus motion. LaViolette then explains how works the B2 developed after Brown's engineering designs: "the positive charges are spread in an umbrella shape in front, and the negative ones are compacted at the rear. The result is an unbalanced force pushing the B2 forward." Interestingly, in SQK, the more the speed, the more the thrust forward.

During an experiment with Eugene Podkletnov on the latter's invention, the *Gravity Impulse Beam generator* (that generates gravity pulses as discharges of very high voltage), LaViolette was able to corroborate that the gravity impulse continues to propagate beyond the anode at the front, as a very coherent beam at the superluminal speed of *64 C* (sixty four times the speed of light). This measurement was at 61 centimeters ahead of the anode, and the beam was still coherent 3 meters ahead. (See LaViolette's book *Secrets of antigravity*, and his Muffon 2011 talk on YouTube.[39])

The deep interconnection between electromagnetic (EM) fields and gravity—that is at the foundation of John Brandenburg's GEM theory (Gravity-Electro-Magnetic unification theory)—is highlighted by an applied technical and experimental research, especially the one focused on antigravity spaceships research. Let's understand first the deep interconnection between static pressure (EM), and gravity pressure, clearly apparent at the scale of our rocky planet Earth. The sum of the thermodynamic or static (EM) pressure (created by the weight of rocks and magma) and of the gravity pressure is always constant, whatever the surface or depth of the measurement inside Earth. At the earth's surface, the gravity field's pressure is millions of atmospheres (we feel it only as our own weight), while the rock pressure is zero. At the center of earth, it's the opposite: having half the weight of earth above us, the thermodynamic (or static EM) pressure is millions of atmospheres, while the gravity pressure is zero. Thus gravity and EM forces appear as perfectly complementary, the gravity field pressure being the force that impedes earth from exploding due to the high pressure in its molten core. (The fact that the intensity of gravity fields was calculated in thermodynamic (EM) pressure units, is what led Sakharov to infer that gravity is equivalent to a radiation pressure of the ZPF, the Zero Point Fluctuations.)

Now, Newton's laws describe gravity as a force *between* celestial bodies (and their mass is calculated as if it were concentrated at their centers). But gravity, as we just saw, is also a force acting on our planet and within it. However, as John Brandenburg (2011, 245) remarks "Force is created by pressure imbalance, not pressure itself. (...) Pressure moves things when imbalances occur."

He explains in "*Beyond Einstein's Unified Field*" that GEM theory brings the equation of an antigravity effect called the "*Vacuum Bernoulli Effect*." In classical aerodynamics, the 'Bernoulli effect' is what creates, on the top of the curved surface of an aircraft wing, a faster airflow and less pressure; below the wings, to the opposite, the airflow is slower and the pressure higher. Moreover, the difference in the speed of flow between the two wing surfaces (above and below) "creates a vortex flow pattern around the wing as it flies." All aircraft designs are conceived to use this vortex flow. Thus the higher air pressure pushing on the below surface (as a vertical upward push) has a lifting effect of the wings. And the two flows (vortex motion and faster airflow on the top of the wing) add to each other to increase the speed of the aircraft. In the hovercraft aerodynamics, the vortex principle is worked out so that the high pressure is confined under the craft in a 'vortex ring,' a rotating torus that makes the stable cushion of air between the water surface and the hovercraft.

Now, Brandenburg's Vacuum Bernoulli Effect is produced by the gravity field pressure, not by the aerodynamic flow of air. In this Vacuum Bernoulli Effect, the relation that remains constant is the difference between EM static pressure and dynamic radiation pressure. John Brandenburg states conclusively: "It is as if the fluid aether has negative mass density, so that its dynamic pressure when it moves is negative."

The design for an 'antigravity' craft envisioned by Brandenburg is to create a "Tesla vortex" underneath the disk, using Tesla's three-phase power to create a vortex (a spin) of Poynting flow and, in essence, "such an electromagnetically propelled craft (...) should be called an 'emfoil' because it shapes EM flow to create a lift" and furthermore, "it bends spacetime itself" and both its mass and its inertia will be so reduced as rendering the craft nearly weightless. He concludes that "the most obvious way to accomplish FTL for human spaceflight is to modify spacetime geometry around a ship in space so that the rest of the cosmos thinks the ship is a tachyon" that is, a faster than light particle that Brandenburg fathoms to be the essence (or 'quintessence' ?) of dark energy. (*Beyond Einstein*, 248-49, and 253-54)

Jack Sarfatti, in his 2006 mind-boggling book *Super Cosmos: Through struggles to the stars,* that tackles the most far-reaching research and concepts in physics, has inserted a chapter on 'UFO phenomenology' developing a novel type of propulsion. Sarfatti calls it "zero point pressure propulsion" and states that it can explain astonishing spacetime anomalies in observed crafts motion, such as an inverse Doppler effect (in normal motion, a shift of frequency toward the blue in the direction of motion, toward the red in the opposite direction). This propulsion is a way to tap into the vacuum's quantum zero point and use the opposite radiation pressures produced by dark energy (negative and with anti-gravity effect) and dark matter (positive and having a gravity effect). Sarfatti states that dark matter generates a positive pressure that causes a gravity redshift at the bow of the craft (i.e. in the direction of motion), whereas dark energy generates a negative pressure that causes a gravity blueshift at the back of the craft, or in the direction opposite to motion).

Now, in the ISS theory, of course we have communication and exchange of information through the 3 metadimensions (CSR), in a way that is necessarily faster-than-light, because it doesn't happen in the time dimension at all. In the cosmic ISS (preceding Planck's wall) we are prior to the birth of space and of time; and in the particle's ISS, we are in the compact and curled up hyperdimension of the CSR, similar (isomorphic) to the cosmic ISS, just like the facet of a hologram, and (in this hyperspace) forming a quasi continuity with the cosmic ISS.

Holographic Principle and Black Hole's Information Paradox

Hawking had modeled the black hole (BH) event horizon as a boundary (a surface) structured as a "light-like geodesics." He stated that the number of geodesics always increases, especially through collision with neighboring BH and their geodesics, so that the horizon surface always increases; this became the second law of BH thermodynamics. In 1974, Stephen Hawking predicted that black holes should emit a black body radiation that came to be called Hawking radiation; Jacob Bekenstein extended this concept and calculated that BH have a finite temperature and entropy. At first, black holes were supposed to have no entropy—just as all EM fields; but Bekenstein showed that if a gas with high entropy (disorder) would fall in a BH, this entropy would be lost, in contradiction with the second law of thermodynamics. The problem became known as the *Black Hole Information Paradox* because it moreover clashed with the Laplacian determinism in science, that stipulates that the state of a system (thus the information about it) always fully determines its future states; a classical physics tenet that found its way into QM in the sense that the total information about a system is encoded in its wavefunction until it collapses, and is still preserved afterwards in its operator, thus allowing to calculate the past state of a system. Bekenstein then modeled BH (in a thermodynamics framework) as random objects with such a huge entropy that it would dwarf the gas' entropy. Already Charles Thorn in 1978, and then Raphael Bousso, had postulated that in a lower-dimensional framework such as a 2D surface (allowed and contained in string theory), gravity would emerge as a discrete holographic and geodesic-like structure.

According to the Wikipedia article (wiki/Holographic_principle), "*The holographic principle* states that the entropy of *ordinary mass* (not just black holes) is also proportional to surface area and not volume; that volume itself is illusory and *the universe is really a hologram which is isomorphic to the information "inscribed" on the surface of its boundary*."

But if a BH really had entropy, pondered Hawking, then it would have to radiate: beyond absorbing photons, it would have to also emit and exchange photons with surrounding gases. Hawking had set himself to show that a BH couldn't radiate; however, he found just the opposite, and, elaborating on Bekenstein's thermodynamics formula, he came up with the ratio that the *entropy of a BH* is one quarter of its horizon area (using Planck length).

According to classical thermodynamics, when heat is increased in a system, its entropy *S* increases proportionally to the increase in its mass-energy divided by the temperature. In brief: the second law of thermodynamics is understood to imply that a system will tend to stabilize in time in its equilibrium state, in a state of maximum entropy or disorder. Yet, interestingly, Boltzmann original equations for thermodynamics were taking care of all possible 'microstates' of the system (e.g. position and momentum of all gas molecules, instead of a global macrostate). Moreover, the most interesting point for us is that *Boltzmann original laws were reversible*: A system, even isolated, could either lose or gain temperature and entropy; and if it lost entropy then it gained order. "The *change* in entropy (ΔS) was originally defined for a thermodynamically reversible process," states the Wikipedia article Entropy.[40] *The concept of an* ***irreversible increase in entropy*** *and disorder in the universe (and predicting its unavoidable ultimate death by entropy) was something added later in the equations,* and then heralded as a scientific paradigm.

But to get back to black holes, the fascinating point about BH thermodynamics is that the maximal entropy was found to be related to the square of the radius and not to the radius cubed—in other words, the information (as entropy) about a system's entropy should be contained on the BH surface (its event horizon), and not in its volume. The entropy is proportional to the logarithm of the number of microstates, and for a BH, the number of states is proportional to the surface of the event horizon.

At this point Nobel laureate Gerard 't Hooft came up with the Holographic Principle and soon after Leonard Susskind modeled this principle within string theory. Using both String and Quantum theories, the Holographic Principle states that "the description of a volume of space can be thought of as encoded on a boundary to the region—preferably a light-like boundary like a gravitational horizon."[41] In other words, the information about a black hole (able to describe all the states of the system, i.e. its state-space), is entirely inscribed on the surface of its event-horizon.

At one point, the holographic principle was extended to a cosmology of a holographic universe, in which all information about this universe would be inscribed on the 2D surface of its cosmological horizon (the diameter of the observable universe is 92 billion light-years, or 92Gly, although some scientists put it at 93Gly).

The two-way flow emitted and absorbed by black holes

Now comes a series of extremely puzzling discoveries along these lines, and you'll understand why I have taken the risk of overflowing my text with details and possibly bore you, my readers. The implications that will follow are of a rare value for us all (and for the ISS model also), because they open stupendous new perspectives.

During his lengthy study, Hawking realized that the radiation emitted by black holes is not correlated to the radiation/matter that they absorb, so that it was very unlikely that the outward radiation would be caused by the scattering of matter that had previously been absorbed. First, there was a location mismatch: the outgoing rays (Hawking radiation) sprang from the edge of the event horizon and circled there a long while; and second, there was a time mismatch: at the time the outward light-rays were darting off (the outgoing radiation), the matter that would be swallowed (the infalling one) was still very far from reaching there. In brief: it seemed that "the infalling and outgoing mass/energy only interact when they cross."[41] Hawking concluded that black holes absorb photons as an integral system-state (the wave-function, as a state-space), whereas what they emit or re-emit much later are different systems of photons in mixed states (precisely "a thermal mixed state described by a density matrix"[41]). In other words, they absorb quantum systems and emit complex intermingled fields, possibly interference patterns. The problem, thought Hawking, was that this possibility (of mixed states) is not allowed by QM. In QM, the collapse of a wave-function gives only one of the superposed states being selected and appearing clearly now in the matter (particle) domain—and having thus lost its wave domain superposed state.

Now, isn't the presence of 'intermingled fields' interesting regarding the X-funnel? The translation of all matter in a previous universe into the CSR manifold gives variegated and intermingled networks of bows that will in-form the birth of numerous viable systems in a new universe. And also pertinent is the fact that the matter-systems in our spacetime send sygons with their snapshot information back to the source (the cosmic ISS), thus setting the two-way constant connection between all systems in QST domain and the Cosmic ISS.

Gerard 't Hooft found a way to model the emitted and infalling matter-energy of black holes, by blending QM (that posed 'quantum fluctuations' on the event-horizon surface) with a particular string theory called 'world-sheet.' Some string theories have used—instead of strings (filaments)—branes (rubber bands) and a model based on superposed 2D sheets (surfaces).

Then Leonard Susskind, who later developed M-Theory and the 'anthropic landscape' of string theory, stated that the use of world-sheet theory ("with long highly excited string states") would give a sound holographic model of black holes, and that the event horizon's quantum oscillations could perfectly well account for the two flows.

But let's see in more details what Gerard 't Hooft really found. Pondering the problem of the two incoming and outflowing matter-energy flows, 't Hooft saw a way in which these flows could interact. The incoming matter-particles of course were accelerated when approaching the BH; but he found that, through their gravitational fields, they were actually deforming the event horizon, in such a way that when exiting, the outgoing Hawking radiation would be modified. As it turned out, tiny bulges were showing on the event horizon's quantum fluctuations, that were the traces of photons either radiating from the BH or emitted by the BH. Says the article "When a particle falls into a black hole, it is boosted relative to an outside observer, and its gravitational field *assumes a universal form*. 't Hooft showed that *this field makes a logarithmic tent-pole shaped bump* on the horizon of a black hole, and like a shadow, the bump is an alternate *description* of the particle's location and mass."[41]

In other words, not only does the information (the descriptors of its states) about matter-energy infalling or exiting the black hole is inscribed on the event-horizon, but moreover, *the incoming energy takes on a conic tepee shape*! But that's not all, for 't Hooft realized that the deformation of the BH surface was very similar to the string theory's 'world-sheet' model for the absorption and emission of particles, and could thus be modeled using world-sheets (superposed layers of string surfaces).

Now this 'world-sheet' model is extremely significant since, in the ISS theory, the spiral staircase is first a conic spiral shape of discrete steps, in which is embedded another cone presenting both an inverted rotation and an inverted global direction (the inner cone

> moving inward toward the X-point of origin; and the outer cone moving outward toward the Planck length and the 4D world). So that we have a the two-way flow emitted and absorbed by the ISS, and both are not identical or even using similar frequencies. Furthermore, the steps (and the Phi logarithm that sets their incrementing or decreasing radii) give the measure of the bow-frequencies; in fact, given they are discrete and connected only via their small inner angle along a vertical inner spiral, each flat upper surface of the step of the staircase can be viewed as a 'sheet' (just like the flat tops of each step in a helicoidal staircase). However, if you remember, I was drawn to give some 'depth' to the ISS steps, however small, so that frequencies could vibrate as the pre-strings they are. Thus, it appears more and more that there are very deep connections between the ISS model and diverse types of string theory, so much so that it seems possible to model the ISS using string theory.

Andrew Strominger and Cumrun Vafa, from Harvard University, made a breakthrough when they used superstring theory to solve the Bekenstein-Hawking entropy. The fascinating model they propose consists of a 5-brane, on which 'open strings' are traveling along a 1-brane filament. The open strings are exchanged with the black hole (itself 1D) as 'closed strings' in a two-way soliton wave, ejected or absorbed by the 5-brane, 5D torus. Vafa also developed a blending of geometry with quantum field theory (F-Theory) and tackled 'topological string theory.'

Information and gravity

The physicist John Wheeler was the first one to stipulate that all matter, at an underlying level, is an informational substrate, that he envisioned as being essentially made of bits—the binary 0 and 1 language of computers. He called this concept 'It from bit' and explained in his 1990 article: "'It from bit' symbolizes the idea that every item of the physical world has at bottom—a very deep bottom, in most instances—an immaterial source and explanation; that which we call reality arises in the last analysis from the posing of yes-or-no questions and the registering of equipment-evoked responses; in short, that *all things physical are information-theoretic in origin and that this is a participatory universe*." (My emphasis.) Jacob Bekenstein

(2003a), one of his brilliant doctoral students, echoes this stand by saying that scientists should "regard the physical world as made of information, with energy and matter as incidentals."

It was in 1948 that the mathematician Shannon postulated his famous concept of information related to entropy, that used the laws of thermodynamics to model the transmission of information in bits. The law being that the more uncertain, random, and out of ordinary the event, the more information it conveys; whereas a redundant or always identical event brings no information. Of course, Shannon's entropy is still widely used for the conception of all transmission and IT technologies; however, as a cognitive scientist myself, and expressing here the general view of this field, we are convinced that Shannon's "information" had everything to do with engineering and computer sciences, and nothing to do with meaningful information, or with significant exchanges, or even with a learning cognitive process. It was far from covering the "active information" that David Bohm, Ervin Laszlo, and myself (with semantic energy) are addressing and modeling. The great difference is that when a human being attributes meaning (via thoughts, feelings, relating, acting, etc.), this process creates organization, order, variety, distinctions, and a meaning-laden social world and natural environment. And this is pure negentropy (as a creation of organization) as opposed to entropy, which is the increase of disorder and uniformity (the uniformized-temperature gas is at maximum entropy: all the molecules are in the same ground state).

Bekenstein solves the conundrum by going back to Boltzmann's original formula that was based on the number of *distinct microstates* within a system being the measure of its entropy; and he makes a parallel with the quantity of information (in bits) necessary to describe all features in a system. In this perspective, we don't look at disorder versus organization, but rather at a measure of complexity and differentiation within a system.

The Black Hole's Firewall Paradox

In the framework of a universe structured like a hologram, all information about a spacetime system is imprinted on its boundary surface; this includes of course the information about matter-energy that had fallen into a BH and indeed the whole information about the universe itself would be encoded on its cosmological horizon. Maldacena, about the BH Information paradox, states that with such a framework "there is no room for information loss."

If you remember, in 1974 Hawking had hypothesized that any radiation coming out of a black hole should be random, and consequently that all information about infalling matter-energy would be lost. But in 2004, Hawking conceded he had been wrong in advocating that no information can be retrieved from matter-energy that has fallen in a BH (and paid the price of his lost bet to John Preskill of Caltech).

In 2013, the "Firewall paradox" ignited the physics community, brought about by Dr. Polchinski of the Kavli Institute and three colleagues from the UC-Santa Barbara, California (namely Ahmed Almheiri, Donald Marolf, and James Sully)—known as the AMPS team, by their initials. The AMPS team, in their February 2013 paper ("Black holes: complementarity or firewalls?"[42]), had calculated that the QM framework made it impossible to view the event horizon as a smooth surface, and that, to the opposite, one should expect a huge discontinuity in the vacuum fluctuations, that would in effect create a sort of "firewall" of high-energy particles, one that would 'fry' any transiting matter-energy, including the poor cosmonaut who would have fallen into the BH (while physicists up to then had thought, based on relativity theory, that he wouldn't even notice he was crossing the event horizon to his own death by the increasing crushing gravity). (*J. of High Energy Physics* 2013(2). arXiv:1207.3123.)

The argument was set as concerning two entangled particles (in a single burst of Hawking radiation)—one quantum in the outgoing radiation, and the other quantum (or particle) falling into the BH. Susskind had proposed that the outgoing particle must be entangled with all previously emitted Hawking radiation, and with the infalling particle. However, in the theory of entanglement, a particle can only be entangled with its paired particle, and that excludes entanglement with other particles. Three possible solutions were aired: (1) the entanglement is instantly broken between the infalling and outgoing particles, and that releases such an immense energy as to create the firewall itself. (2) There's no entanglement between the two particles, therefore there's an information loss. (3) There's a gradual loosening of the entanglement, and thus the energy release, more gradual, doesn't create a firewall.

Raphael Bousso (of UC-Berkeley) declared that the Firewall theory, if correct, would present to physicists a "menu from hell" with, as sole dishes to select from, the obligation to either discard the equivalence principle in Relativity Theory, as demanded by solution 1, or discard the QM Unitarity principle (the base of a theory's consistency and the possibility to infer the past state of any system), as demanded by solution 2), or else to correct the quantum field theory in order to allow for a gradual energy release, as necessary in solution 3. As for Polchinski, he felt no qualms about Relativity: "We've known for years that spacetime is not fundamental. Spacetime is emergent. Gravity is emergent. Maybe sometimes it doesn't always emerge."[43] As he noted succinctly in his 2011 seminar at the University of California Davis: "Space, and perhaps time, are probably emergent, not fundamental. (...) The idea that physics must (be) holographic, that *the fundamental variables are nonlocal* and locality is emergent, is a complete change from previous experience, and we have much to understand."[44]

In his ArXiv 2014 article titled "Information preservation and weather forecasting for black holes," Hawking now discards the term 'black hole' and suggests instead to use the term "grey hole." The grey hole would have much more of a transient structure, as opposed to a stable one. The grey holes' boundaries would only be "apparent horizons" and the matter and light composing them would be only temporary. This would allow an escape of energy and information from the grey hole.

The implications of the Grey Hole hypothesis, is that a quantum could escape a GH from any point of the 'apparent horizon,' and likewise, information could be released from any point. However the information would end up in such a randomized and scrambled state that, in a practical sense, it would be as impossible to decode it as to forecast weather. But many scientists, relying on one of quantum theory's tenets, believe that information can't be lost.

The enormous stake of the Information and Firewall paradoxes are well described by Raphael Bousso: "The firewall paradox tells us that the conceptual cost of getting information back out of a black hole is even more revolutionary than most of us had believed."[43] In other words, only a revolutionary theory could solve the problem of the universe's past information release, and that of the emergence of spacetime and gravity. I would add that in ISST the Big Bang is nothing but a universe-scale white hole issued from the previous universe final black hole.

Micro black holes and the trans-Planckian problem

Micro black holes (MBH) have been predicted to exist due to the fact that black holes lose mass and energy through the Hawking radiation they emit. MBH are allowed and posited in a large variety of models, including the ones that posit (following Hawking), an array of them right at the origin, called primordial black holes.

Yakov Zel'dovich, working on the vacuum in synergy with Sakharov in the USSR, was the first to postulate (after WWII), that micro black holes of opposite charges would form in the vacuum at the origin, at the exact Planck length and would immediately annihilate each other (according to the particle-antiparticle annihilation posed by Dirac). This would create an immense turbulence in the ZPF and turn it into a foam. These MBH would *tunnel out of their event horizon*. Zel'dovich understood that the Planck length was a cut-off, a threshold before which no quantum particle or EM wave could exist. The annihilation concept could get rid of the infinities before Planck cut-off, but not of the excess of matter over antimatter in our universe bubble at least, that allows us to exist on a planet.

MBH would emit a much larger Hawking radiation than large BH. The CERN's Large Hadron Collider (LHC) has been trying to detect these MBH through their Hawking radiation. And the Fermi space telescope, launched by NASA in 2008, is searching the skies for these gamma rays emitted from primordial BH, that would be nearing the end of their process of evaporation.

Wavelengths smaller than Planck length: The Trans-Planckian problem

It has been calculated by Hawking that near a black hole horizon, particles should have wavelengths smaller than the Planck length, that is, they should cross over the Planck constant toward infinite smallness—whereas no particle is allowed before the first quantum of energy, the wavelength of which is the Planck length. Hence its name:

the Trans-Planckian problem. Says the Wikipedia article on Hawking Radiation:[45] "A particle emitted from a black hole with a finite frequency, if traced back to the horizon, must have had an infinite frequency there and a trans-Planckian wavelength." In fact, Hawking posed that the same problem applies to photons outgoing of black holes and matter-energy falling into a white hole (both radiations being in opposite directions to the main stream of matter-energy swallowed by the sink in case of a black hole, or ejected in case of a white hole). As Hawking modeled both singularities as if time was stopped at the horizon, but with a finite beginning, all BH-outgoing radiation has its mathematical source in a singular point, and all WH-infalling radiations have no future (inside the singularity) but will contract in a singular point also. From the viewpoint of an outside observer, both would exhibit singularity not only in gravity but in frequencies also. If we trace back in time the BH-outgoing photon: "as it gets closer to the horizon, (... the latter) requires the wavelength of the photon to 'scrunch up' infinitely at the horizon of the black hole."

This 'scrunching up' of the wavelength of a particle below Planck scale and toward the infinity is of course a capstone of the Spiral Staircase dynamical topology, but the ISST proposes in addition that it does so following the logarithm of Phi in integer numbers (thus modeled as vibrating pre-strings. The sub-Planckian particles are the FTL sygons, that will become particles (strings) only after tunneling through the Higgs field/ vacuum.

The very interesting point (for the ISST) in Hawking's modeling, is the two unusual directions of particles (opposite to the main flow of matter), as outgoing of a black hole or infalling in a white hole. Both unusual trajectories demand that these particles' wavelengths be not only smaller than the Planck length, but also tending toward the infinite.

9
ISS: SELF-ORGANIZED CONSCIOUSNESS AT THE ORIGIN

"It will remain remarkable, in what ever way our future concepts may develop, that the very study of the external world led to the scientific conclusion that the content of the consciousness is the ultimate universal reality." Eugene Wigner, Nobel Prize in Physics.

In this chapter, I'll develop the cosmological stand of the ISS theory, but let me first summarize the model currently held as the most likely, concerning the emergence of the matter-universe, that of the 'symmetry-breaking'.

Symmetry-breaking versus ISS scenario

Most physicists agree that at the Planck scale happens a *phase transition*: reaching the size of the first quantum of energy, the universe will now turn into the energy-matter universe of the Standard Model. At first starts the *'radiation-dominated era,'* with energy particles such as the gauge bosons, and then light quarks, electrons, photons, and the heavy W and Z particles. An infinitesimal fraction of a second later, at 10^{-36} second (or else 10^{-34} to 10^{-32} second), the size of the primordial universe is multiplied 10^{50} times, setting the *inflation phase*, or the big-bang itself. Several physicists emphasize that this gigantic sudden expansion of the universe displayed a billion times faster-than-light speed.

At the first second the decoupling of neutrinos from the quark-gluon plasma happens, that will produce the cosmic neutrino background. An immense time later, within the second minute, the radiation era ends (at 10^2 s) and the *matter era* starts: quarks assemble by three to form protons, the nuclei of atoms. Between 10^2 and 10^{13} seconds (for some physicists) starts the decoupling of matter from radiation, that allows the photons to now stream freely with an enormous surge of energy, this rendering the universe luminous; its 'afterglow image' or relic radiation is the Cosmic Microwave Background (CMB) that we have mapped at about 370,000 years after the Big Bang.

During the *nucleosynthesis* phase started within the second minute (also at 10^2 s), the magical first couple electron-proton forms the hydrogen atom, and an immense number of them; a part of them fusions into helium (4 protons, 2 electrons). The first twenty minutes see the *formation of atoms* that make the majority of the atoms that *are still with us* at the present time. Then the matter universe will be born, unfolding in ever greater complexity of atoms, and then structures, leading to the formation of galaxies hardly 420 million years after the Big Bang. The Planck time, at 10^{-43} second, is also the beginning of the Grand Unification epoch (or GUT epoch): all forces are unified and symmetry reigns; matter particles (such as electrons) and antimatter particles (such as positrons, the positive antiparticles of the negative electrons) are in equal numbers (and in some models are supposed to annihilate each other), until 10^{-34} second when a *symmetry-breaking* occurs, as highlighted by Grand Unified Theories (GUT). At 10^2 seconds, according to Alan Guth, whose input was seminal in elaborating the Inflationary Universe theory, most antimatter particles (such as positrons) have disappeared, leaving the matter universe to unfold—thus favoring matter (over antimatter) and our universe hospitable to the formation of galaxies and to life.

However, there are two well-known (and hotly debated) problems with the currently accepted symmetry-breaking scenario. The first one is that we still don't know what made the *baryogenesis*, that is, the excess of baryons (matter particles) that had set the matter universe. According to Paul Dirac, the discoverer of antimatter, both matter and antimatter particles should have been produced in equal numbers at the origin, and since he posited the law that, as soon as they meet, particles and their paired antiparticles cannot not annihilate each other, in the end positrons and electrons should have all disappeared. Yet this is not the case and we are left with an immense number of electrons (compared to their antiparticles or positrons) that are going to be essential in the formation of atoms and in electromagnetic fields. And the second problem is that there should be a great number of monopoles (imagine magnetic rods with just one polar charge) and these haven't been found yet.

ISS scenario for a matter-friendly universe

From the immaterial Center-Syg-Rhythm hyperdimension at the point-scale (or Bindu scale), the universe will develop, after Planck length, via the medium and large sygons, into an *apparently* matter-dominated or at least matter-friendly universe. (Let's remember that baryonic matter—all stars and the 4D matter around us—makes only less than 5% of all energy of the universe.) Meanwhile, the first Free sygons, of extremely high frequency, to have been issued after the X-point of origin from the unfolding Infinite Spiral Staircase, will create the hyperdimensional bulk that will enclose our spacetime region.

Large sygons becoming particles in spacetime domain

Let's review what we have seen of the early phase of the ISS scenario. From the origin, the Infinite Spiral Staircase (the ISS) kept forming itself and growing, in a topological hyperspace, driven by the Phi ratio, the natural logarithm that incremented the radius of the golden spiral in sequence by multiples of Phi; conjointly the number Pi would instantly, for each new radius, create a quarter circle drawing the next 'bow' delimiting the large end of a new step of the spiral staircase; *each bow or step of the ISS has a specific and unique frequency, that sets the virtual string-particle (sub-Planckian) called a sygon.* While the whole cosmic ISS unfolds its spiral in a blinding instant, it creates its own hyper-space and hyper-time (that we could also call pre-space and pre-time). The hyperspace is a spiraling expanding vortex, thus a discrete suite of radii and circles; however the cosmic ISS contains a double rotation with two embedded spirals, one inward and one outward, thus setting two opposite charges and directions of flow. The hypertime (Metadim Rhythm) is the rhythmic and complex vibration of the whole spiral, due to the frequencies of all bows, and the cyclical rotation (angular momentum) on a quarter of circle and on a virtual full circle. Thus this hypertime (before our spacetime sprang to existence at Planck length) is a cyclical, yet evolving and expanding, pretime. (It concurs somehow with the myths about a cyclical time at the origin, but doesn't regard it as an 'eternal return,' as if the universe was on a loop time.)

Now, the sygon is a complex entity, that has the shape of an arc (the bow of ¼ circle) and is also a frequency as a vibrating string. However, it is set in the virtual full circle (drawn from the quarter circle of this spiral's bow). It is still unclear for me if the sygon as a virtual string should be considered as having the shape of an arc + a bow-frequency (thus computing the sygons as quarter-circles virtual strings), or if they should better be computed as virtual circles or virtual 'closed strings,' containing 4 bows and 4 wavelength nodes and vibrating segments (thus as a square in a circle), and each vibrating at a unique frequency four times the frequency of the quarter of circle. However, it seems that the torsion waves and the creation of loops in the vacuum would favor a closed circle-string. But most importantly, the sygon is alive information and semantic energy and dynamics.

The important point is that when the sygons were still attached to the cosmic ISS, as quarter-circles on the unfolding spiral, they had formed *networks of links* that had memorized the informations and systems organization from a parent universe, and thus coded for specific systems and beings. These sygons-networks will retain their systemic information, links, and *organizational closure* (what makes their identity as a system) even after the large sygons becoming particles in our 4D manifold. For example, the sygons-networks can code and organize specific matter-systems and bio-systems, but nevertheless, as they are self-organized, they will adapt, evolve, and transform themselves.

Metadim Center-Circle as triggering self-organization

I should stress again here why, in systems theory, the concept of *organizational closure* is of paramount importance; it is what sets the identity and boundary of a complex system, its relative autonomy compared to other systems, even those who are linked to it or coupled with it. This is also what triggers the self-organization of a complex system, its ability to modify itself internally and to adapt to a changing environment. So that when we deal with biosystems and intelligent complex systems, such as minds or a cultural group, the identity of a system is what allows self-reference or self-reflection, the 1st person perspective. And this self-organization dynamics—a basic tenet of systems theory, chaos theory, Varela's auto-poietic systems, and my own semantic fields theory—, is what allows emergent processes and intentional changes. The dynamics of both organizational closure and self-organization have definitely to be integrated in a deep theory of physics that intends to model consciousness as a hyperdimension of the universe that is blended and fusioned with matter-energy.

This is why metadim Center-Circle is such an essential dimension, with dynamics that effect changes and that works all the way within matter-systems: it sets the self-reflection of the boundary (the circle) over its own center (the identity or self), and allows the self to enact and know its wholeness (its system's semantic field). Without Metadim C, the mind would not know its body, nor its own mental potentials, and an animal would not be able to react as a whole body. Metadim Center is the dynamics of the soma-significance, a concept elaborated by physicist David Bohm and to which Josephson and Pallikari refer, as a biosystem having a sense of one's own body and self.

Metadim Center is an example of how we can introduce in the core of physics a quasi-mental dynamics that accounts for the existence in our universe of self-referent biosystems such as minds or complex ecological systems.

Large Sygons at post-Planck scale

Each bow, as soon as it is formed, emits a sygon that has its own specific frequency—a virtual particle/string. Sygons are ejected with enormous strength as torsion waves forming a large cone open in the direction of flow of the spiral staircase (its quasi Poynting drift). However the sygons' velocity is immensely higher that the ISS. And while the ISS quickly enlarges, the sygons ejected have a progressively greater wavelength and lower frequency. Sygons with medium and low frequency (and large wavelength) will create the Higgs field (we'll see how shortly) and as the field will become more and more dense, the sygons will now have to cross it or to tunnel their way through it.

But the Planck length happens to be a stupendous gate indeed for the sygons, each bearing the information of a melodic group of bows, of networks of links back in the pre-space.

As if the whole ISS were a hologram, most sygons, while crossing the Higgs field at or just above Planck length, become wrapped in a cloud of charges and acquire a relative mass; these sygons become matter-energy, that is, immensely bigger strings, the wave-particles of the standard model that are populating our 4D spacetime universe. However, the ISS sygons themselves now form the (curled up) core of all particles and atoms, their compact 5th dimension. All particles and atoms retain in their core, within the curled up sygons, the replica of our cosmic DNA, with the information of the whole universe-bubble inscribed into it. In all particles, this compact 5th dimension opens at the Planck length as a hyperdimension—the triune Center-Syg-Rhythm manifold, that is, prespace-consciousness-pretime. It can definitely be modeled as a Kaluza-Klein compact 5th dimension (or hyperspace), or as a 5D Randall-Sundrum model with a compact hyperspace, both models featuring a 4D spacetime plus a compact, sub-Planckian, hyperspace.

The Higgs field in the standard model

The field (or ocean) of Higgs opens after the inflation phase at 10^{-12} and lasts up to 10^{-10}. It is described as a sort of quark-gluon plasma with virtual particles jumping out of it and being attracted back into it, until the time when the photons will have the right energy to set themselves free at the time of symmetry-breaking (or decoupling of radiation-matter) happening at 10^{13} second.

The Higgs field is now understood as an energy field pervading the universe. Says Linda Randall in her 2012 book *Higgs Discovery*: “The Higgs field, on the other hand, is everywhere. It is spread throughout the universe. The field isn’t made up of actual particles. In a sense, it involves something like a type of charge spread everywhere through empty space. Particles that experience the weak force (...) interact with that “Higgs charge” and thereby acquire mass.” (12) With the Higgs field, underlines Randall, “A single field both permeates the vacuum—empty space—(...) and is also responsible for particle creation.” (13).

As for the famous Higgs bosons, there’s still some perplexity among scientists. Higgs bosons are postulated to be a fundamental particle of a *scalar type* that acts as the medium of interaction with all other particles. However, according to Randall (2012), if the Higgs field is well established, the Higgs boson (as it was postulated) may not be the perfect, or even the only, candidate to be the "Higgs particle," i.e. the particle associated with the Higgs field. Indeed, a perplexing fact with the near certain 'discovery' of the Higgs particle (announced by the LHC's Atlas and CMS experimenters, on July 4, 2012) is that this particle is more heavy than the Standard Model had predicted for the Higgs boson. According to Randall, this is "beginning to make ordinary vanilla supersymmetric models look increasingly unnatural. (...) The theory (...) just doesn't want the Higgs boson to be this heavy." (Randall 2012, 43)

We know since March 14, 2013 that the experiments with the LHC (Large Hadron Collider) at CERN, Geneva, corroborate the existence of the Higgs boson and the role of the Higgs field in giving their mass to particles as a function of their interaction with the field. This is called the 'Higgs mechanism' by the original team (Higgs, Brout, Englert, *et al*) and it shows how, through the symmetry-breaking of the electroweak symmetric force into EM and weak forces, the $W^{\pm}$ and Z bosons thereafter implied in the weak force will acquire their mass. Similarly all other gauge bosons (gauge refers to their role in the four 'forces'), responsible for three out of four fundamental forces (apart from gravity)—such as the gluons for the strong force, and the photons for the EM force—will get their mass through their interaction with the Higgs field (and very probably the Higgs bosons). The process is described in the Wikipedia article on Higgs Bosons as such: at the beginning the Higgs field is *potential energy* that, when gauge bosons interact with the Higgs field, will be converted into mass-energy. "In Higgs-based theories, the property of 'mass' is a manifestation of *potential energy* transferred to particles when they interact ("couple") with the Higgs field, which had contained that mass in the form of energy."[46] Lisa Randall, while explaining how the Higgs field gives masses to particles, adds to the mechanism, at its very start, a necessary 'jiggle' of the field to get the initial energy. Talking about the putative Higgs boson, she states: "Although not a field itself, the [Higgs] particle is indeed associated with the Higgs field. Essentially, *when you jiggle the Higgs field—add a bit of energy*—you can create an actual particle." (12-3, my emphasis.) This initial energy is absolutely necessary, of course, not only to launch the process, but also for the law of conservation of energy to be preserved; Randall emphasizes the point: "*With enough energy*, a Higgs field can create these particles, and they have properties specific to the role of the Higgs field in allowing for particle masses." (Randall, 2012, 13)

However how the Higgs field and this potential energy happen to be there in the first place, or *who or what is jiggling the field*, all this remains an enigma.

Here is the precursor dynamic that the ISS model lends to:

The X-funnel is a ***two-way translation mechanism*** *from HD energy to spacetime energy and the reverse. The X-funnel translates all masses (in the big crunch) from a previous universe into its CSR hyperdimension, and, on the other (white hole) side of the X-funnel, a new translation occurs, from the CSR syg-energy of the parent universe toward the new universe-bubble, within both its CSR bulk and its Quantum-spacetime (QST) region.*

The creation of the Dirac Sea and Vacuum

Three models have been proposed to explain processes happening just after Planck scale:

- *as Dirac sea*: a sea of negative energy posited by Dirac, and in which dwell all the antiparticles, with a sort of *flat* surface-boundary impeding the particles in spacetime to be attracted by their paired antiparticles in the sea; the sea is thus a sort of immense negative medium the other side of the boundary. (All antiparticles have the same mass as their paired particles; most antiparticles are negative, a few are positive, such as the positron.) What is the nature of the boundary that impedes the contact and annihilation of matter and antimatter is a mystery.
- *as the Higgs field*, at or after Planck scale, in which the Higgs bosons (the first virtual particles to appear at 10^{-10} second) dwell, and populated by lots of particles (mostly quarks and electrons) that, by interacting with the field will get more or less mass.

- *as the vacuum* (the "false" one), precisely *the zero point fluctuations (ZPF) field*, a medium akin to a superfluid, existing at the Planck (or quantum) scale, and in which virtual particles create enormous fluctuations and vortices, or "networks of loops," making it a very turbulent medium (sometimes compared to a "firewall"). This vacuum at the quantum scale is of course part of the basic 4D spacetime manifold included in the standard 10-Dimensional M-theory, whose extra 6D are compact hyperdimensions. This quantum vacuum pervades the universe at (or just above) Planck scale. Particles are hypothesized to interact with it and, as first proposed by H. Puthoff and B. Haisch, would have an impact on gravity, mass, and inertia; according to H. Müller, pressure and torsion waves could propagate through this superdense medium and, through resonances, influence the configuration of matter and matter-systems.

Sygons creating the ZPF and bulk: ISST hypothesis

The spiral staircase allows us to fathom that all bows that are activated at the origin (in the cosmic ISS) are, as we saw, sygons vibrating in an excited state.

Giant coherent field of sygons before Planck scale

Now you have to visualize that these sygons are attached to any step or bow of the ISS near-infinite spiral, each sygon having a specific frequency (and its harmonics) and bearing an active information. Given the enormous (rotational and *torsional*) momentum of the unfolding of the ISS, each bow-frequency launches a sygon torsion wave with enormous strength (mirroring this bow's set of frequencies in wave packets). Torsion waves are defined as being *polarized*, and having a *spin* and a *rotational frequency*. The sygons torsion waves exhibit: (1) a spiral torsion, (2) a spin or cyclical rotation, (3) a lefthanded or righthanded polarization according to which (inward or outward) spiral they belong, a direction of flow (a quasi-Poynting drift) either *inward* toward the X-point or *outward* toward the Planck length and vacuum, and (4) a specific frequency.

Each sygon torsion-wave would tend to be ejected from its bow (on the spiral staircase) perpendicular to the bow. However the global direction of flow of the informing cosmic ISS brings coherence to their individual direction of flow, and thus the primordial sygons' torsion waves espouse the ISS enormous drift outward. As a result, all initial sygons waves create a giant conical array of rays that are fanning out, *a coherent global field of information and sygons' networks with their CSR parameters.* This is a coherent CSR field that nevertheless contains networks of dynamical links issued from the original ISS spiral staircase.

Active sygons from parent DNA: the X-funnel BigCrunch–BigBang

Each sygon wave in the budding ISS is ejected with immensely faster than light velocity. The birth of the novel ISS gets its unfathomable energy thrust from the conversion of all remaining matter (in the previous universe) into pure CSR information, this happening in the final hyper-massive black hole. Thus matter is translated or sublimed from the QST manifold to the CSR manifold within the anti-ISS of the parent universe. At the X-point, the whole sygons system (the parent cosmic DNA) flows through the X-funnel exerting an enormous radiation pressure that sets the novel ISS deployment in the new universe-bubble.

At that point, sygon waves are ejected only by the bows (or bow-networks) that were activated or in an excited state at the origin—those bearing the active information of the realizations of our ancestor universes—the Cosmic DNA or Akashic records of our parent universe-bubbles.

Thus the ISST hypothesizes that a universe-bubble has a beginning with an 'ordered' or informational Big Bang—precisely the white hole side of the X-point, that is, pointing toward the Planck scale and our spacetime; and it has an end in a cosmic black hole, which is in effect the other side of the X-point of the next universe (in this particular collar of universe-bubbles).

This similarity between the Big Bang and black holes is the foundation of Lee Smolin (1992) *Fecund Universes* theory in which he proposes a *cosmic natural selection*, that is, at the cosmic scale. As we saw, the mathematics of the Big Bang and the black hole are one and the same: the theorems of singularity, developed by Hawking and Penrose in 1970. The Big Bang is modeled as a white hole, with just the opposite direction of flow, matter being either sucked in (in case of a black hole) or ejected out of it (in the Big Bang). Smolin proposes that massive black holes could be the seed of budding universes, and would retain the parameters selected and reinforced by a parent universe. Those are very resonant concepts indeed with the ISST, while I had no knowledge of Smolin's theory when I conceived the ISST.

In ISST, when a universe-bubble arrives at the stage of a cosmic black hole or Big Crunch, *all its remaining matter systems will progressively be translated into complex syg-networks in its hyperdimension—its CSR field of information—so that no information is ever lost.* In other words, all matter will be annihilated and only its syg information will remain in the cosmic syg dimension. All networks of links of the sygons actualized in a previous cosmic ISS will thus be imprinted in its anti-ISS, that is, the event-horizon and sink (the singularity) of its final cosmic black hole setting its Big Crunch.

A Big Crunch is, in effect, the anti-ISS leading to the X-point of translation BigCrunch–BigBang. The anti-ISS is the black hole on the other side of the X-point of origin, that translated itself into our own cosmic ISS, the two conic structures meeting at the X-point (forming the shape of an hourglass or *X funnel*) and translating their active syg information from one sinking ISS (a Big Crunch) to the budding and springing ISS.

I have modeled in *The Sacred Network* such a *X-funnel* allowing access to the syg dimension; in this book, I had applied it to a translocation of a personal consciousness (or semantic field) from one location in Brazil to another one in India (272-6). The cosmic X-point is a *X-funnel* that will translate and resolve all matter systems into their sole CSR manifold. For a universe-bubble, the anti-ISS contains: a field of active information + self-organizational dynamics + all realized and manifested systems and beings + information on all the historical developments of natural worlds and intelligent civilizations in this universe-bubble. A dying universe's CSR manifold is thus imprinted via dynamical networks of vibrating sygons (and <u>not</u> as a digital code) on its anti-ISS, and at the X-point, it will be transferred to the budding ISS springing to life.

The cosmic DNA resembles our own human DNA in the ovum, in the sense that only specific sets of this immense double-helix code for genes and are activated, the rest deemed 'junk DNA' by the late twentieth century biologists. At the very instant of the unfolding of the original cosmic DNA, only already activated sygons and networks of sygons (coding for realized systems and potentials of the previous universe bubble) are set free as torsion waves carrying their bow-frequency and syg information. All the while that our universe bubble will start implementing new potentialities and explore its mental, creative, and relational capacities, and as new worlds and civilizations will come to be, then all the beings and their realizations will activate new networks of bows in our cosmic DNA, thus modifying its dynamical and self-organized field of syg information (its ISS at the origin).

Creation of the ISS lattice and Higgs field with bubbling vortices

While the ISS unfolds (like a horn) at blinking speed, and the bows' wavelength increases with the logarithm of phi, the sygons become larger and larger, and more numerous. Very soon, *the created sygon waves have so large wavelengths that they start interfering and sort of foaming, creating a medium of denser and denser interference patterns*.

Soon this medium (the vacuum and ZPF to be) becomes superdense yet fluid and in hypervibrational mode, with the diverse sygons spins creating vortices surrounded by clouds of charge, and lots of turbulence in a fluid medium or 'sea' *that is none other than the Higgs field.* This medium densifies perpendicular to the ISS horn's mouth, thus creating a nearly-flat 'wall' or lattice, *the ISS lattice*. While this lattice forms, it gets a higher density in front of the central quasi-Poynting drift of the ISS, thus creating a *bulge*, the lattice being thicker there. (So that the whole ISS lattice resembles the disk of our galaxy seen from the side, with the central bulge, and is orthogonal to the ISS outward direction of flow, thus parallel to white-hole event-horizon). (See Figure 9.1)

The superdense foaming medium is a turbulent bubbling surface and it presents myriads of spinning vortices bathing in clouds of charges (for an observer in 4D universe).

The lattice is none other than the Dirac sea (with its holes), and the Higgs field (creating mass through interaction). Soon it will become the vacuum's Zero Point Fluctuations (with maximum quantum turbulence and virtual particles), and Sarfatti's loop vortices in the vacuum.

However, one specific prediction of the ISST is that the rotational and torsion field of the whole ISS and its enormous radiation pressure, as well as that of the torsion waves of all sygons, *create a huge rotation of the whole Higgs field or vacuum that will imprint a curvature to it.* These torsion fields (global ISS and networks of sygons) also build up the gigantic radiation pressure of the lattice and vacuum, and create its bubbles of polarized vortices, its 'loops'.

- Despite the thickness and density of the lattice of crisscrossing and interfering sygons waves, each sygon (or network of sygons) sustains its own frequency (or set of frequencies) from the original inherited sygons.

- Each previously activated sygon (or sygon-network), after being ejected from the bows with a fantastic force as a torsion wave, maintains its direction of flow and sustains its velocity, thus being in effect a torsion wave with a Poynting force.

- Each sygon's wave (+ specific wavelength and frequency, + active information and network of links with other sygons), will cross through the dense medium, the Higgs field, at enormous velocity, creating a sort of wormhole through it, yet this crossing will modify some of its parameters, such as the velocity.

- The modification of the sygons' velocity, and their acquiring mass depend on their wavelength and amplitude, because the longer the wavelength and the wider the amplitude and the bigger the 'quasi-surface' of its interference. This quasi-surface of interference creates an inertial effect that slows down the now string-particle, and part of its kinetic energy is transferred into mass.

- Additionally, these interferences from an immense number of sygons produce a sort of agitated foam in the medium and clouds of negative charges. However, we are now looking only at the first thrust from the origin and through the Higgs field or vacuum, toward our spacetime region. But when the matter universe will be formed, each particle and system will send back active information via its own individual ISS (nested within its compact CSR hyperdimension), and this will create an opposite flow of positive charge and torsion waves of inverse spin crossing the vacuum in the opposite direction, that will generate, in the negative foam, clouds and pockets of spinning positive charges. The two flows of opposite spins and charges will create more foam and turbulence, thus polarizing the vacuum. Moreover, this reverse flow will set the reverse inward spiral of the ISS.

- When sygons become particles (strings) within the lattice and Higgs field, each sygon will carry in itself (as a facet contains the whole information of the parent hologram) a replica of the cosmic ISS. In this *individual ISS*, it will keep its own information-network excited. Meanwhile, through their interaction with the lattice, the particles acquire a cloud of charge and mass that constellate around the core sygon. The core sygon's replicate ISS (called the 'individual ISS') now becomes the core of a massive and immensely bigger particle. It becomes the compact CSR hyperdimension of all particles, each with its own frequency (Rhythm metadim), radius or wavelength (Center metadim) and active information (Syg metadim), but nevertheless bearing the information of the whole and being in 'holographic' instant communication with the cosmic ISS.

- During their crossing of the horn's lattice, the sygons (depending on their basic frequency and syg-energy-information) will become specific particles within whole families of different particles such as quarks or electrons. Then the quarks will regroup by three to form a proton, etc. Doing so, they acquire progressively more mass, and mass tends to take a spherical form (due to Hilbert Action). The sygon's original bow-frequencies as well as its compact individual ISS will adjust themselves at the center of mass.

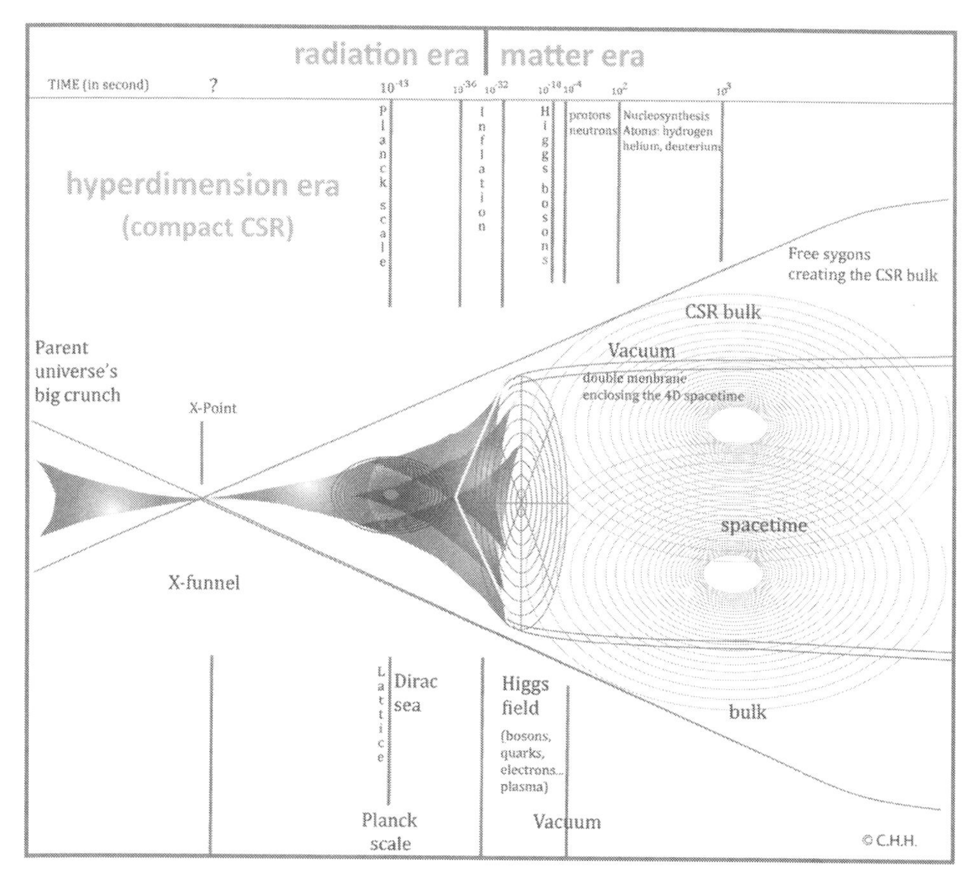

Fig. 9.1. The ISS X-funnel at origin and the lattice (Higgs field). To the left: the X-point of origin and X-funnel. Then the ISS cone, and the ISS lattice (eye-shaped) creating the Higgs field and later the vacuum. The vacuum as an extended membrane surrounding the spacetime manifold. Free Sygons exiting from the early ISS, setting the CSR manifold and creating the larger hyperdimensional CSR bulk. The smaller and cylindrical 4D spacetime bubble is encompassed by the bulk. (Time ratio not preserved). (Concept & Artwork: Chris H. Hardy)

A sort of central black hole or wormhole of the particle will be the opening on its individual CSR hyperdimension of negative charge and negative energy. Thus we end up, as posited by the LHC experiments at CERN, with particles getting the more mass (and the less velocity), the more they interact with the Higgs field.

But the tricky point was not to be in agreement with the experimental findings, but rather to explain where from the energy and the original virtual particles had come, and what was their nature before they crossed the Higgs field and acquired their specific mass, as well as how the Higgs field ever came to be.

The Free Sygons creating the HD bulk

We have seen how the ISS lattice took form and densified progressively by the interferences of the sygon waves in front of the ISS mouth, thus creating the Higgs field.

But it took some virtual 'time' for this turbulent medium to densify into a wall or a sea that any sygon wave had now to cross, getting some mass in the process. And before it did, I propose that a stupendous event happened: that of an advanced wave of sygons passing freely at highly superluminous speed, and creating the vaster region of hyperspace, an extended thin grid of the CSR hyper-dimension, bearing the information and potentials of the new universe-bubble.

A tachyonic foreground

How could this happen? Think of the extremely high frequencies and ultra-tiny wavelengths of the bows situated at the very beginning of the ISS spiral unfolding, just coming out of the X-funnel through the X-point.

We have seen how the direction of flow of the global ISS creates a coherent field that takes the form of a cone that will set the wide angle of the inflation phase (just after Planck scale, from 10-36 second up to 10-32s), generating the 1050 increase in size of the universe in a split instant. Without the thrust from the X-funnel and the ISS' direction of flow, the universe-bubble would have formed as (1) a sphere around the central ISS, and (2) with a huge expansion rate. We have seen also that at the end of this inflation phase, the universe suddenly shifted to a slower expansion rate, at the moment the first particles acquired mass by crossing the Higgs field.

Given that these extremely-high-frequency (EHF) sygons are the first ones to be set free or 'ejected' from the bows, the orthogonal ISS horn's lattice has not formed yet, and the torsion waves pass unimpeded and unobtrusively, acquiring no mass, no cloud of charge, and keeping up their rotational momentum and velocity. I fathom them to be the quasi instant, beyond spacetime, harmonic tuning of Rhythm metadim. These first sygons are hyper-energy, hyper-frequency massless strings, possibly gravitational waves. As they are the first waves to be spinning out of the cosmic ISS, they are the ones to create the universe's CSR dimension as a pervasive medium of active information. In this model, the particles of the first thrust should bear the imprint of the Rhythm metadim, and that means they should be highly sensitive to frequencies and resonances, and we should find them associated with resonance waves at the galactic and cosmic scale, as well as with the CSR compact hyperdimension. They may have set patterns of organization for the future matter universe; in contrast, they should neither be influenced by 4D+ spacetime, nor being stopped or slowed down by interacting with matter and EM fields.

The bulk and dark energy

The first sygons being ejected by the unfolding ISS have extremely high frequencies, and blinding velocity, enormously higher than the speed of light. They are tachyonic torsion waves belonging to the CSR manifold and having active information, pure CSR sygon frequencies with negative energy. All sygons are unique, since they follow Pauli's exclusion Principle. (Look again at Fig. 9.1, p. 313.)

The Free Sygons instantaneously spread out, creating the large conic shape of the inflation area. While spreading out, they extend the CSR hyperdimension in the form of a large cone of hyperspace, in the process setting the range of the dark energy or Quintessence field, in other words, the extended bulk of the hyperdimensional universe.

I meet here John Brandenburg's hypothesis of dark energy being filled with negative-energy tachyons. The FLT sygons, with negative energy pressure, are separated from the particles in 4D universe by the vacuum membrane. These extremely high-energy tachyonic sygons are the source of the immense negative pressure of dark energy, and of the limitless energy thought to be in the vacuum. According to ISST, once the vacuum is formed, the Free Sygons will just tunnel in and out of the vacuum membrane, through wormholes.

As they have passed and keep passing freely, the lattice begins to densify, and sygons of much greater wavelengths will start interfering between them and create clouds of charge, and vortices in a foam-like medium. The sygon torsion waves ejected at a later time will now be slowed down by the progressively denser medium of the Higgs field that will eventually become the vacuum's ZPF. As the Higgs field densifies, and the lattice gets denser and more turbulent, the flow of Large Sygons follows a less large cone, more in keeping with the ISS direction of flow—the matter-universe cone we have mapped (See Fig. 2.3, p. 69). Thus, in this slimmer cone looking more like a cylinder, where the horn's mouth (just above Planck length) is now blocked by an orthogonal slice that is the Higgs Field or vacuum, the inflation area ends abruptly and the rate of expansion decreases enormously to be just above flat.

Meanwhile, the Free Sygons keep on their course and keep dashing and expanding as a large cone (probably isomorphic to that of the inflation phase), and create the CSR hyperdimensional bulk, i.e. the quintessence field of negative energy pressure, filled with the pure CSR energy of the FTL Free Sygons. A sort of hyperspace bubble or *region*, immensely larger than the matter universe (and fully encompassing it), but still somehow conic, is thus *preceding* the expansion of the 4D matter universe we are inhabiting. In this scenario, the point is not that our universe-bubble 'chose matter over antimatter,' but rather that we live in the matter, inner, sub-region of a larger dark energy region (negative energy and negative pressure). This pure CSR manifold region would be separated from the matter-sub-region, by the type of double membrane that Sarfatti (2006) describes in his model, bordering the Dirac Sea/vacuum—one whose outer side would be made of dark matter and anti-particles. (We'll come back to it.) This type of organization could be why we find dark matter as halos surrounding the galaxies; and we'll certainly find similar halos at all scales, such as around galactic clusters.

Consequently, because all particles will have an hyper-dimensional opening on the CSR manifold (their compact CSR or triune 5thD), and all systems at all scales likewise, this explains the two facets of the CSR hyperdimension:

- a hyperdimensional bulk enclosing our matter universe-region with a much larger quintessence field, and separated from it by a double boundary membrane, an organizational closure (the Center-Circle metadim), and
- a curled up and compact individualized 5th dimension opening at Planck scale on all matter systems (extremely small compared to the radius of even a quark). Yet these

individual ISSs are being ONE and the same manifold with the cosmic ISS at the X-point, and the CSR cosmic bulk.

Soon, the extended hyperdimension of tachyonic sygons, by being itself a cone and curved, will set (or impose) the future curvature of the 4D spacetime. The bulk is definitely dark energy modeled as a new type of energy, Quintessence; the ISST predicts it will be filled (within the bulk, not the membrane itself) with a field of tachyonic Free Sygons, thus torsion waves expressing the CSR manifold's syg energy of cosmic consciousness, self-organizing and self-actualizing.

All in all, we end up with a Quintessence being a CSR field, forming an immensely larger conic field than the 4D cone of the radiation-matter universe.

The difference in the range of frequencies and in velocity makes that the CSR field (Quintessence, dark energy) is:

1. enclosing and containing the 4D cone,
2. yet pervading all points in 4D spacetime,
3. setting an advanced wave (and field) for our universe-bubble that contains all the syg information about our universe's potentialities in a dynamical and self-organizing triune energy (hyperspace-consciousness-hypertime). This Quintessence-CSR field has, embedded in its Center-Syg-Rhythm metadims, all the active information for favoring intelligent life and for adjusting all constants and parameters in order to birth life and biosystems.
4. this CSR hyperdimension is for its greatest part in our future, immensely ahead of the unfolding of our matter-universe in its quantum-spacetime (QST) cone (the latter shown in the 'Timeline of the universe' famed figure). Of course the CSR bulk, being beyond spacetime, pervades not only our future but our past and present as well.

The bold stand of ISST is that the CSR hyperdimension consists of both an inner compact hyperdimension, and an extended enclosing bulk. It is accessible from the compact triune 5th dimension curled up at the core of all particles, systems, and beings, or else from the hyperspace bulk—dark energy as a quintessence field.

I hypothesize that the advanced wave (field) of Free Sygons at the origin, should have left a relic radiation that pervades not only our 4D spacetime (QST cone), but also the CSR cone, or HD bulk—the larger region of tachyonic sygons outside our universe's matter or light cone: the quintessence or dark energy that is also the Elsewhere of Minkowski's cone. But why would this relic radiation of HD sygons be detectable also inside our QST cone? Because at the (hyper)time when the Free Sygons launched their tachyonic course and created the HD bulk, the Higgs field and vacuum had not formed yet (as the horn's lattice) and so these HD sygons covered the entire wide angle of their CSR cone (this angle still represented in the inflation cone). (See again Figure 9.1, p. 313.)

This primordial CSR cone included of course the horn's mouth where the radiation density of the larger sygons will create the interference patterns leading to the formation of the lattice and Higgs field. So that when our quantum-spacetime universe, much later (possibly between 102 and 1013 seconds), launched its own advanced wave of photons, these photons spread as a thinner cone WITHIN the boundary set by the earlier tachyonic sygons field (the CSR bulk).

The vacuum surface becoming our spacetime region's boundary

The ISST hypothesizes that the vacuum and Zero Point field, in its actual state, has evolved from the ISS lattice and Higgs field (or Dirac sea) of the origin. Given Hilbert Action law, the lattice has spread as a curved surface surrounding our QST cone, this spreading in sync with the spacetime itself spreading with the photons' advanced wave. The vacuum, then, is the membrane acting as a boundary between the 4D spacetime and the hyper-dimensional CSR bulk (quintessence). This would be why the vacuum inherits its own paradoxical topology from the CSR hyperdimension: it is accessible both at the quantum scale within all particles, and as a cosmological vacuum (an extended surface or brane). Hence the law stipulating that any quantum system rolling on a hypersphere surface will take the frequency of that surface, can be adapted to: any vacuum surface embedded within a CSR hyper-bulk will be in sync with the range of frequencies and CSR properties inherent to that HD-bulk.

Consequently, our QST matter universe region has been molded by the CSR hyperdimension, not only in terms of its curved spacetime geometry and of metadim Rhythm's resonant patterns, but also by the organizational influence of the syg hyperdimension of cosmic collective consciousness. In brief: Our matter universe (region) had been organized by the Cosmic DNA being spread and scattered, or sowed like seeds, that is, by the information-laden tachyonic sygons (and sygon-networks).

As a result, the resonant patterns with the proton, duly demonstrated by Müller's Global Scaling theory, are due to the proton being itself in resonance with the Free Sygons that created the hyperdimensional universe bubble. Follow the proton resonant rhythm (in the inverse direction) toward the infinitely small, and you will find the original signature of the Free Sygons exiting from the X-funnel.

I surmise that the relic radiation from these first extremely high frequency (EHF) Free Sygons (the sygon torsion waves from the very origin), would have kept the tachyonic velocity of metadim Rhythm, would be only waves, would become the syg-energy and all the universe's syg-fields (laden with self-organizing information); they would moreover imprint their specific information on the topology (metadim Center) of the individual particles' ISS, and act as the subtle nonlocal links between all individual ISSs and the cosmic ISS. Could there be a link with the newly discovered (March 2014) gravitational waves showing a torsion wave grid underlying the CMB? To the neutrino background?

- The sygons torsion waves are also an active information on themselves and on their original network-links to other bows. They act as two-way links between the cosmic ISS and the individual ISSs.
- All activated bows are still constantly emitting sygons torsion waves. New bows are activated and new networks of bow-sygons are formed within the cosmic ISS, through the individual ISSs being in sync with the cosmic ISS.
- When the disk-shape of the vacuum will be densified, it will still be enormously less dense and more cold on its spiraling edge. And it means that the (WMAP and PLANCK) map of the cosmic microwave background radiation (CMB) reflects the configuration of an earlier state of the ISS lattice, that was the ocean of Higgs and Dirac, and then became the ZPF of the vacuum.

Crossing the vacuum boundary between the CSR bulk and the 4D spacetime

- All sygons, crossing or tunneling through the vacuum as torsion waves, have left and still leave a print of their passage in the form of vortices of specific wavelength and charge, as well as their frequency signature, on the surface of the lattice (the disk and its bulge facing the ISS), which could also be viewed, in its original state, as the ISS white-hole's event-horizon.
- To an observer in our 4D space, that is, on our side of the lattice and looking at the dense and bulging vacuum surface boundary actually blocking the ISS horn's mouth: The point of exit from the ISS and entry into our 4D spacetime, of an immense number of sygons (now strings-particles, with more or less mass), should show numerous vortices, each one with a central hole the radius of the particle, and an event-horizon which is its cloud of charge. These vortices at the exit mouth of the wormhole (where stands our observer on the 4D side of the boundary), should be negative with (dark energy) negative pressure. Inversely, when sygons flow in sync from individual ISSs toward the cosmic ISS, they enter the vacuum through a wormhole (the size of the triune 5th dimension), and their entry point should present a wormhole mouth of positive energy. (Could they be appearing as what we call dark matter then, as Sarfatti postulates?)

Of course, a wormhole-mouth can be equated to a micro black hole, each negative energy-pressure mouth to a negative micro-black hole, and inversely. So that we meet Zel'dovich's concept of the vacuum pierced with micro-black holes of opposite charge or spin. We meet also Dirac's idea that the (Dirac) sea of negative energy in which all anti-particles reside (behind a boundary impeding them to meet and annihilate with their paired particles) appears as a multitude of 'holes.' As for Klein and Alfvén, they proposed that matter and antimatter are in balance in the universe, separated by a boundary consisting of two layers of plasma of opposite charge, an idea developed further by Sarfatti.

All in all, (for a 4D observer) the vacuum as a quasi surface or brane presents the texture of (take your pick): innumerable vortices (of opposite spin and charge) surrounded by clouds of opposite charge, or the micro black holes proposed by Hawking and earlier by Zel'dovich; or else the entries of wormholes posited by Jack Sarfatti (2006), the ISST being the nearest to the latter.

For Sarfatti, the vacuum consists in a foamy two-dimensional surface boundary that is in essence "a neutral ionized plasma" where phase transitions happen. One side of the surface is Dirac sea (dark energy, of negative energy pressure) containing the antiparticles such as a positron, and where the antiparticle will have its (negative pressure) wormhole mouth. The wormhole crosses one layer of the surface and its other mouth (of positive pressure) dwells in a thin layer of dark matter, where stands its paired matter particle (the electron). All anti-particles (antimatter) reside in the Dirac sea or dark energy, and are thus linked through wormholes to their matter particles in spacetime, thus forming an immense number of such "stable bound pairs," lots of them being network-linked. (Sarfatti, 2006, 150-2) Therefore these "virtual pairs" are part of "networks of loops" (the loops being the closed branes forming the mouths' vortices) and appear in the macro-quantum and spacetime side of the surface as *psi wavefunctions with a bulge*, because they contain an immense number of states of these virtual pairs.

But let's see at a finer grain the process with the vacuum first forming as the cosmic ISS horn's lattice, and dark energy being the HD bulk of the CSR hyperdimension. The thing is, the sygons vortices are serving as two-way passages (through the vacuum membrane) to and from the cosmic ISS, in or out of *our 4D (or nD) spacetime*, and they are the traces of the sygons passage. The sygons vortices have all the above-mentioned characteristics, being microscopic black or white holes with an event horizon and a trace-information on it (such as the sygon's frequency, radius length, and charge). However, we should not expect to catch the real sygons' syg-information, because they are only traces. What we could find, though, are the networks of sygons, because they would erase some of the CSR-hyperdimension's nonlocal links and shrink to form a system in 3D-space that holds together, so that they should appear on the vacuum surface as networks or groups of wormholes' mouths.

The vacuum as boundary extended to enclose the spacetime region and to shield it from the HD bulk

The ISST then proposes that the vacuum membrane, soon after the lattice densified, tended to spread like that of a balloon, and finally enclosed our whole matter-region. The vacuum becomes thus a boundary between the matter-region (the quantum-spacetime or QST region) and the much larger tachyonic sygons-region that is in effect the bulk of the triune Center-Syg-Rhythm or CSR hyperdimension. (In topology, this dimension enclosing another one that can't be contracted is said to be 'multiply-connected.')

Antiparticles self-morphing into their own paired particles

Let's see now how ISST proposes a somewhat original perspective on the 'pairing' of particles with their anti-particles, and also on the boundary separating the region in which antimatter and dark energy would reside, thus letting our quantum-spacetime region to be one to favor a matter universe filled with stars and galaxies, allowing the existence of bountiful living organisms and intelligent civilizations.

ISST has proposed a dynamics for the formation of the Higgs field (the ISS' orthogonal lattice) and how it will, once densified, become a barrier that large sygons will now have to cross or to tunnel through it. At the very moment that the large-wavelength sygons start interfering and raising clouds of charge, creating a sort of foam that will densify to become the lattice, the faster-than-light Free Sygons have already dashed by, propelled by their density pressure within the whole angle of the inflation cone. Thus they have already created the bulk of the CSR hyperdimension, spreading and seeding the hyperspace with syg-energy and active information stemming from the cosmic DNA (issued from preceding universe-bubbles)—information about the ISS, the syg-dimension and its organizational dynamics.

But a moment later, the lower frequency sygons ejected from the larger part of the Spiral Staircase (where the wavelengths are the larger, up to Planck length) start meeting a quickly densifying quasi-liquid medium. They keep their basic frequency, each one, intact, but create a lot of foam and charged clouds around each of their sygon's vortex, and their network of related sygons' vortices.

Each sygon (negative energy) is now wrapped in a cloud of positive charge, its kinetic energy slows down and is transferred into density of charge and mass; and thus the large CSR sygons acquire mass and charge according to their fundamental frequency and wavelength. And still they tunnel at great velocity through the dense Higgs field, spinning wildly, and finally emerge the other side of this boundary as particles spinning in their vortices, with a reduced velocity (equal or lower than C), more or less mass, and their overall charge is now opposite to the one they had before crossing. In brief: they have become the particles paired with the anti-particles they had been.

As we have seen, each original sygon (necessarily of sub-Planckian wavelength) remains at its original scale, and serves now as the compact 5th dim residing at the core of the new particle. This particle is now 'clothed' in a lower harmonic or resonant wavelength above Planck length, and has acquired charge and mass. This new particle, through its core compact CSR hyperdimension, and its individual replicate of the ISS hologram in it, will remain entangled with its original sygon inside itself, of course, but also with the bow it occupied (as well as its network of links) in the original cosmic ISS.

Taking this perspective, I may offer a hypothesis:

What is called 'paired particles,' of opposite spin and charge and identical frequency, would only be *a unique system*, comprising:

- at its core a 5^{th}D sygon as a pure vibratory wave (a pre-Planckian string) with either one direction of spin or two opposite spins; and
- a (post-Planckian) large massive particle enclosing it, of harmonic but lower frequency, with more or less mass or massless, with opposite spin or no spin, and with a charge or neutral.

Then how could we explain what we see (in the EPR-type experiments) as two separate particles, still paired and entangled, in nonlocal correlation or communication? Let's not forget the main prerequisite for the experiment to be successful: the two particles must have been issued from the same source—such as a particle smashed by a collision that sent the two particles, due to their opposite spins, in opposite directions. Then could this collision have simply broken the nut open and force the sygon at the core to duplicate (with inverse symmetry) to maintain its identity and information intact?

Predictions derived from the ISS model, on the CMB anisotropies:

- The CMB anisotropies have formed due to the foaming and bubbling of interference patterns of the large sygons' torsion waves crossing through the ISS lattice or Higgs field.
- The vacuum should still present a much higher density where lies the lattice's 'bulge.' This bulge would point to the direction of the cosmic ISS, situated behind this bulge and covered, in terms of signals, by its more dense and intense ZPF fluctuations 'noise.' The point opposite to the bulge (on the universe event-horizon) would point the direction of flow of our QST cone, and thus our future.

Two-way information flow

From the side of intelligent beings living in a 4D world: Each being or biosystem will have instantaneous two-way communication at all times (unconsciously or consciously) with their individual ISS, and through it, with the cosmic ISS. For us human beings, the individual ISS is our Self (soul, atman) with whom our ego (the conscious 'I') keeps constant links. These links can be reinforced to the point of becoming quasi-stable bridges; it's as if we could be network-linked to some melodies, experiences, capacities, beings, and future events, via our strong ISS connections. So that I do believe that the genius Ramanujan was really obtaining complex and precise mathematical information directly from the cosmic ISS, and similarly Pauli, Poincaré, Tesla, and Kekule, and so many other inspired scientists and artists...

This is the ontological difference with a god-creator: no command, no judgment, no restricted domain of inquiry, no imposed worldviews, values, behaviors, and then some... running in the unconscious. And this is the ontological difference with a materialistic theory: this organizational and dynamical model of the cosmos is alive, is a network of collective intelligence, is self-organizing and generative of dynamical knowledge, and I feel I'm, while developing it, part of this alive network of minds-psyches in a shared hyperdimensional mindspace that feels wide open on many worlds that are not only random numbers but real paths of exploration being trodden or tried by other beings.

Thus, there is a two-way flow between the ISS and the collective cosmic consciousness. The Free Sygons' torsion waves bear also an active information on themselves and on their original network-links to other bows, and they were, as we said, the first ones to deploy, well in advance of the spiraling unfolding ISS. When the ISS duplicates become the internal hyperdimension of all particles and systems at all scales, including the Self of intelligent beings, they will act as the two-way link with the Cosmic ISS.

The two-way sygon information flow (carrying the organization, evolution, and network of links of a system) sets in our spacetime region a retrocausal or back-propagation flow (a future→past arrow), as well as synchronistic transversal influences on the evolution of linked events. (See again Figure 6.3, 213.) This creates a ***retrocausal attractor*** working at both the QST and the CSR levels of a wavefunction. With this RC-Attractor, the beyond-time consciousness (Syg metadim), as well as the Rhythm metadim of the hyperdimension can retro-influence past events and systems or modify the course of actual events nonlocally, as I have modeled it within a systems theory framework in *Multilevel webs stretched Across time* (Hardy, 2003 [48]).

This information within the individual CSR dimension will not only be 'imprinted' on the collective consciousness field (as in the Akashic model) but *it will also influence it*—the more intense the syg energy, the stronger the influence. Thus the human minds and all sentient and intelligent civilizations in our universe-bubble (at the very least) will be in 'conversation' and inter-influence with our cosmic consciousness field (ISS, Tao, brahman); thus we can expect collective cosmic waves of change spreading through the galaxy at least, and intensifying a galactic mental phase-transition to another type of mind-body-psyche intelligence (as we may feel is happening right now).

Anti-ISS and the X-funnel: A collar of universes

ISST posits, at the X-point of origin, a X-funnel that is a double translation Black-Hole–White-Hole, or ISS–Anti-ISS. The anti-ISS and the ISS are forming a hourglass shape, or X-shape, called a X-funnel. Each ISS (like the one at our origin) is the white hole directly issued from the previous universe-bubble's anti-ISS black hole.

A fascinating model featuring such a double system [black hole + white hole], facing each other as a X-funnel shape, is that of physicist Roy P. Kerr. For Kerr, each singularity has two event-horizons. In his 1963 article in the *Physical Revue Letters*[47] he postulates:

- *The properties of space-time between the two event horizons allow objects to move only toward the singularity.*
- *But the properties of space-time within the inner event horizon allow objects to move away from the singularity, pass through another set of inner and outer event horizons, and emerge out of the black hole into another universe or another part of this universe without traveling faster than the speed of light.*
- *Passing through the ring-shaped singularity may allow entry to a negative-gravity universe.*

The anti-ISS is a black hole, big crunch, and a sink, into which the whole universe bubble will one day transfer all its matter-energy into the CSR consciousness manifold. It is a subliming phase transition, from matter to the hyperdimension. The ISS of the origin is generative, it is a birthing phase-transition. The Cosmic DNA of a universe's ISS contains the active melodic information of previous parent universes as on a collar, in which the nearest parent universe-bubble is of course the most active information at birth. Yet, there's a collective cosmic consciousness implied in the birthing process who may decide to smother parts of the melody and give more weight to other more promising parts. When a spacetime universe has become enriched with intelligent life, then the mental and biological processes of a multitude of evolved civilizations will now be the trigger for choices and therefore the weighting of preferred possibilities and paradigms. We'll observe then a whole gamut of extremely variegated collective choices made by an immense number of such intelligent civilizations.

This resembles what we see (at a smaller scale) on Earth with various cultures: a vast process of inter-influence that tends toward harmonization and political federation. Such political federation is bound to happen in a universe-bubble teeming with intelligent civilizations in a vast number of pockets at the galactic scale. It will set a trend for harmonization and cross-fertilizing of life and of the diverse civilizations, and will secure the sheer survival of the worlds involved. Therefore, some sort of loose holistic or holographic resonance of shared paradigms and aims between most evolved worlds (within a galactic federation at least) is likely to occur and take precedence. In the opposite case of a continuous warring and competitive and destructive modes of intra-or inter-galactic relationships, then the worlds implicated would destroy themselves and disappear from the matter universe, leaving only their alive memory in the CSR manifold.

So what happens to stars swallowed by a central galactic black hole? Such black hole is an anti-ISS at the galactic scale. In such a BH, the stars and all matter resolve into pure CSR systems (information, consciousness, knowledge, organization), a melody memorized in this galactic anti-ISS. At the universe scale, the anti-ISS is the attractor standing in the future of our QST matter-region. Far from representing an immutable and deterministic future, it consists in the *probability lines* of our universe. These probability lines are created, modified, and reinforced while each galaxy (as a collective syg system comprising all intelligent civilizations and evolved systems) generates new intentions, paradigms, values, behaviors, and thus novel probable futures. The anti-ISS has thus to be modeled as a *Retrocausal Attractor* (RCA), and similarly the whole X-funnel [ISS–anti-ISS]. The RCA logic leads to understand that we create (at all individual and collective scales) our future as we walk along and make choices, when we think, feel, and create, when we invent new ways of living and expressing ourselves.

A collar of universes: Penrose's Aeons and the ISS

Roger Penrose, in his recent *Conformal Cyclic Cosmology* (CCC) postulates a chain of universes or "aeons" whose 'big bang' (or point of origin) is issued from a previous universe/aeon (each aeon lasting 10^{100} years). Penrose (2014)[49] develops a "gravitization of quantum theory" that discards both the initial singularity with its high entropy (within a White Hole modelization) and the inflation phase, postulating instead that "the initial singularity (the Big Bang) was an exceptionally *low*-entropy singularity." Penrose concludes to a real "'black-hole information loss' namely that Hawking, in his powerful original analysis [in 1975][50] was correct in arguing that there is indeed a loss of information—or, as I would prefer to put it, a *loss of degrees of freedom*—in the black-hole evaporation process." (This information loss entails a violation of the 'Unitarity' principle, the possibility to deduce a system's past states.) With the "gravitational degrees of freedom hugely suppressed," and once the final black holes in a universe have all disappeared, there is a "re-setting of the zero of entropy" at the start of each new aeon, that allows the new aeon to show accrued entropy with time. Penrose views this zero entropy as a "transcendence" rather than a violation of thermodynamics' 2d law. He estimates "The extraordinary specialness of the Big-Bang state, as opposed to the extremely high entropy potential possibilities in black (or white) holes," to have a probability of $10^{10\wedge 124}$ (with two exponents) and concludes "This figure makes clear how enormously far from random the initial state must have been." Moreover, the phase-states of the end and the beginning of each aeon are isomorphic (similar in form, i.e. "conformal"): "there was actually a conformal *continuation* of our own Big Bang *B* to a previous universe phase *prior to B*, whose conformal infinity joined smoothly to *B*."

We have several points of agreements here between CCC and ISST: the anti-ISS at the end of a UB is modeled as the reverse of the ISS at the origin (with a conformal or isomorphic spiral organization), and the ISS is as far as possible from a random and high entropy system.

Let's note an interesting perspective lending some ground to the ISS theory. Penrose states that "the main good reason" (for which "most cosmologists appear to believe that inflation necessarily *did* take place") has to do with a "somewhat remarkable feature of the generation of the tiny spatial variations in the cosmic microwave background (CMB), namely that they were (very closely) *scale invariant* and this could be explained by the basically self-similar exponential expansion of an early inflationary phase." (Penrose, 2014) Now, such "self-similar exponential expansion" could fit perfectly a golden spiral whose phi ratio of expansion is logarithmic, and therefore "scale-invariant." This leads me to make some further predictions, namely that our whole (nearly 14^{10} years old) UB is still spinning as a cosmic spiral, this *forma* imprinting an aggregation (and clumping) of matter (thus a higher matter density) on the crest of these spiral arms. This would explain different properties of our universe, namely, (1) global scalings and their resonance patterns, (2) the Great Walls as the crest of these spiral arms (the nearer to the crests the higher the speed). Therefore, (3) the Dark Flow (and Great Attractor) would just be the spiral arm we are on, thus explaining its much higher velocity, that, by being relatively ours also (depending on our remoteness to the crest) is (4) deepening our speed-difference with the 'sides' of the UB horizon (the lower speed being at 90° or orthogonal to our direction of flow) and thus adding to the *apparent* acceleration of expansion (higher receding speed) in any direction but our flow (front and back). Then a derived property would be (5) rotational wave patterns and harmonic resonance within (and due to) cosmic-scale gravitational waves.

X-funnels, spiraling and layered structures in astronomy

Now, the interesting thing is that we have such X-funnels in astronomical objects.

One is a *Herbig-Haro object* catalogued HH-30.

The X-funnel is formed by two protoplanetary accretion disks (that is, the accretion disk that will slowly form the ancestor of a planet). (See Figure 9.2.) Moreover, a reddish thin double jet of gas is ejected from the forming star.

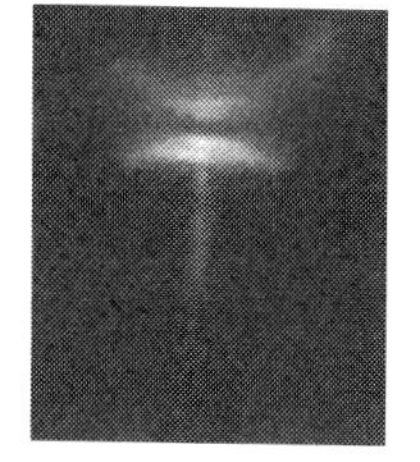

Fig. 9.2. Herbig–Haro object: showing a X-funnel structure, by a type of protoplanetary accretion disc. A greenish jet of gas emanates from the forming star HH-30, surrounded by a protoplanetary accretion disc. (Credit: NASA/ESA 1995. Public domain.) http://hubblesite.org/newscenter/archive/releases/1995/24/image/e/

The most revealing X-funnel is the Red Rectangle nebula *(*HD 44179, on Figure 9.3a*).* It shows a X-funnel or syg-funnel structure (a double cone) with distinct layers of ejected material forming embedded isomorphic structures. These embedded structures seem to be straight sides of embedded pyramids, but in fact (see 9.3b) they are conical, like embedded wine glasses, whose rims are vortices forming a torus. Says the NASA article: "Hubble's sharp pictures show that the Red Rectangle is not really rectangular, but has an overall *X-shaped structure*, which the astronomers involved in the study interpret as arising from outflows of gas and dust from the star in the center. The outflows are ejected from the star in two opposing directions, producing a shape like *two ice-cream cones touching at their tips*. Also remarkable are straight features that appear like rungs on a ladder, making the Red Rectangle look similar to a *spider web*, a shape unlike that of any other known nebula in the sky. These rungs may have arisen in episodes of mass ejection from the star occurring every few hundred years. They could represent a *series of nested, expanding structures* similar in shape to wine glasses, seen exactly edge-on so that their rims appear as straight lines from our vantage point. (...) The disk has funneled subsequent outflows in the directions perpendicular to the disk, forming the bizarre **bi-conical structure** we see as the Red Rectangle. The reasons for the periodic ejections of more gas and dust, which are producing the 'rungs' revealed in the Hubble image, remain unknown." [51] Note on the 9.3b NASA/Feild illustration that the vortices are in fact creating the 'rungs' (the rims of the embedded glasses), the rims thus being tori, each of them having an eight-shape dynamic flow that springs from inside each 'glass.'

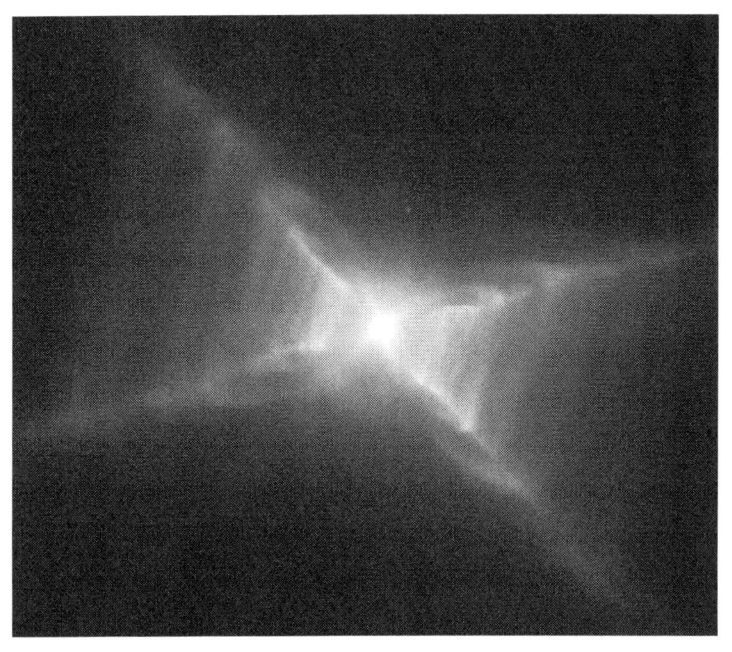

Fig. 9.3a. X-funnel with layering in the Red Rectangle nebula. Two layered cones touching by their tips around the dying two-star system (at the center), itself creating a curved X. Credit: NASA/ESA; H. Van Winckel and M. Cohen. Public domain.) Source: http://hubblesite.org/newscenter/archive/releases/2004/11/image/a/

Such eight-shape or hourglass galaxies and nebulae seem frequent. See for example the quasar on Figure 9.4, that shows a fine mesh structure in the huge outflow ejected from it. If you visualize the 3D structure from above, you will see sorts of undulating glass rims.

Finally, many galaxies and nebulae show a fine mesh structure, definitely layered, circular, curved, or spiraling (such as the ones on Figures 9.5 and 9.6).

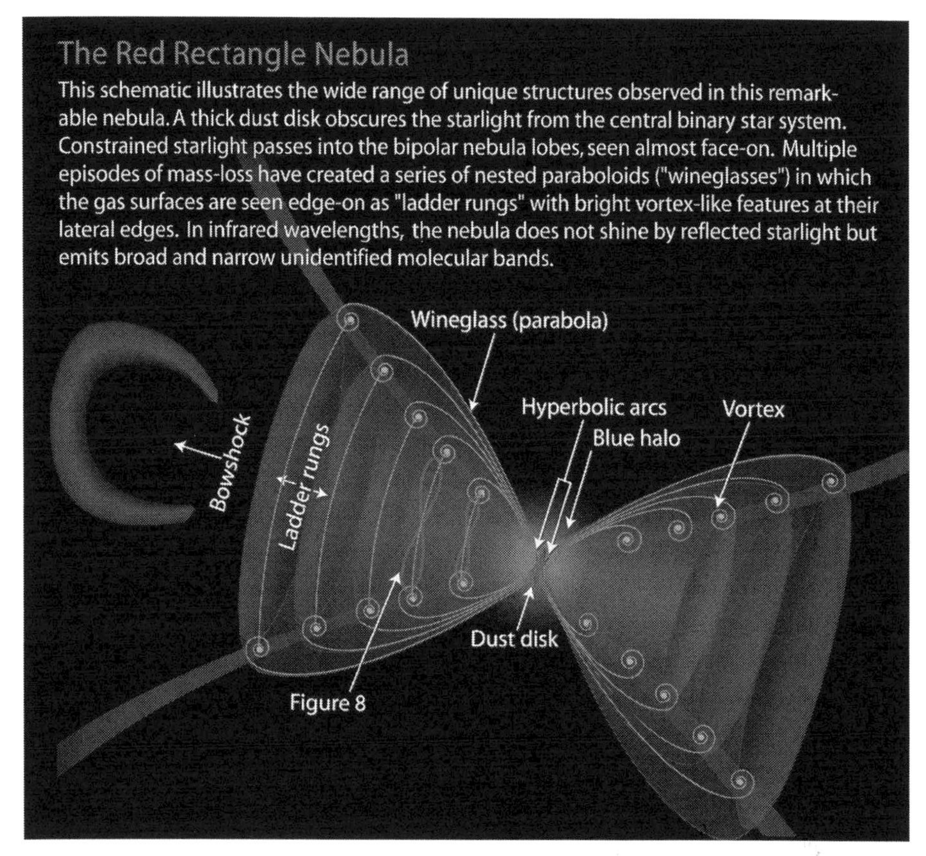

Fig. 9.3b. Red Rectangle's embedded wine-glasses with vortices forming the 'rungs.' Hubble Site. Illustration Credit: NASA and A. Feild (STScI) http://hubblesite.org/newscenter/archive/releases/2004/11/image/b/

Fig. 9.4. Hourglass shape with fine mesh structure in the huge outflow ejected by a quasar (SDSS J1106+1939) from the region around its central supermassive black hole (in white) and that spreads 1000 light-years out into its host galaxy. (Image ESO.org, 2012; Artist's rendering by L. Calçada/ESO. Wikimedia Commons)

The finest structure, presenting spokes and orbs, could be the *Cat's Eye nebula* (NGC 6543), discovered in 1786 by William Herschel (Figure 9.5). It is a planetary nebula, the first one to have been studied with a spectroscope that revealed it was made of rarefied gases forming concentric shells. Recent observations of the *Cat's Eye* using the Hubble telescope show also jets of gas as well as "knots of gas." It has either one brilliant star at its center, or a binary star system. In an earlier phase, it ejected spherical shells. It is also surrounded by an immense halo of great beauty (on the right).

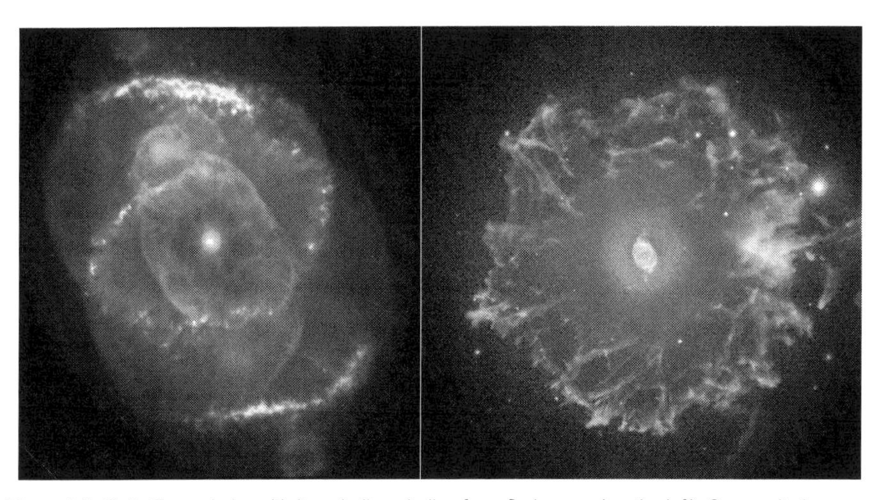

Figure 9.5. Cat's Eye nebula, with its spiraling shells of rarefied gases (*on the left*). Composite image using optical images from the HST and X-ray data from the Chandra X-ray Observatory. Credit NASA & ESA (Public domain). *On the right:* The Cat's Eye surrounding halo. (Public domain) Nordic Optical Telescope and Romano Corradi - http://www.spacetelescope.org/images/heic0414b/ Find them at: http://en.wikipedia.org/wiki/Cat%27s_Eye_Nebula

Another planetary nebula, bipolar, the Egg nebula, presents definite undulating arcs and bows, placed in concentric circles (Figure 9.6). Also called CRL 2688, the Egg nebula was discovered in 1996.

Orthogonal arcs or bows on torus-shape spires are visible in galaxy NGC 4725 (Figure 9.7, left), that belongs to the "barred spiral galaxy" category. NGC 4725 has a supermassive black hole at its center (an active galactic nucleus) and it lies in the Coma Berenices constellation, some 40 million light-years away.

As for the Hoag's object (quite mysterious), it shows also transversal rings on a huge torus, that evoke the keys of my futuristic piano (Figure 9.7, right). The Hoag's object is an atypical ring galaxy, discovered in 1950 by astronomer Art Hoag.

Fig. 9.6. Egg nebula: concentric layers and two crisscrossing jets. Credit: NASA and The Hubble Heritage Team (STScI/AURA); W. Sparks (STScI) and R. Sahai (JPL). (hubblesite.org).

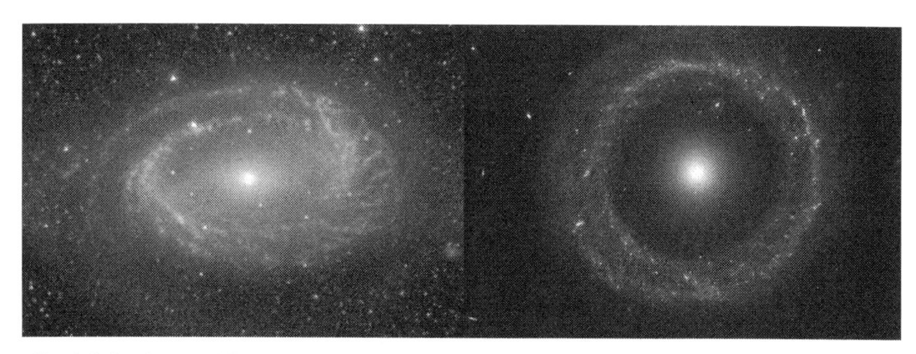

Fig. 9.7. A galaxy and Hoag's object showing transversal arcs set in a torus shape. (Left:) Galaxy ngc 4725 (A mid-infrared image of NGC 4725 taken by Spitzer/SST. (Public domain. Credit: Spitzer Infrared Nearby Galaxies Survey/SST/NASA). (Right:): Hoag's object. (Public domain. credit: NASA/ESA) http://en.wikipedia.org/wiki/Hoag%27s_Object

The central body consists of yellow stars (predominant), while its outer ring is mostly made of blue stars. It belongs to the Serpens constellation, and stands about 600 million light-years away.

Thus it seems that the astronomical systems (not only galaxies but also nebulae) are much more structured than we had anticipated, and I believe that what appeared at first to be random gas clouds will reveal more and more their fine-grain organization.

10
THE ONTOLOGICAL ARGUMENT
FOR A HYPERDIMENSIONAL UNIVERSE

"Most people who haven't been trained in physics probably think of what physicists do as a question of incredibly complicated calculations, but that's not really the essence of it. The essence of it is that physics is about concepts, wanting to understand the concepts, the principles by which the world works." Edward Witten, physicist.

What follows is the argument concerning the ontological ground and logic pertaining to a universe (or universe-bubble) composed of two regions: a hyperdimension (HD) and a spacetime region—that is, a HD existing beyond the quantum-spacetime manifold or QST.

The term ontology derives from *ens* meaning entity, being, in Greek, and was, as a central theme of metaphysics ('meta physica' or 'after or beyond matter sciences'), the crown of all sciences in Greek antiquity, the domain of the ultimate debate on the nature of Being and reality. Thus, ontology is the philosophy of Being in its essence and/or fundamental existence (Heidegger's *dasein* as both an openness to Being and a questioning about Being). In my view, ontology poses the question of Being, and of the fundamental nature of reality, in its wholeness or system's organization. Not only should it tackle cosmology (another branch of metaphysics) and the universe, but the latter's global organization and dynamics as well, as a suite of postulates.

The ontological argument developed here adopts the principles of systems theory as exposed by Ludwig von Bertalanffy, in his landmark 1968 book *General system theory*. The body of the argument suits any type of two-region [hyperdimension + spacetime] universe(-bubble), each region possibly N-dimensional, and this, whether or not within a multiverse-type framework. As in our actual physics, the quantum domain of the Standard Model particles is integrated to spacetime, to form the quantum-spacetime manifold or QST.

The last postulate (§ 12) is specifically developed using the present Infinite Spiral Staircase Theory (ISST), with its HD region being the Center-Syg-Rhythm manifold/hyperdimension or CSR HD, and its spacetime region being the quantum-spacetime manifold or QST.

❖ 1. Given that consciousness exists in our universe (-bubble) and that it cannot be accounted for by any field or force dependent on space, time or mass—that is, the 4D spacetime manifold, or any N-dimensional Quantum-Spacetime manifold (QST),

➢ a. Then it has by necessity to be accounted for via a hyperdimension of consciousness (let's call it a *syg-hyperdimension*) with its own parameters and dynamics.

❖ 2. Given that consciousness is able to *effect an action* on the physical world (on matter systems, bio-systems, and on the distribution of randomness as shown in bio-PK, micro-PK and Global Consciousness (GC) effects in experiments (see Nelson, Radin),

➢ a. Then by definition it is an energy (let's call it *syg energy*) and

➢ b. It has to be treated as such in equations (e.g. it is quantifiable by the force of its action and its dynamics).

❖ 3. Given that consciousness operates beyond space, time, and mass (even if part of it is strongly coupled with biological layers of organization in biosystems),

➢ a. Then it is not constrained by any constant or law linked to spacetime parameters, e.g. neither by C nor by the inverse square law of EM fields.

➢ b. However, *some* of its *effects on matter* systems may be quantifiable using the units of matter parameters (such as the translation of a psychokinetic (PK) effect in terms of the force in Joules needed to simulate it).

➢ c. Statistics (as presently used in laboratory double or triple blind experiments to show a deviation from the baseline of randomness), can only point to the reality of consciousness events, such as diverse forms of psi, and can only help to sort out the influence of psycho-social factors on these psi events. Yet, it doesn't explain (1) the nature of consciousness, (2) the syg-energy of consciousness, nor (3) the modes and dynamics of action of this syg-energy.

❖ 4. Given that a syg-hyperdimension of consciousness exists at some point in the evolution of a universe (or universe-bubble in a multiverse framework)

➢ a. Then this syg-hyperdimension has to be considered a fundamental part/ aspect/ quality/ potentiality of the cosmos and of reality.

❖ 5. Given that even in an emergent framework, only phase-transitions happen, not Creation Ex Nihilo,

➢ a. Therefore the syg-hyperdimension exists at all times, whether as virtual, potential, or manifested reality;

➢ b. Conversely, even before Planck's threshold, (where happens the phase transition to matter and particles in spacetime, i.e. to the QST manifold), this 4D matter reality existed as potential or virtual;

➢ c. Therefore also, spacetime and matter *laws* exist as potentials or seeds in a hyperdimension such as the sub-Planckian universe.

➢ d. So that the seed of the QST manifold exists in the syg-HD, and inversely the seed of the syg-HD exists in the QST manifold.

❖ 6. From the previous postulate, it follows that *the whole universe or multiverse has a self-consistent reality*, that is, a system's self-consistency, and that some laws may apply at different scales, even if some of their parameters are wildly different and some of their constraints are violated.

➢ a. The speed of light C is definitely a barrier only within the manifold of spacetime (QST).

➢ b. Tachyons and other forms of faster-than-light particles/ waves/ processes are highly probable in hyperdimensions and at dimensional boundaries or phase-transition scales (such as the ISS and anti-ISS in ISS theory).

❖ 7. Any syg-hyperdimension beyond space and time (e.g. the CSR HD),

➢ a. exists at all points of space and in particles, meaning that all points of space and all particles are open onto the syg-HD.

➢ b. Therefore all systems in the QST region, at all scales (from particles to universe-bubbles), exist also within the syg-HD region, as *syg-HD systems.*

➢ c. From 7b, it follows that the syg-HD region is not a uniform indefinite stuff, but rather contains, as individual systems, the syg information and reflections of all individual systems existing within the QST region. (These are called syg-fields in ISST)

❖ 8. This creates *the Point-Circle Paradox*: A hyperdimension is called compact when it is curled up as a point (near Planck scale). However, given that (according to 7a) a HD must exist at all points of space and within particles, then it is pervading the whole universe and is quasi-spatial and extended as a virtual hyperdimensional surface-layer.

➢ a. So that any compact HD can also be treated as a quasi-spatial *bulk*, especially for an observer standing within this HD itself.

❖ 9. There are two definitions of 'being alive':

➢ a. *Being alive as a growing, interacting, evolving biosystem*, that is, the living in spacetime

➢ b. *Being conscious, that is, alive in spirit*, as a sensing, self-referential, self-conscious and a thinking and learning cognitive system, in syg-HD; and that is the definition of intelligence, as the capacity to develop strategies and to evolve based on learning and thinking.

❖ 10. Due to *the Point-Circle Paradox, any hyperdimension is pervading the universe by definition*—whether as a quasi-spatial surface in case of a compact HD, or whether as an embedding system in case of a 'bulk-type' HD. Using systems theory's framework, any hyperdimension is axiomatically embedding the potentialities and the information of its component dimensions, just as any embedding system will vis-à-vis its component systems. Since consciousness is by necessity a hyperdimension (as stated in §1),

➢ a. Then, according to 9b, the whole cosmos (as hyperdimensional and a hypersystem) is conscious, intelligent and alive;

➢ b. Then, all matter-spacetime systems, having part of their existence in the syg-HD, are entities, that is, individualities endowed with a proto-consciousness at the very least.

➢ c. Then consciousness exists at all coordinates of time and space of the embedded 4D spacetime universe.

❖ 11. If consciousness (given 10c) exists at all time and space coordinates of the 4D matter-universe, and given that the cosmos is a system, a whole entity,

➢ a. Then, according to systems theory, all components of the global system are in mutual exchange and inter-influence; no sub-system can be modified without the embedding system 'knowing' about it, and being modified itself. Inversely, the global system cannot be modified without the information of this modification in-forming all the sub-systems, and without these sub-systems being themselves modified.

➢ b. Therefore consciousness (the syg-HD region) and the matter universe evolve in mutual interaction and exchange;

➢ c. And consequently, all matter-systems we observe are imbued and infused with syg-energy, and their systemic organization is the product of their constant interaction with syg-energy.

Given all the above, the ISS theory poses as postulates that:

❖ 12.1. The universe as a whole is organized by:

➢ a. A collective intelligence of the cosmos (the syg-HD), who is perpetually self-creating and complexifying.

➢ b. Therefore all intelligent civilizations in the cosmos are co-evolving.

➢ c. This syg-metadimension is interlocked in a triune braid with Center metadim and Rhythm metadim, to form the CSR hyperdimension.

❖ 12.2. The 'seed' of the ISS or *Cosmic DNA* in our universe-bubble (UB) came from parent or previous UBs, and there are very probably other such 'collars' of UBs.

➢ a. So that our UB, at the origin, had the condensed knowledge of a chain of previous UBs.

➢ b. The cosmic ISS is the cosmic DNA bearing the information of our collar of UBs, the latest potentialities that have been expressed and experienced by our parent UB being more active and strong at the birth of our own cosmic DNA, than that of the previous UBs in our collar.

➢ c. The cosmic ISS, as cosmic DNA, has myriads of potentialities of possible states, yet only some of them are active at birth. The collective consciousness of our own UB will choose its own paths of exploration and development, being largely free to tread new paths and change the 'network-weights' and trajectories of past UBs' experiences and realizations.

❖ 12.3. All UBs are striving for a more harmonious evolution and overall interactions among intelligent civilizations, because it insures their survival.

❖ 12.4. All matter systems (including biosystems and quantum systems), at all scales, have a CSR hyperdimension:

➢ a. As a curled-up, compact dimension, yet also enmeshed in an extended bulk,

➢ b. starting at a sub-Planckian radius (below Planck length), and

➢ c. spiraling, as the ISS, toward the near-infinite X-point;

➢ d. This individual ISS in all matter systems is both a facet of the cosmic ISS (cosmic DNA), and an individualized syg-field.

CONCLUSION

Irrespectively of the worldline of this specific ISS theory, the research I've been involved with along the years both on ISST and SFT, has highlighted priorities regarding the development of science in this 21st century. Allow me to list the ones I deem most needed and essential to insure the well-being of humanity.

- We need a full-blown theory of **nonlocal** consciousness grounded in cognitive sciences and consciousness studies, able to account for the beyond-spacetime processes of the mind and consciousness—such as high meditative states, peak states, Aha! or Eureka! states, creative experiences, flow states, fusion states, and psi phenomena including the influence of mind over matter.
- We need to lay the ground for the '**physics of synchronicity**,' that is, (1) a physics reaching beyond the Einsteinian spacetime as well as quantum indeterminacy, and also beyond strict causality as it is grounded on linear time; (2) able to sort out the laws of acausal, two-way, inter-influences, and (3) that integrates consciousness as the self-organizing and negentropic force of the universe.
- We need to research the domain of the **merging of mind and matter** (along the visionary hints given us by Pauli and Jung), or more precisely, of consciousness-as-energy, and to sort out the laws of their interaction, communication, and inter-influence, as well as the specific domains of application of these laws.
- And finally: an integral theory has to integrate consciousness as a set of specific dimensions (a N-dimensional hyperdimension)—at both the individual mind level, and at the cosmic level.

The sad point is that by not recognizing the action of consciousness on the world of matter, and not being aware that it is a force in itself (working within metadimensions of its own and behaving according to another set of laws than matter-energy), we keep enchaining humanity and keeping it hostage to a blind paradigm. This purely materialistic worldview of the science of the last three centuries, that has for the most part complied blindly with the politics of control of the powers-that-be, has been the rationale for all worldwide or domestic interest groups to pursue their petty aims. It has led our societies to disregard protecting and nurturing the beauty of cultures, of original individualities, of divergence and creativity, and of the natural harmony with Earth as a living entity—Gaia.

I'm deeply confident that science, in this 21st century, is bound to transform itself into a new paradigm, in which the keywords are consciousness and mind potentials, depth of vision, harmony, being in sync, soft inter-influence, respect of divergence, creativity, flow states. And scientific research will put a high priority on the harmonious development of the potentials of human beings. In fact, as any keen observer and mind in sync can experience it, science is already mutating at exponential speed, and its aims and values are shifting toward a more human-friendly worldview. This has been long due!

Three intelligent dimensions/forces at the origin?

There's still a last point I want to make: an open question for us to ponder. While re-reading the book prior to publishing it, I'm struck by an idea that didn't really cross my mind while developing the theory. It's one of those ideas that can only emerge when a synthetic comprehension of the whole system is achieved, as a second phase.

Early on, I had assimilated the semantic fields of individuals to their Self (or soul), and the planetary semantic field to the collective unconscious, the Tao, Brahman... that is, collective consciousness.

With the ISS, I now pose, as the source of our universe, a cosmic consciousness that is also a force, syg energy, pervading our universe at all times and all scales, and thus implied and activated in any thought process and cognitive act. I had an insight when I pondered anew the ISS in each person (their Self) being in constant two-way conversation and interaction with the cosmic ISS, the Source.

As the person's ISS has its own CSR (3-metadim) manifold, thus acquiring the status of an intelligent and autonomous entity dwelling in a beyond-spacetime dimension (which is exactly what Jung posed with the Self, as the subject of the unconscious), it appears that the ISS theory gets a step nearer to clarifying the nature of the classically called 'immaterial soul.' The 'soul' is indeed a persistent concept in all cultures, even when presented in a non-religious philosophy such as the Advaita Vedanta in India (*advaita* means non-dualist, and *vedanta* means philosophy), or Taoism in China (another philosophy of Oneness), and also as the *monad* in Leibniz's philosophy. The 'soul' in Christian lore is believed to be 'near' to God, and the saints themselves are thought to be in a specific semantic space: the 'communions of the saints' in which they are in constant empathy, conversation and divine energy sharing (in SFT's terms: connected between them via their resonant semantic fields).

We also saw many qualities traditionally attributed to God or a deity, being expressed and instantiated by the Cosmic ISS, notably:

- the *'omnipotence'* attribute, or immense power (= syg energy);
- the *'ubiquity'* attribute: the *deity's essence pervading the universe* and creating it (syg dimension);
- the *'ordering principle'*: the universe seen as the 'Grand Architect' by the Freemasons, thus hinting at a *sacred geometry*, handed down from Pythagoras; the enigmatic letter G, omnipresent, central, and sacred for them, in my opinion stands for the *Geomatria* of the Greek philosophers (letter Gamma in Greek, translated as the G of our Latin alphabet). This sacred geometry was thought to be the foundation of the order of the cosmos, in-forming matter by reflecting the First Principles;
- the *'omniscience'* capacity, that can be understood readily if one speaks about a collective consciousness of cosmic proportion, or cosmic consciousness (and in contrast, is rather questionable when a personalized entity is invoked, such as a creator god as a person).

As for the resonances of the archetype of a primordial intelligence and/or of a Creator image, in the 'creatures' and the 'manifested world' at large:

- *the concept of a soul* who belongs to the divine realm, and is of similar essence (despite the lessening of its 'light'), or at least in communication with it. A basic dogma of Christian religions, the soul is also a central concept in Egypt (with the Ka), and as well in the philosophies of Oneness (also badly termed 'negative philosophies') such as Taoism, Advaita, and in Plotinus' system. For example, in the *Upanishads* (expounding the Advaita Vedanta), the individual ego (*jiva*) may, by meditation and samadhi states, rise to connect and get harmonized with his/her *Atman* (or Self, soul), and in that state gets spontaneously in communion with the cosmic consciousness, the beyond-duality *Purusha/Brahman*;

- *the concept of Akasha* (Vedic, Hindu, and Tibetan Mahayana Buddhist religions), meaning that all beings and events in the universe are memorized in a sort of central databank often referred to as the 'Akashic Records, or Library.' Some ancient texts attribute this spontaneous 'imprinting' process to the force of *prana*. Prana is cosmic consciousness as energy or force, and as such it pervades the universe and is also replete in the atmosphere. The yogi's 'prana yoga' is a set of techniques for inhaling this spiritual and healing force through breathing. The Akasha (just as the prana) has several levels, the most basic one being a world's memory-field. At its global level, the Akasha is a highly spiritual ether defined in somewhat identical terms with the brahman or cosmic consciousness. Brahman, Tao, and Prana could be the religious-philosophical concepts the nearest to our actual scientific term of 'hyperdimension.' This is the meaning retained by Nikola Tesla and also by Ervin Laszlo. The latter quotes Swami Vivekananda stating in his book *Raja Yoga* that at the beginning of creation as well as at the end of 'a creation' there's only the Akasha. In this latent state, all forces are called Prana; and all things emerge from the Akasha for the next creation. (Laszlo, 2013, Chapter 9);
- the Golden ratio Phi, viewed by Pythagoras as the 'measure' of the whole universe from the smallest scale to the cosmic scale.

The most striking similarity is the resonance of number 3 (as Center-Syg-Rhythm manifold), with triune divine realms or principles, notably:

- the Christian divine Trinity (or plerome) of three masculine essences, including the enigmatic Holy Spirit, with the Father as Creator, and the Son (Jesus) incarnated as Savior;
- the trinity of gods in Hinduism—Brahma (masculine in Sanskrit, as distinct from Brahman neutral, the cosmic consciousness), Shiva, and Vishnu (Krishna), each one with his feminine consort;
- the three cosmic essences (the One, the Intelligence, the Universal Soul) as 'first causes' in Plotinus, each with their counterpart in the individual human being;

- the three 'ideas' in Plato as archetypes existing at the source of the universe, and all events in the world being their dim 'reflection;'
- the three divine concentric circles (or cosmic dimensions) in Celtic religion (Druidism), with the dimension of The One at the center. Sometimes also represented as a triune spiral, as in the Triscel, showing the dynamic interaction between these dimensions;
- the three active principles of the mantra AUM, the most sacred mantra of Hinduism, representing in sound-frequencies the three dimensions of conscious existence: Brahman/Purusha (cosmic consciousness), Atman (the dimension of Selfs), and the ego-matter layer. This mantra is used for creating the harmonization of a mind with the whole.
- the three cosmic forces in Hinduism: Sat-Chit-Ananda, where *Sat* is the essence, *Chit* is cosmic consciousness, and *Ananda* is bliss;
- the isosceles triangle (3 equal sides, and 3 angles of 60°) as sacred for Pythagoras; its complex 2D structure: the Sri Yantra or Star of David (comprising 6 isosceles triangles); and also its two complex 3D structure: the tetrahedron (4 isosceles triangles), and the pyramid (4 isosceles triangles and a square base);
- the double spiral + central axis, featuring two entwined snakes, in Hermes' Caduceus (symbol of science and wisdom), already existing in Sumer, through his father Enki's double-snake symbol (see Hardy, *DNA of the Gods*, p. 212-3).

The above was just a small glimpse at the occurrences of the number 3 regarded as a creating and ordering principle among the first causes/principles —in religions, metaphysics, and cosmogonies.

So, given the preponderance of number 3 in so many cultures and religions, symbolizing the creative principle existing at the origin, here is an intriguing question: What if the three masculine and relatively humanized deities of the divine trinity (in both Christian and Hindu religions) had been, in a long forgotten past, *three neutral entities or intelligent forces at the origin?* We have the clear case of Brahman (neutral in Sanskrit) as the cosmic consciousness, closely associated with the god of creation, no less, Brahma (masculine in Sanskrit). This could also explain the injunction, in some religions, not to represent in humanized images the One deity—although the personalization is nevertheless pronounced via specific volitions, commands and actions. Furthermore, in some languages, for example in French, the neutral subject is mostly expressed by the masculine subject 'he' ('*il*'). This is of course a sound and educated guess, since our Western sciences and

philosophies are derived from the Greek ones, and given that the Greek sciences all issued in Egypt, while in turn the Egyptian ones are rooted in the Sumerian civilization (see my book *DNA of the Gods* for specific examples).

A very interesting point to consider, especially in the light of what we saw about the qualities presented by the ISS and found as classical attributes of the deity, is the fact that the ISS is devoid of moralistic judgments. It doesn't show the moral flaws of the gods of old, relatively projected out of our own human psychological profile—such as wrath, jealousy, despotic and domineering tendencies, warlike drives, etc. So that, if there's anything like a law of karma (as a balancing effect of good/bad actions), it won't be rooted at the moralistic level, and we'll rather have to look for dynamics of psychological balance and of projection-introjection (as in SFT), or of compensation as expressed by Carl Jung.

The most unexpected development of the ISS theory and a very heartening one, is to integrate a very ancient Hermetic knowledge with a state-of-the-art scientific knowledge about the origin and the pre-space, and to do so in a way that prods our scientific paradigm to move beyond the materialistic worldview so as to welcome a meaning-laden universe—one in which participation and co-creation, as well as self-organization, are the core dynamics.

The self-creating ISS is neither towering, nor despotic, nor overshadowing; it leaves free rein to our creativity, inventiveness, and autonomy. The ISS multilayered universe is indeed a very beautiful place to live in, exquisitely complex and variegated, yet gracefully ordered... It is the cosmic piano on which we may play and create our individual and planetary Grand Opus, while connecting with intelligent civilizations in our galaxy.

It is also, I believe, how we'll one day get to know the universe-bubbles that have preceded our own and left their print on our own cosmic DNA, and how we'll discover or deduce the existence of other such collars of pearls making up the grand tapestry of a non-random yet non-deterministic, relatively structured, pluriverse—one that offers us enough slack for inventiveness, art and beauty.

NOTES

1. http://en.wikipedia.org/wiki/EPR_paradox)
2. http://en.wikipedia.org/wiki/Wilkinson_Microwave_Anisotropy_Probe
3. Tesla, "The new art of projecting concentrated non-dispersive energy through natural media." http://www.tfcbooks.com/tesla/1935-00-00.htm
4. http://science.nasa.gov/astrophysics/focus-areas/what-is-dark-energy/
5. T. Beardsley, *Scientific American* 267:6 (1992), p. 19; see also http://www.ias.ac.in/jarch/jaa/18/455-463.pdf
6. Citation from http://www.khouse.org/enews_article/2013/2045/print
7. http://en.wikipedia.org/wiki/Redshift_quantization
8. Dr. Emily Baldwin "Cosmic dark flow mystery deepens." *Astronomy Now,* March 12, 2010. http://www.astronomynow.com/news/n1003/12darkflow/
9. http://www.nasa.gov/centers/goddard/news/releases/2010/10-023.html. See also *Astrophysical Journal Letters*, March 20, 2010, at http://arxiv.org/abs/0910.4958;
10. www.nasa.gov/centers/goddard/universe/black_hole_sound.html
11. www.global-scaling-institute.de/files/gscompv18_en.pdf
12. "The Problem of Increasing Human Energy." Find it at: http://www.tfcbooks.com/mall/more/330prob.htm
13. C. Hardy. Synchronicity: Interconnection through a Semantic Dimension. See also Chris H. Hardy, *The Sacred Network,* p. 324-7.
14. See also http://en.wikipedia.org/wiki/Polar_jet http://en.wikipedia.org/wiki/Black_hole; http://simple.wikipedia.org/wiki/Black_hole
15. Schödel, R. *et al* (2002). "A star in a 15.2-year orbit around the supermassive black hole at the centre of the Milky Way". *Nature 419* (6908): 694–696. doi:10.1038/nature01121. PMID 12384690. (Last checked: 5/5/2013) And also: Henderson, Mark (2008). "Astronomers confirm black hole at the heart of the Milky Way." London: *Times Online*. http://www.timesonline.co.uk/tol/news/uk/science/article5316001.ece.
16. See the Wikipedia's article called 'Hawking Radiation': http://en.wikipedia.org/wiki/Hawking_radiation)
17. Plotinus, Enneads 2, Liv. 3, Google Books, p. 134; in: Plotin (1857). *Les Ennéades*. Traduction M.N. Bouillet. Paris: Hachette. See the scanned text at http://books.google.com
18. Cited by I. & G. Bogdanov, *Le Visage de Dieu*, Paris: J'ai lu, 2011; Grasset 2010; p. 259.
19. http://en.wikipedia.org/wiki/Cosmic_microwave_background_radiation (last acc. 4/2/2013)
20. http://www.michaelglickmanoncropcircles.com/blog/why-pi/

21. http://chris-h-hardy-dna-of-the-gods.blogspot.fr/2014/03/embedment-of-pi-number-within-barbury.html
22. Here is how to calculate the Fibonacci sequence: 1 + 1 = 2; 2 + 1 = 3; 3 + 2 = 5, and so forth. The beginning of the sequence is thus 1, 2, 3, 5, 8, 13, 21, 34, 55, 89, 144, 233, 377, 610, 987... As the sequence progresses toward large numbers, the proportion between two sequential numbers keep approaching the golden proportion, or phi (1.618, also referred to by the Greek letter φ). One way to show this is to divide a number in the sequence by the previous number in the sequence. Thus, 8/5 = 1.6000, 13/8 = 1.6250, and 21/13 = 1.6153. Beginning at 55/34 = 1.6176, we are very near to the value of phi. At the proportion 233/144 = 1.618055, we are exact at four digits beyond the decimal point (that is, exactly 1.6180). Continuing to divide according to the rules of Fibonacci results in 6,765/4,181 = 1.6180339; that is, the more we go in the sequence, the more the numbers approach the phi value.
23. Nikola Tesla, "The Problem of Increasing Human Energy." http://www.tfcbooks.com/mall/more/330prob.htm
24. This is part of a poem I composed at a very young age. Christine Hardy (1983). *Paysages d'infini*. Paris: Ed. L'Originel. (p. 41)
25. Carr, 2003, http://www.tcm.phy.cam.ac.uk/~bdj10/psi/carr2003.html
26. Penrose, Roger & Hameroff, Stuart (2014). "Reply to Seven Commentaries on Consciousness in the Universe: Review of the 'Orch Or' Theory.'" *Physics of Life Reviews,* 11(1), March 2014 (94–100). See also :
    - Penrose, R & Hameroff, S. (2011) "Consciousness in the Universe: Neuro-science, Quantum Space-Time Geometry and Orch OR Theory." *Journal of Cosmology,* 14. http://journalofcosmology.com/Consciousness160.html
    - Hameroff, S. (1996). "Did Consciousness Cause the Cambrian Evolutionary Explosion?" In: *Toward a Science of Consciousness II*. Hameroff, S., Kaszniak, A.W., & Scott, A.C., (Eds.) Cambridge, Ma: MIT Press/Bradford.
    - Hameroff, S.R., & Penrose, R. (1996). "Orchestrated reduction of Quantum Coherence." In S.R. Hameroff, A.W. Kaszniak, & A.C. Scott (Eds.), *Toward a science of consciousness*. Cambridge, Ma: MIT Press/Bradford.
27. Ervin Laszlo (2014) "Consciousness in the cosmos." http://clubofbudapest.org/ervinlaszlo
28. Towler, M. Lecture on Pilot-wave theory, Bohmian metaphysics. Cambridge Univ. UK; 2009. www.tcm.phy.cam.ac.uk/~mdt26/PWT/lectures/bohm1.pdf)
29. John Bush (2015 in print). See the PDF text at : http://math.mit.edu/~bush/wordpress/wp-content/uploads/2014/09/Bush-ARFM-2015.pdf
30. Natalie Wolchover, *Quanta Magazine,* June 2014: (http://www.wired.com/2014/06/the-new-quantum-reality)
31. Hardy, C. (2003) "Multilevel Webs Stretched across Time: Retroactive and Proactive Inter-influences." *Systems Research and Behavioral Science* 20, no. 2 (2003): 201–15.
32. Read and download Chris Hardy's articles on Semantic Fields Theory at http://cosmic-dna.blogspot.fr.

33. Daryl Bem (2011). "Feeling the Future: Experimental evidence for anomalous retroactive influences on cognition and affect." *J. of Personality and Social Psychology, 100*, (407-25). (http://dbem.ws/)
34. YouTube, *Unwrapping a tesseract*, https://www.youtube.com/watch?v=BVo2igbFSPE)
35. See Wikipedia/ Gauge Theory, and also Compact Set.
36. http://en.wikipedia.org/wiki/Kaluza%E2%80%93Klein_theory
37. Tesla, A Machine to End War, *Liberty*; http://www.tfcbooks.com/tesla/1935-00-00.htm)
38. Susskind, L. (2003) *The Anthropic Landscape of String Theory*: (arXiv:hep-th/0302219v1 - 27 Feb 2003)
39. Paul LaViolette, *Genesis of the Cosmos* (2004), *Secrets of antigravity (2008)*. Paper at Muffon 2011. www.youtube.com/watch?v=ifEgGMFK-VU
40. http://en.wikipedia.org/wiki/Entropy (*l.a.* 1/25/2015)
41. http://en.wikipedia.org/wiki/Holographic_principle (*l.a.* 1/25/2015)
42. Polchinski et al. "Black holes: complementarity or firewalls?" In *J. of High Energy Physics* 2013(2). arXiv:1207.3123.)
43. http://www.nytimes.com/2013/08/13/science/space/a-black-hole-mystery-wrapped-in-a-firewall-paradox.html?
44. http://particle.physics.ucdavis.edu/seminars/data/media/2011/may/polchinski.p
45. http://en.wikipedia.org/wiki/Hawking_radiation (*l.a.* 01/25/2015)
46. http://en.wikipedia.org/wiki/Higgs_boson (last accessed 01/21/2015)
47. Roy P. Kerr. In his 1963 article in *Physical Revue Letters* (11, 237); citation in http://en.wikipedia.org/wiki/Black_hole (last accessed Jan 2011)
48. Hardy. C. Multilevel Webs Stretched Across Time. (2003)
49. Penrose, R. (2014). "On the Gravitization of Quantum Mechanics 2: Conformal Cyclic Cosmology." *Foundations of Physics* 44 (8) (873-890). PDF at http://link.springer.com/journal/10701/44/8/. DOI 10.1007/s10701-013-9763-z.
50. Hawking, S.W. (1975)"Particle creation by black holes." *Commun. Math. Phys.* **43**, 199–220.
51. http://hubblesite.org/newscenter/archive/releases/2004/11 (last acc. 01/21/2015)

BIBLIOGRAPHY

Abraham, F. (1993). Book review: Chaos in brain function, edited by E. Basar, in *World Futures.* 37 (41–58).

Abraham, F., Abraham, R., & Shaw, C. (1990). *A visual introduction to dynamical systems theory for psychology*. Santa Cruz, CA: Aerial Press.

Abraham, F.D., & Gilgen, A.R. (Eds.) (1995). *Chaos theory in psychology.* Westport, CT: Praeger Publishers.

Alfvén, H. *Worlds-Antiworlds*. New York: W.H. Freeman/Macmillan, 1966.

Almheiri, A., Marolf, D., Polchinski, J., & Sully, J. (2013). "Black holes: complementarity or firewalls?" *J. of High Energy Physics* 2013 (2). arXiv:1207.3123.

Anderson, P. W. (1972). "More is Different." *Science 177*, 4047, August 4, 1972.

Atmanspacher, H. & Primas, H. (1996). The hidden side of Wolfgang Pauli, *Journal of Consciousness Studies, 3* (2).

Barrett, T. & Grimes, D. (Eds.) *Advanced Electromagnetism: Foundations, Theory and Applications.* Singapore: World Scientific, 1995.

Barrow, J.D. & Tipler, F.J. *The Anthropic Cosmological Principle*. Oxford, UK: Oxford Univ. Press, 1988.

Bateson, G. *Steps to an Ecology of Mind*. Chicago, Il.: Univ. Of Chicago Press. 2000.

Bekenstein, J.D. (2003a). Black holes and information theory. *Physics Archives*: arXivquant-ph/0311049.

———. (2003b). Information in the holographic universe. *Scientific American*, August 2003, 59–65.

Bem, D. J. (2011). "Feeling the Future: Experimental evidence for anomalous retroactive influences on cognition and affect." *J. of Personality and Social Psychology, 100*, (407-25). (http://dbem.ws/)

Bergson, H. *Creative Evolution*. CreateSpace. 1911/2012.

Bierman, D.J. (2010). "Consciousness Induced Restoration of Time Symmetry (CIRTS): A psychophysical theoretical perspective." Journal of Parapsychology 74, (273-99).

Bierman, D.J., & Scholte, H.S. (2002). "Anomalous anticipatory brain activation preceding exposure of emotional and neutral pictures." *Proceedings of the Parapsychological Association 45th Annual Convention*, Paris, France.

Bogdanov, I. & G. *Avant le Big Bang*. Paris: Grasset Poche, 2004.

———. *L'esprit de Dieu.* Paris: Grasset, 2012.

———. *Le visage de Dieu*. Paris: Grasset poche, 2010.

Bohm, D. (1986). "A new theory of the relationship of mind and matter." *J. of the American Society for Psychical Research*, 80, 113–36.

———. *Wholeness and the Implicate Order*. London: Routledge & Kegan Paul, 1980.

Bohm, D. & Hiley, B.J. *The Undivided Universe: an Ontological Interpretation of Quantum Theory*. London, UK: Routledge, 1993.

Bousso, R. (2002). "The holographic principle." *Reviews in Modern Physics.* 74: 825-74.

Brandenberger, R. & Vafa, C. (1989). "Superstrings in the early universe". *Nuclear Physics B* 316(2) (391–410). Bibcode:1989NuPhB.316..391B. doi:10.1016/0550-3213(89)90037-0.

Brandenburg, J. *Beyond Einstein's Unified Field. Gravity and Electro-magnetism Redefined*. Kempton, ILL: Adventures Unlimited Press, 2011.

———. (1995). "A Model Cosmology Based on Gravity-Electro-magnetism Unification," *Astrophysics and Space Science*, 227 (133)

———. (2010). "A Combined Kaluza-Klein Sakharov Model of Baryo-Genesis. *APS meeting proceedings*," Flint Mich, Ohio.

Brillouin, L. *La science et la théorie de l'information*. Paris: J.Gabay, 1988.

Broughton, R. *The Controversial Science*. New York: Ballantine, 1991.

Bunnell, P. (1999). "Attributing nature with justifications." *Proceedings* of the 43rd annual conf. of the ISSS, Asilomar, CA.

Burns, Jean (2002). "Quantum fluctuations and the action of the mind," *Noetic J.*, 3(4) (312-17).

———. (2003). "What is beyond the edge of the known world?" in J. Alcock, J.E. Burns, & A. Freeman (Eds.), *Psi Wars.* Exeter, UK: Imprint Academic. Reprinted in R.M. Schoch & L. Yonavjak (Eds.), *The Parapsychology Revolution.* New York: Tarcher, 2008.

Bush, JWM (2015)."Pilot-Wave Hydrodynamics." *Annual Review of Fluid Mechanics*, Vol 47, 01-2015. http://math.mit.edu/~bush/wordpress/wp-content/uploads/2014/09/Bush-ARFM-2015.pdf

Carr, B. *Universe or multiverse*. Cambridge, UK: Cambridge Univ. Press, 2009.

———. (2010)."Seeking a New Paradigm of Matter, Mind and Spirit." *Network Review*, Spring, Summer 2010.

———. (2014)."Hyperspatial models of matter and mind." In E. Kelly, A. Crabtree & P. Marshall (Eds.). *Beyond Physicalism: Towards Reconciliation of Science and Spirituality.* Lanham, MD: Rowman & Littlefield.

Carter, B. (1974). "Large number coincidences and the Anthropic Principle in cosmology." In: *Confrontation of cosmological theories with observational data*. Boston, Ma: Reidel.

Casti, J. *Lost paradigms.* New York: William Morrow, 1989.

Chalmers, D. (1996). "Facing up to the problem of consciousness." In S.R. Hameroff, A.W. Kaszniak, & A.C. Scott (Eds.), *Toward a science of consciousness*. Cambridge, Ma: MIT Press/Bradford Book.

Chalmers, D.J. *The Conscious Mind: In Search of a Fundamental Theory*. New York: Oxford University Press, 1996.

Changeux, J.P. *Neuronal Man: The Biology of Mind*. Princeton, NJ: Princeton Univ. Press, 1983-1997.

Combs, A. *The Radiance of Being. Complexity, Chaos and the Evolution of Consciousness*. St Paul, MN: Paragon House, 1996.

———. (1995). "Psychology, chaos, and the process nature of consciousness." In F.D. Abraham & A.R. Gilgen (Eds.), *Chaos theory in psychology*. Westport, CT: Praeger Publishers.

Combs, A. & Holland, M. *Synchronicity: Science, Myth, and the Trickster*. New York: Marlowe, 1995.

Costa de Beauregard, O. *Le second principe de la science du temps*. Paris: Seuil, 1963.

———. (1975). "Quantum paradoxes and Aristotle's twofold information concept." In *Quantum physics and parapsychology*. New York: Parapsychology Foundation.

———. (1985). "On some frequent but controversial statements concerning the Einstein-Podolsky-Rosen Correlations." *Foundations of Physics,* 15, (8), 871-887.

Couder, Y. & Fort, E. (2012). "Probabilities and trajectories in a classical wave-particle duality." *J. Phys. Conf. Ser.* 361:012001

Couder, Y., Protière, S., Fort, E. & Boudaoud, A. (2005). "Walking and orbiting droplets." *Nature,* 437:208.

Csikszentmihalyi, M. *Flow: The Psychology of Optimal Experience*. New York: Harper & Row, 1990.

Davies, P. *The Goldilocks Enigma*. New York: Allen Lane/ Penguin, 2006.

Davies, P. & Gregersen, N.H. (Eds.) *Information and the Nature of Reality*. Cambridge, UK: Cambridge Univ. Press, 2010.

Deardorff, J., Haisch, B., Maccabee, B., & Puthoff, H.E. (2005). "Inflation Theory; Implications for Extraterrestrial Visitation." *J. of the British Interplanetary Society (JBIS), Vol. 58, pp. 43-50.*

De Broglie, L. *Matter and Light: The New Physics*. Mineola, NY: Dover Publ., 1st Publ. 1939.

———. *Physics and microphysics. New York*: Pantheon, 1st Publ. 1955.

Deleuze, G. & Guattari, F. *A Thousand Plateaus*. New York: Continuum, 2004.

Dennard, Linda (1996). "The New Paradigm in Science and Public Administration." *Public Administration Review* 56 (15): 495–99.

Dossey, Larry. *Recovering the Soul: A Scientific and Spiritual Approach*. New York: Bantam New Age Books, 1989.

Feinstein, D. & Krippner, S. *The Mythic Path.* New York: Tarcher/Putnam, 1997.

Feynman R.P. (1969). "Very High-Energy Collisions of Hadrons." *Phys. Rev. Lett. 23 (1969), p. 1415.*

———. *Surely You're Joking, Mr. Feynman!* New York: W.W. Norton & Co., reprint 1997.

Fredkin, E. (1992). "Finite Nature." *Proceedings of the XXVIIth Rencontre de Moriond.* 1992.

Freeman, W. (1999). "Consciousness, intentionality and causality." *J. of Consciousness Studies*. 6:143-172. (Nov/Dec).

———. *Societies of brains: A study in the neurosciences of love and hate.* Hillsdale, NJ: Lawrence Erlbaum, 1995.

Gell-Mann, M. *The quark and the jaguar.* New York: W.H. Freeman & co, 1994.

Germine, M. (2007). "The Holographic Principle Theory of Mind." *Dynapsych* (online J.) http://goertzel.org/dynapsyc/2007/holomind.htm

Gibson, J.J. (1986). *An ecological approach to visual perception*. Hillsdale, NJ: Lawrence Erlbaum. (First printing: 1979).

Gleick, James. *Chaos*. New York, NY.: Viking Press, 1987.

Gödel, K. *On formally undecidable propositions of principia mathematica and related systems.* Mineola, NY: Dover Publ., 1992.

Goerner, S. *After the clockwork universe*. Edinburgh, Scotland: Floris Books, 1999.

Greene, B. *The Elegant Universe - Superstrings, Hidden Dimensions, and the Quest for the Ultimate Theory*. New York: W.W.Norton & Co., 2010.

———. *The Hidden Reality: Parallel Universes and the Deep Laws of the Cosmos*. New York: Vintage/Random, 2011.

Gribbin, J. *In Search of the Multiverse: Parallel Worlds, Hidden Dimensions, and the Ultimate Quest for the Frontiers of Reality.* NY: John Wiley, 2010.

———. *In Search of Schrödinger's Cat*. Toronto, Canada: Bantam Books, 1984.

Guastello, S. *Chaos, catastrophe, and human affairs*. Mahwah, NJ.: Lawrence Erlbaum As., 1995.

Guth, A.H. *The Inflationary Universe*. Reading, Ms: Perseus Books, 1997.

Haisch, B. *The God Theory: Universes, Zero-point Fields, And What's Behind It All*. York Beach, ME: Red Wheel/Weiser Books, 2006.

Hameroff, S. (1996). "Did Consciousness Cause the Cambrian Evolutionary Explosion?" In: *Toward a Science of Consciousness II*. Hameroff, S., Kaszniak, A.W., & Scott, A.C., (Eds.) Cambridge, Ma: MIT Press/Bradford.

Hameroff, S.R., & Penrose, R. (1996). "Orchestrated reduction of Quantum Coherence." In S.R. Hameroff, A.W. Kaszniak, & A.C. Scott (Eds.), *Toward a science of consciousness*. Cambridge, Ma: MIT Press/Bradford.

Hardy, Chris H. *Diverging Views. On Our Way to the Galactic Club.* Delhi, India: Terra Futura, 2008.

———. *DNA of the Gods: The Anunnaki Creation of Eve and the Alien Battle for Humanity*. Rochester, Vt: Bear and Co., 2014.

———. *The Sacred Network*. Rochester, Vt: Inner Traditions, 2011.

Hardy, Christine (2000). *Two types of memory-as-process in a dynamical networks' view of the mind.* Paper, Society for Chaos Theory in Psychology and Life Sciences, Annual conf., Philadelphia, NJ.

———. (2001). "Self-organization, Self-reference and Inter-influences in Multilevel Webs: Beyond Causality and Determinism." *J. of Cybernetics and Human Knowing 8*, no. 3 (July 2001): 35–59.

———. (2003). "Complex intuitive dynamics in a systemic cognitive framework." *Proceedings* (Cd) Inter. Soc. for the Systems Sciences, Crete.

———. (2003). "Multilevel Webs Stretched across Time: Retroactive and Proactive Inter-influences." *Systems Research and Behavioral Science* 20, no. 2 (201–15).

———. (2004). "Synchronicity: Interconnection through a Semantic Dimension." Presentation (and *Proceedings*) at Second Psi Meeting, April 21–26, Curitiba, Brazil.

———. *La prédiction de Jung: la métamorphose de la Terre*. Paris, France: Dervy/Trédaniel, 2012.

———. *Le vécu de la transe*. Paris: Editions du Dauphin, 1995.

———. *Networks of Meaning: A Bridge between Mind and Matter*. Westport, Conn.: Praeger, 1998.

Hawking, S.W. *A Brief History of Time*. New York: Bantam Books, 1988.

———. (2003). "Cosmology from the Top Down." Paper given at Davis Inflation Meeting, 2003. http://arxiv.org/abs/astro-ph/0305562

———. (2014). "Information Preservation and Weather Forecasting for Black Holes." arXiv:1401.5761v1 [hep-th] 22 Jan 2014.

Hebert, J. "The Higgs Boson and the Big Bang." *Acts & Facts*. 41 (9): 11–13. 2012.

Heidegger, M. *The Principle of Reason*. Bloomington, In.: Indiana Univ. Press, 1996.

Herbert, N. *Quantum Reality: Beyond the New Physics.* NY: Doubleday, 1985.

Hiley, B.J. (2010). "The Bohm Approach Re-Assessed." http://www.bbk.ac.uk/tpru/RecentPublications.html

Hiley, B. & Peat, D. *Quantum Implications*. NY: Routledge, 1991.

Jahn, R.G. (Ed.) *The Role of Consciousness in the Physical World*. Boulder, CO: Westview, 1981.

Jahn, R.G. & Dunne, B. *Margins of Reality: The Role of Consciousness in the Physical World.* Princeton, NJ: ICRL Press, 2009.

———. *Quirks and the Quantum Mind.* Princeton, NJ: ICRL Press, 2012.

Jourdan, J-P. (2000). "Just an extra dimension." Paris, Fr.: *Les Cahiers de IANDS-France*.

Jordan, M.J. "An introduction to linear algebra in parallel distributed processing." In D.E. Rumelhart & J.L. McClelland, *Explorations in the microstructure of cognition; Vol1: Foundations*. Chapter 9. Cambridge, Ma: MIT Press/Bradford Books, 1986.

Josephson, Brian D. (2003). "String Theory, Universal Mind, and the Paranormal." *http://arxiv.org/html/physics/0312012v3*

Josephson, B.D., & Pallikari-Viras, F. (1991). "Biological Utilisation of Quantum NonLocality." *Foundations of Physics*, 21: 197–207. http://www.tcm.phy.cam.ac.uk/~bdj10/papers/bell.html

Josephson, B.D. and Carpenter, T. (1996). 'What can music tell us about the nature of the mind? A Platonic Model.' In S.R. Hameroff, A.W. Kaszniak and A.C. Scott (Eds.) *Toward a Science of Consciousness*, Cambridge, Ma: MIT Press (691–4).

Jung, C. Alchemical Studies. In *The Collected Works of C.G. Jung,* Bollingen Series, vol 13. Adler, G., and R. F. Hull (Eds.) Princeton, N.J.: Princeton University Press, 1968.

———. *Answer to Job*. New York: Routledge and Kegan Paul, 1954.

———. Introduction, In Wilhelm, R. *The Secret of the Golden Flower: A Chinese Book of Life*. Mariner Books, 1962.$

———. *Man and his Symbols*. Garden City, N.Y.: Windfall Books/DoubleDay, 1964.

———. *Memories, Dreams, Reflections*. New York: Vintage/Random, 1965.

———. Psychology and Alchemy. In *The Collected Works of C. G. Jung,* Bollingen Series, vol. 12. Adler, G., & R.F. Hull (Eds.) Princeton, N.J.: Princeton University Press, 1968.

———. Synchronicity: an Acausal Connecting Principle. In *The Collected Works of C. G. Jung,* Bollingen Series, vol. 8. Adler, G., and R. F. Hull (Eds.) Princeton, N.J.: Princeton University Press, 1960.

———. *The Red Book*. (Ed.) S. Shamdasani. Philemon Series, The Philemon Foundation. New York: W.W. Norton & Co. 2009.

———. Two Essays on Analytical Psychology. In *The Collected Works of C.G. Jung,* Bollingen Series, vol. 7. Adler, G., & R.F. Hull (Eds.) Princeton, N.J.: Princeton University Press, 1966.

Jung, C. & Pauli, W. *The Interpretation of Nature and the Psyche*. New York: Pantheon Books, 1955.

Kafatos, M. and Nadeau, R. *The Conscious Universe: Part and Whole in Modern Physical Theory.* New York: Springer-Verlag, 1990.

Kaku, M. *Hyperspace: A Scientific Odyssey Through Parallel Universes, Time Warps, and the 10th Dimension.* New York: Anchor, 1994.

———. *Parallel Worlds: A Journey Through Creation, Higher Dimensions, and the Future of the Cosmos.* NY: Anchor, 2006.

Kauffman, S. *At home in the Universe: The search for the laws of self-organization and complexity*. Oxford, NY: Oxford University Press, 1995.

Kerr, Roy P. (1963). "Gravitational Field of a Spinning Mass as an Example of Algebraically Special Metrics." *Physical Review Letters* 11(5): 237–238.

Koestler, A. *The act of creation*. NYC: Penguin, 1989.

Kuhn T. *The structure of scientific revolutions*. Chicago, Il.: Univ. of Chicago press, 1970.

Laszlo, E. *The Akashic Experience: Science and the Cosmic Memory Field.* Rochester, Vt: Inner Traditions, 2009.

———. *Science and the Akashic Field: an integral theory of everything*. Rochester, Vt: Inner Traditions, 2004.

———. *The Self-Actualizing Cosmos*. Rochester, Vt: Inner Traditions, 2014.

LaViolette, P. *Genesis of the Cosmos*. Rochester, Vt: Bear & Co., 2004.

———. *Secrets of Antigravity Propulsion.* Rochester, Vt: Bear & Co, 2008.

Leibniz, G.W. *Discourse on Metaphysics and the Monadology*. NY: Prometheus Books, 1992.

Lewin, R. *Complexity: Life at the edge of chaos*. Chicago: Univ. of Chicago Press, 1999.

Libet, B., Wright E.W. Jr., Feinstein, B., & Pearl, D.K. (1979). "Subjective referral of the timing for a conscious sensory experience," *Brain 102,* 193–224.

Linde, A. (1994). "The Self-Reproducing Inflationary Universe." *Scientific American*, 271 (5), 48-55. http://web.stanford.edu/~alinde/1032226.pdf

Lorenz E. *The Essence of Chaos*. Seattle: Univ. of Washington Press, 1993.

Lovelock, J. *Gaia: A new look at life on earth.* Oxford, UK: Oxford Univ. Press, 1979-2000.

Lovelock, J., & Thomas, L. *The ages of Gaia*. NYC: Bantam books, 1990.

McMoneagle, J. & Tart, C. *Mind Trek: Exploring Consciousness, Time, Space Through Remote Viewing.* Charlottesville, Va.: Hampton Roads, 1993.

McRae, R. (1984). *Mind Wars: The true story of government research into the military potential of psychic weapons.* New York: St. Martin's.

McTaggart, L. *The Field: The Quest for the Secret Force of the Universe*. NYC: Harper Perennial, 2012.

Margulis, L. & D. Sagan. *Microcosmos. Four Billion Years of Evolution from Our Microbial Ancestors*. NYC: Simon and Shuster/Summit Books, 1986.

Mather, J. & Boslough, J. *The Very First Light: A Scientific Journey Back to the Dawn of the Universe.* New York: Penguin, 1998.

Maturana, H. & Varela, F. *Autopoiesis and cognition*. Boston: Reidel, 1980.

Maturana, H. (1999). "Autopoiesis, structural coupling and cognition." *Proceedings* of the 43rd annual conf. of the ISSS, Asilomar, CA.

May, E.C., Utts J.M., & Spottiswoode, S.J.P. (1995). "Decision augmentation theory: applications to the random number generator database." *J. Scientific Exploration,* 9:453-488.

Minsky, M. *The society of mind*. New York: Simon & Schuster, 1985.

Mishlove, Jeffrey. *The Roots of Consciousness*. New York: Marlowe & Co., 1997.

Mitchell, Edgar R. *The way of the explorer*. New York: Putnam, 1996.

Mongan, T.R. (2007). "Holography and non-locality in a closed vacuum-dominated Universe." *Intern. J. of Theoretical Physics,* 46. (399–404).

Morin, E. *Homeland Earth*. Cresskill, NJ: Hampton Press, 1998.

———. *Method: Toward a study of humankind; The nature of nature.* American Univ. Studies (Series V, Philosophy, Vol. 1.), 1992.

Müller, H. (1987). "The General Theory of Stability and evolutional trends of technology." In *Evolutional trends of technology and CAD applications*. Volgograd Institute of Technology, 1987. (In Russian)

Narby, J. *The Cosmic Serpent: DNA and the Origins of Knowledge*. New York: Tarcher/Putnam, 1999.

Nelson, R.D., Bradish, G.J., Dobyns, Y.H., Dunne, B.J., & Jahn, R.G. (1996). "FieldREG anomalies in group situations." *J. of Scientific Exploration,* 10(1), 111–41.

Neppe, V.M. & Close, E.R. (2012). Reality Begins with Consciousness: A Paradigm Shift That Works. E-Book: brainvoyage.com

Nunn, C.H.M., Clarke, C.J.S., & Blott, B.H. (1994). "Collapse of a quantum field may affect brain function." *J. of Consciousness Studies, 1* (1), Thorverton, UK.

Pagels, Heinz. *Perfect Symmetry: The Search for the Beginning of Time*. New York: Bantam, 1995.

Pauli, W., "The Influence of Archetypical Ideas on the Scientific Theories of Kepler," in C.G. Jung and W. Pauli, *The Interpretation of Nature and Psyche*, New York: Ishi Press, 2012. (Bollingen Series L1, 1955).

Pauli, W. & Jung, C.G. *Atom and Archetype. The Pauli/Jung letters, 1932-1958.* Princeton, NJ: Princeton University Press, 2014.

Peat, F.D. *Synchronicity: the Bridge between Matter and Mind.* NY: Bantam Books, 1987.

Penrose, R & Hameroff, S. (2011) "Consciousness in the Universe: Neuro-science, Quantum Space-Time Geometry and Orch OR Theory." *Journal of Cosmology,* 14. http://journalofcosmology.com/Consciousness160.html

———. (2014a). "Reply to Seven Commentaries on 'Consciousness in the Universe: Review of the 'Orch Or' Theory.'" *Physics of Life Reviews,* 11(1), March 2014 (94–100).

———. (2014b). "On the Gravitization of Quantum Mechanics 2: Conformal Cyclic Cosmology." *Foundations of Physics* 44 (8) (873-890).

———. *Cycles of Time.* Oxford, UK: Oxford Univ. Pr., 2010.

———. *Shadows of the Mind.* Oxford, UK: Oxford Univ. Pr., 1994.

Plato. *The Dialogues of Plato.* New York: Theommes Press, 1977.

Plotinus. *The Enneads.* LP Classic Reprint Series. 1992.

Poincaré, H. *Science and method.* NY: Dover Publications, 1952.

Popp, F.A., Chang J.J., Herzog, A., Yan, Z., & Yan, Y. (2002). "Evidence of non-classical (squeezed) light in biological systems." *Physics Letters A,* 293(1-2): 98-102.

Pribram, K.H. *Brain and perception: Holonomy and structure in figural processing.* Hillsdale, NJ: Lawrence Erlbaum, 1991.

———. (1997). "The deep and surface structure of memory and conscious learning: Toward a 21st-century model." In R.L. Solso (Ed.) *Mind and brain sciences in the 21st century.* Cambridge, Ma: MIT Press, 1997.

Prigogine, I., & Stengers, I. *Order out of chaos.* New York: Bantam Books, 1984.

Puthoff, H.E. (1989, 1993). "Gravity as a zero-point-fluctuation force." *Phys. Rev. A,* 39, 5, 2333; *PRA* 47, 4, 3454. http://earthtech.org/publications/PRAv39_2333.pdf

———. (1998). "Can the Vacuum Be Engineered for Space Flight Applications?" *J. of Scientific Exploration,* 12 (1) (pp. 295–302). (http://earthtech.org/publications/JSEv12_295.pdf)

Puthoff, H.E. & Ibison, M. (2003). "Polarizable Vacuum 'Metric Engineering' Approach to GR-Type Effects," *MITRE Conf.*, McLean, VA, 5/8/ 2003. (http://earthtech.org/publications/Mitre%20Conference.pdf)

Puthoff, H.E., Little, S.R. & Ibison, M. (2001). "Engineering the Zero-Point Field and Polarizable Vacuum For Interstellar Flight." *Institute for Advanced Studies*, Austin, Texas, January 2001.

Radin, D. *Entangled Minds.* NY: Paraview Pocket Books, 2006.

———. *The Conscious Universe.* New York: Ballantine, 1997.

Radin, D., Michel, L., Galdamez, K., Wendland, P., Rickenbach, R., & Delorme, A. (2012). "Consciousness and the double-slit interference pattern: Six experiments." *Physics Essays* 25, 2 (157-71) [DOI: 10.4006/0836-1398-25.2.157] http://media.noetic.org/uploads/files/PhysicsEssays-Radin-DoubleSlit-2012.pdf

Radin, D. & Nelson, R. (1989). "Evidence for consciousness-related anomalies in random physical systems." *Foundations of Physics, 19,* (12), 1499–514.

Radin, D., Rebman, J., & Cross, M. (1996). "Anomalous organization of random events by group consciousness. Two exploratory experiments." *J. of Scientific Exploration, 10,* 143–68.

Ramon, C. & Rauscher, E.A. (1980) "Superluminal transformations in complex Minkowski space." *Foundations of Physics* 10, (661-69).

Randall, L. *Higgs Discovery: The Power of Empty Space*. New York: Bodley head/ Random House, 2012.

———. *Warped Passages: Unraveling the Mysteries of the Universe's Hidden Dimensions*. New York: HarperCollins, 2005.

Randall, L. & Sundrum, R. (1999)."An alternative to compactification." *Physical Review Letters* 83 (4690-93).

Ray, P. & Anderson, S. *The Cultural Creative*. Three Rivers Press. 2001.

Reber, A. S. *Implicit learning and tacit knowledge*. New York: Oxford Univ. Press, 1993.

Ricoeur, P. *Interpretation Theory*. Fort Worth, Tx: Texas Christian Univ. Press. 1976.

Ricoeur, P. The Hermeneutical Function of Distantiation. In *From Text to Action: Essays II*, 75–88. Evantson, Il: Northwestern Press. 1991.

Robertson, Robin. *Your Shadow*. Virginia Beach: A.R.E. Press, 1997.

———. *Jungian Archetypes: Jung, Gödel, and the History of Archetypes*. York Beach, Me.: Nicholas-Hays, 1995.

Ruelle, D. *Chance and chaos*. Princeton, NJ: Princeton Univ. Pr., 1993.

Sarfatti, J. *Super Cosmos; Through struggles to the stars. (Space-Time and Beyond III)*. Bloomington, In.: Author House, 2006.

Schlitz, M.M. (2006). "A Study of Experimenter Effects in Psi Research." *Shift*. No. 9:40-41, Dec. 2005-Feb. 2006.

Schlitz, M., Amorok, T. & Micozzi, M. *Consciousness and Healing: Integral Approaches to Mind-Body Medicine.* London, UK: Churchill Livingstone, 2004.

Schwartz, Stephan. *Open on the Infinite*. Nemoseen Media, 2007.

———. *The Secret Vaults of Time*. Charlottesville, Va.: Hampton Roads, 2005.

Schwartz, S. & Dossey, L. (2010). "Nonlocality, Intention, and Observer Effects In Healing Studies: Laying A Foundation For The Future." *Explore 2010, vol 6* (p. 295–307). Published by Elsevier Inc.

Seligman, M. *Learned Optimism: How to Change Your Mind and Your Life*. New York: Vintage/Random, 2006.

Sheldrake, R. *Dogs that know when their owner is coming home.* New York: Broadway Books/Random, (revised ed.), 2011.

———. *Morphic Resonance. The nature of formative causation*. Rochester, Vt.: Park Street Press. 2009.

———. *Science set free*. New York: Deepak Chopra/ Crown /Random, 2012.

Sirag, S-P. "Consciousness: A hyperspace view." In J. Mishlove (Ed.) *The Roots of Consciousness*. New York: Marlowe & Co., 1993-1997.

Smythies, J.R. *Analysis of Perception*. London, UK: Routledge & Kegan Paul, 1956.

Smythies, J.R. (2003)."Space, time and consciousness." *J. of Consciousness Studies* 10 (47-56).

Smolin, Lee. *The life of the cosmos*. New York: Oxford Univ. Press, 1997.
———. *The Trouble with Physics*. Boston, Ms: Houghton Mifflin Harcourt, 2006.
Smoot, G. & Davidson K. *Wrinkles in Time; Witness to the Birth of the Universe*. New York: Harper Perennial, 1994-2007.
Stapp, H. *Mind, Matter And Quantum Mechanics*, Heidelberg, Germany: Springer (The Frontiers Collection), 2009.
———. *Mindful Universe: Quantum Mechanics And The Participating Observer*. Heidelberg, Germany: Springer, 2011.
Susskind, L. (2003) "The Anthropic Landscape of String Theory." *arXiv:hep-th/0302219*
Swanson, C. *The Synchronized Universe*. Tucson, AZ: Poseidia Press, 2003.
Talbot, Michael. *The Holographic Universe: The Revolutionary Theory of Reality*. New York: Harper Perennial, 2011.
Targ, R. & Puthoff, H. *Mind-Reach: Scientists Look At Psychic Abilities*. Charlottesville, Va.: Hampton Roads, 2005.
Targ, R., Puthoff, H. & May, E. "Direct perception of remote geographic locations." In C.T. Tart *et al.* (Eds.) *Mind at Large*. New York: Praeger, 1979, 78-106.
Tart, C. *States of consciousness*. New York: Dutton, 1975.
Tart, C. (Ed.) *Altered states of consciousness*. New York: John Wiley & Sons, 1969.
Teilhard de Chardin, P. *Phenomenon of Man*. NY: Harper Torch Book, 1965.
Teodorani, M. (2014) "A strategic 'viewfinder' for SETI research." *Acta Astronautica* 105 (512–516 & 547–552)
———. *Bohm: La physique de l'infini.* Cesena, Italia: MacroEditions, 2001.
———. *Entanglement*. Cesena, Italia: MacroEditions, 2007.
———. *Synchronicité: Le rapport entre physique et psyché, de Pauli et Jung à Chopra*. Cesena, Italia: MacroEditions, 2010.
't Hooft, G. (2009). "Entangled quantum states in a local deterministic theory." arXiv:0908.3408 [quant-ph].
Varela, F. & Shear, J. "First-person methodologies: What, Why, How?" In F. Varela. & J. Shear (Eds.), The view from within. *J. of consciousness studies.* 6(1,2), 1999.
Varela, F., Thompson, E., & Rosch, E. *The embodied mind.* Cambridge, Ma: MIT Press, 1991.
Vivekananda, Swami. *Raja Yoga*. Calcutta, India: Advaita Ashrama, 1982.
von Bertalanffy, L. *General system theory*. New York: G. Braziller, 1968.
———. *Robots, men and mind*. New York: G. Braziller, 1967.
von Lucadou, W. (1983). "On the limitations of psi: A system-theoretic approach," in W. Roll, J. Beloff & R. White (Eds.), *Research In Parapsychology 1982.* Metuchen, NJ: Scarecrow Press.

———. (1987). "The model of pragmatic information (MPI)," *Proceedings of the 30th Annual Convention of the Parapsychological Association*. Edinburgh, Scotland: Edinburgh University.

Waite, A.E. (Ed.) *Hermetic and Alchemical Writings of Paracelsus*. Boston: Shambhala, 1976.

Walker, E.H. "Consciousness and Quantum Theory." In *Psychic Exploration*, (Ed.) J. White. New York: Putnam, 1974.

Walker, E.H. *The Physics of Consciousness: The Quantum Mind and the Meaning of Life*. Cambridge, MA: Perseus Books, 2000.

———. (1975). "Foundations of paraphysical and parapsychological phenomena," in L. Oteri (Ed.), *Quantum physics and parapsychology*. New York: *Parapsychology Foundation.*

———. (1984). "A review of criticisms of the quantum mechanical theory of psi phenomena," *J. of Parapsychology, 48,* 277–332.

Weinberg, Steven. *Dreams of a Final Theory*. London: Hutchinson Radius, 1993.

Wesson, Paul. *Five-Dimensional Physics: Classical and Quantum Consequences of Kaluza-Klein Cosmology*. Singapore: World Scientific, 2006.

Wheeler, J.A. *Geons, Black Holes, and Quantum Foam: A Life in Physics.* New York: W.W. Norton & Co., 1998.

———. (1990). "Information, physics, quantum: The search for links." In W. Zurek, *Complexity, Entropy, and the Physics of Information*. Redwood City, Ca: Addison-Wesley.

Witten, E. (1981). "Search for a realistic Kaluza–Klein theory." *Nuclear ¨Physics B 186*(3) (412–28). Bibcode: 1981NuPhB.186..412W. doi:10.1016/0550-3213(81)90021-3.

———. (1995). "String theory dynamics in various dimensions." *Nuclear Physics* B 443 (85–126).

Wilson, C. *From Atlantis to the Sphinx*. New York: Virgin Books, 1997.

Wolchover, N. (2014)."Have We Been Interpreting Quantum Mechanics Wrong This Whole Time?" *Quanta Magazine*, 06.30.14. http://www.wired.com/2014/06/the-new-quantum-reality

Zohar, D. & Marshall, I. *The Quantum Society*. NY: William Morrow, 1994.

INDEX

f = figure
11(n1) = page 11, note 1
¤ = definition or central

THE AUTHOR

Cognitive systems scientist, Ph.D. in ethno-psychology and former researcher at Princeton's Psychophysical Research Laboratories, Chris H. Hardy has spent the past two decades investigating nonlocal consciousness and thought-provoking mind potentials.

Author of more than fifty papers and fifteen books on these subjects, she is an authority in the domain both in scientific terms and as an author and workshop facilitator.

In her book *Networks of Meaning: A Bridge Between Mind and Matter* she developed a cognitive theory (Semantic Fields Theory) posing a nonlocal consciousness, of which professor and author Allan Combs said, "This book may well be the first step to an entirely new and deeply human understanding of the mind." In the present ISS theory, she has broadened her previous theory to fathom a collective consciousness and a field of active information at the origin of the universe, a Cosmic DNA boosting its birth and its self-organization.

Dr. Hardy presents her research regularly at various international conferences and is a member of several scientific societies exploring systems theory, chaos theory, parapsychology, and consciousness studies.

For more information (e.g. new research papers and presentations, media appearances, and recent physical and cosmological discoveries supporting the Infinite Spiral Staircase theory, etc.), visit her blog at:

http://cosmic-dna.blogspot.fr

54485087R00206

Made in the USA
San Bernardino,
CA